Environmental History

Volume 19

Series Editor

Mauro Agnoletti, Florence, Italy

The series intends to act as a link for ongoing researches concerning the historical interrelationships between man and the natural world, with special regard to the modern and contemporary ages. The main commitment should be to bring together different areas of expertise in both the natural and the social sciences to help them find a common language and a common perspective. Interdisciplinarity and trans-disciplinarity are needed for more and better understanding of the environment and its history, with new epistemological frameworks and methodological practices. The links between human activities and flora, fauna, water, soil, are examples of the most debated topics in EH, while established discipline like forest history, agricultural history and urban history are also dealing with it. The human impacts on ecosystems and landscapes over time, the preservation of cultural heritage, studies of historical trajectories in pattern and processes, as well as applied research on historical use and management of landscapes and ecosystems, are also taken into account. Other important topics relate to the history of environmental ideas and movements, policies, laws, regulations, conservation, the history of immaterial heritage, such as traditional knowledge related to the environment.

Mauro Agnoletti

History of the Italian Forest Landscape

unesco
Chair

Springer

Mauro Agnoletti
UNESCO Chair, Agricultrural Heritage
Landscape
University of Florence
Florence, Italy

Translated by Mauro Agnoletti

ISSN 2211-9019 ISSN 2211-9027 (electronic)
Environmental History
ISBN 978-3-031-81630-7 ISBN 978-3-031-81631-4 (eBook)
https://doi.org/10.1007/978-3-031-81631-4

Translation from the Italian language edition: "Storia del Bosco. Il paesaggio forestale italiano" by Mauro Agnoletti, © Laterza 2018. Published by Laterza. All Rights Reserved.

This Springer imprint is published by the registered company Springer Nature Switzerland AG
The registered company address is: Gewerbestrasse 11, 6330 Cham, Switzerland

[...] trees and forests were considered the greatest gift given to man. From these, he originally drew his own food, their leaves made the caves softer and from their barks, clothes were made.

(Gaius Pliny the Elder, Natural History, Book XII, I)

Contents

About the Author

Prof. Mauro Agnoletti is the Holder of the UNESCO Chair Agricultural Heritage Landscapes at the University of Florence, Italy. He is Vice President of the International Association of Landscape Ecology (IALE) and Editor of the book series on Environmental History by Springer. He has been the Coordinator of the Research Group "Forest History and Traditional Forest Knowledge" of the International Union of Forest Research Organizations (IUFRO), Vice President of the European Society for Environmental History (ESEH), chair of the Scientific Advisory Group of the FAO program "Globally Important Agricultural Heritage Systems". In Italy he has coordinated the institution of the National Register of Historical Rural Landscapes and was appointed Chair of the committee for the implementation of the law on the preservation of ancient forests. In 2024 he received the IUFRO Distinguished Service Award and was included in the Elsevier and Stanford University top 2% scientists list.

Chapter 1
Introduction

An attempt to interpret the historical dynamics of the Italian forest landscape has not yet been undertaken, while many local or thematic essays have been published. The reasons for this lack are many but justified by the difficulties of being able to represent all the complexity linked to some millennia of social and economic events that have deeply influenced the natural characteristics of the forests, to which is added the great diversity of environmental and socioeconomic conditions that characterize our peninsula. However, the time may have come to attempt a synthesis, albeit partial, for several reasons including the risk of seeing completely lost the knowledge of the fundamental contribution of the forest to the social and economic development of our country and the definition of important aspects of our landscape. Like the agricultural landscape, forests can be considered a man-made object, having been entirely shaped by human work in our country. In this sense, the volume intends to offer a contribution to the recognition of the forest as a product of centuries of human influence, no different from other elements of the Italian landscape. Italian forests belong to the category of cultural landscapes as defined by the geographer Carl Sauer in 2025: "The cultural landscape is fashioned from a natural landscape by a culture group. Culture is the agent, the natural area the medium, the cultural landscape the result". However, the importance of the cultural values of the Italian forest landscape has also been recognized by scientific research (Tieskens et al. 2017; Agnoletti et al. 2022).

Given the difficulty of taking into consideration the forests in all the different parts of the peninsula in various historical periods, I have chosen to document some activities useful for understanding the main historical process in some particularly significant areas of the country, accompanied by some case studies. It is a work of synthesis and connection between different themes, using the results of the research carried out in the past decades in various areas of Italy, integrated by contributions taken from some scholars who have investigated specific themes. For reasons of space, I unfortunately had to neglect many important research, while Tuscany, where

M. Agnoletti, *History of the Italian Forest Landscape*, Environmental History 19,
https://doi.org/10.1007/978-3-031-81631-4_1

I carried out an important part of my research activity, has received more attention compared to other regions.

From a historiographical point of view, the analysis of the landscape requires expanding the perspective compared to a literature that for decades has addressed the history of the forest with a narrative predominantly aimed at highlighting the reduction of the forest cover by man, or the degradation resulting from the distancing of its characteristics from an ideal state of naturalness (Williams 2002). The theme is rather the deepening of the forms assumed by the forests over time, in function of the various ways in which they have accompanied the socioeconomic dynamics. In this sense, the historical persistence and diversification of vegetation structures associated with grazing, the production of construction timber, firewood and charcoal, shipbuilding, or fruit chestnut groves, are fundamental processes for interpreting much of the history of Italian forest landscape. However, the need for synthesis has prevented me from addressing many forms of use linked to a vast heritage of traditional forest practices, which have often been the focus of ethno-anthropological studies but have rarely been formalised in scientific texts. This book has also omitted to describe the situation of the woods before the period in which humans began to modify their natural features.

A brief clarification concerns the terms ‘wood’ and ‘forest’, often used as synonyms in scientific texts and in many linguistic registers, both in spoken and literary language. As illustrated in the second chapter, ‘forest’ (“foresta” in Italian language) is probably a term of Germanic origin, but perhaps further from the characteristics of the Mediterranean wood that dominates much of Italy, while the term ‘wood’ (bosco) has Greek roots, with the literal meaning of pasture, even though such semantic variability exists in many European languages. Some associate the different use of ‘wood’ and ‘forest’ with some scientific approaches, such as when forest ecosystems are assigned tasks such as the regulation of water flows, or climate, functions for which it is often necessary to consider extensive and homogeneous tree formations, with a high degree of structural maturity, useful for maximising some environmental functions. Even in this, the specificity of the Italian landscape is perceived, in which the intertwining with agricultural and pastoral activities has diversified the original characters of the woods. The term ‘wood’ has been synonymous with pasture for many centuries and only in the last hundred years has animal grazing been considered an enemy of the wood. This marks an important difference compared to other parts of the world, where vast, dense, forests relatively less influenced by man dominate entire regions. However, in this English translation of the original Italian text, the use of the terms wood and forest, is not related to the different meaning of the two terms described above.

1.1 Woodland History and Landscape History

In the last decades of the last century, studies of forest history and landscape history often intersect their paths. I have already given an overview of such relationships in other essays to which we refer for any further study (Agnoletti 2000), but I would like to recall here a few references useful for better framing the content of the volume in the context of such studies. Without forgetting the monumental 'Natural History' by Pliny the Elder, in addition to other Latin authors to whom I have frequently referred, the history of the wood sees in the 'Studies of Forest Archaeology' by Adolfo Di Bérenger, published around 1864–65, the first important work appeared in Italy. Forest archaeology was a subject of study within the Royal Forestry Institute, founded at the Abbey of Vallombrosa in Tuscany in 1869, of which Di Bérenger was the first director. He was born in Munich (Germany) by French parents. After graduating in forestry in Wien he was sent to manage the forests of the north east of Italy, a territory belonging to the Austrian empire in the first half of the nineteenth century, but was later hired by the Italian government, after the unification of the country, to direct the forest school. His work was the result of a conception that according to the author indicated in history the "fundamental stone" missing "to the current edifice of forest science", considering the subject an integral part of the preparation of forestry technicians. This is still a very modern vision, as I cannot say that the teaching of history has really taken root in the forestry sector at university level in Italy, a problem resulting from the rigid separation between humanities and scientific disciplines existing today in this academic field, which does not favour the often invoked "interdisciplinarity" let alone "transdisciplinarity". The forester produced by the university in Italy is often no longer a manager, an expert with a comprehensive view of several aspects that are useful to understand and manage a forest system, but someone who has acquired a sectorial knowledge of distinct topics (biology, ecology, pathology, genetic, chemistry, remote sensing etc.), with little skills in the understanding of the historical factors affecting the features of a forests. A past director of a well-known Italian National Park standing on an old charcoal kiln platform, explaining the naturality of the surrounding beech forests to some international colleagues, although the forest was clearly the result of a conversion of an old coppice into a high stand, it is an example that I frequently offer to students.

It is worth reporting that in the same years of Di Berenger's activity, George Perkins Marsh visited Vallombrosa. Marsh was the United States ambassador to Italy for about twenty years and died in Vallombrosa in 1882, writing during this period his famous book "Man and Nature" (1867), considered a milestone of North American environmentalism and often cited in environmental history studies. The book essentially deals with the degradation of nature carried out by man, an idea shared by other American authors in those times like David Thoreau and John Muir. The title initially proposed to the publisher was indeed "Man: the disturber of nature's harmony". When publisher Charles Scribner asked him if man did not instead operate in harmony with nature, Marsh simply replied: "no". He was a politically influential figure who actively participated in the debate on the first Italian forest laws. However, it is worth dwelling on the different approaches to the role of man according to Di Bérenger and Marsh. The former aims to enhance the history of the use of the forest, the latter the degradation due to anthropogenic activities, making a clear break from the often-enthusiastic descriptions of the Italian landscape by many Grand Tour travelers, from Michel de Montaigne to Wolfgang Goethe, from Stendhal to Edward Lear. The Grand Tour was a long journey through continental Europe undertaken by the wealthy of the European aristocracy from the sixteenth century, intended to perfect their cultural background. It had an undefined duration and usually had Italy as its destination resulting also in the production of hundreds of essays and travel reports about the country. Marsh does not exalt the landscape shaped by man, let alone the Italian one described by previous travellers, on the contrary, he highlights the elements of degradation. It is a vision that will progressively come to dominate the following century and that develops in parallel with the study of ecology and nature conservation, which will provide a different interpretative key to the Italian rural landscape compared to the Grand Tour. Investigations on forest history continued and evolved in the post-war period: in 1951 von Hornstein defines the concepts of Forstgeschichte and Waldgeschichte, to separate the history of forestry sciences from the social history of the forest. In 1961 the International Union of Forest Research Organization (IUFRO) created a research group on forest history confirming the growing interest for this topic in forestry.

We must wait until the early 60s to observe a direct interest in the landscape in Italy. The History of the Italian agricultural landscape, by Emilio Sereni, published in 1961, but written several years earlier, proposed not only a new perspective but also original interpretative models in the context of national historical studies. His approach took into consideration the forms impressed by the man to the natural landscape, considering traditional practices. Sereni's interest in the forest is reflected in some of his writings (Sereni 1981) on the "debbio", a traditional fire regime and the "macchia", a vegetation type made of shrubs formations, and in some of his unpublished works recently printed which I have drawn on in the first chapter (Agnoletti 2017). In this historical context, the growing interest in the transformations of the Tuscan Forest landscape was the subject of a first work published by a forester, Antonio Gabbrielli, in 1964 (Gabbrielli 1964). Two years later, the journal of the Italian Academy of Forestry Sciences published an article by Professor Pietro Piussi dedicated to the history of forest utilisation and silvicultural treatments in the north-eastern Alps (Piussi 1966). In the case of Piussi, historical investigation is applied to ecological studies, with an approach that will characterise the teaching and research carried out at the Institute of Silviculture of the University of Florence between the late 60s and the 90s. With these studies, the forest began to be analysed as the result of prolonged interaction between man and nature, a line of study that has continued to this day.

One of the most innovative research perspectives that have influenced studies in the sector is that proposed by the English botanist Oliver Rackham in the mid-70s, with several studies on the English landscape (Rackham 1976) and later the Mediterranean. In his working method, field work assumes a priority role as a source of investigation, with a continuous shift from the local scale to the broader scale and vice versa, challenging traditional historiographical approaches based solely on printed sources and archive documents and including field evidence among historical sources. His 1980 book "Ancient Woodland, its History, Vegetation and Uses in England" led to the formal recognition of such forests by the Forestry Commission and in planning legislation. Although it is not automatically the case that any given ancient woodland is protected.

It is within this framework that the development of the collaboration between the group coordinated by Diego Moreno of the Institute of Modern and Contemporary History at the University of Genoa and that of Pietro Piussi, at the Institute of Silviculture of the University of Florence, can be interpreted. They developed investigations particularly sensitive to this method of work. In the issue of the journal "Quaderni Storici" edited by Diego Moreno titled "Forests: history and archaeology", there is explicit talk of "history and archaeology of a particular ecosystem", to affirm the need for a change in the way of studying the forest (Moreno et al 1982). The effectiveness of such investigations is evident, among other studies, in the case of alder cultivation in Liguria, which I described in the third chapter. These are themes that even the rediscovery of peasant civilization and material culture that took place in the 60s had not deepened, especially with regard to the processes of landscape construction. In the last twenty years the history of the forest has been absorbed into the broader context of environmental history, a discipline that has developed mainly in Northern Europe and North America, a field in which, however, degradationist perspectives have played a priority role, especially in the initial decades, and in which the landscape plays a marginal role. More effective, in this regard, is the approach proposed by historical ecology and the studies of Marc Antrop on traditional landscapes. These research activities and the recognition of the importance of a landscape approach in rural policies led to the establishment in 2012, at the Ministry of Agricultural, Food and Forestry Policies, of the "National Register of Historical Rural Landscapes, Knowledge and Traditional Agricultural Practices" (Agnoletti 2012), the national initiative specifically dedicated to the protection of agricultural landscapes. While we publish this volume, a new law establishing the protection of forests having cultural, historical and landscape value is going to enter action. This law, finally recognize the need to preserve woodlands whose values are the results of the influence of man on the natural environment. This law completes the framework of forest protection in Italy which now takes into consideration both natural and cultural forests (Fig. 1.1).

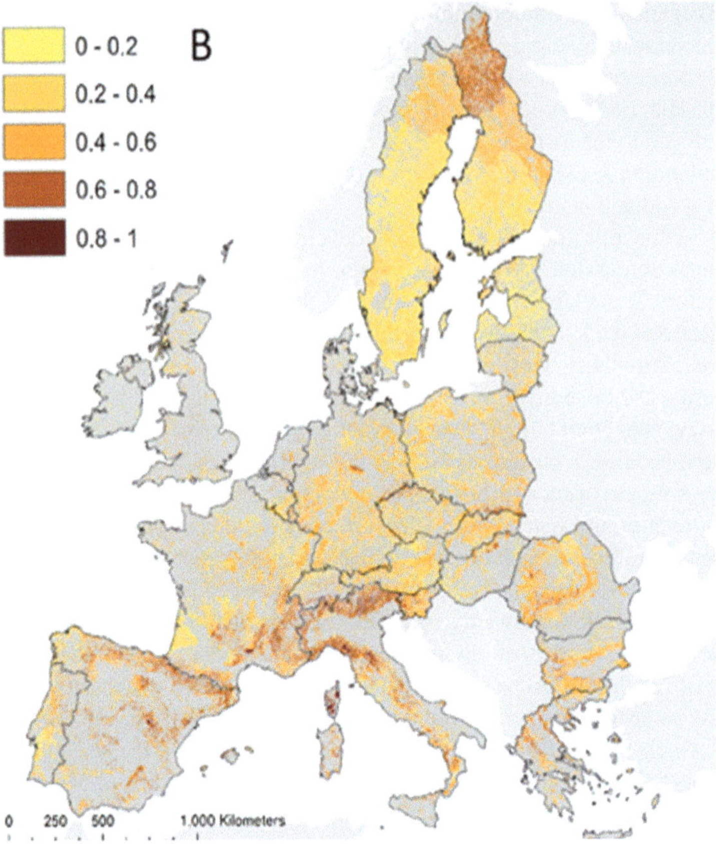

Fig. 1.1 Cultural index of European forests (Tieskens et al 2017)

1.2 Interpreting Forest Landscape

Even though Italian forests have more than doubled in size in the last century due to abandonment of farmed land, no increase in surface area will give to wooded landscape a particular significance in terms of quality.[1] The importance and quality of the landscape does not increase proportionally to the forested area, on the contrary, beyond certain limits, this phenomenon often leads to landscape degradation. In Italy, the main values of forests are found in other characteristics, also because their vitality allows them to recover even after disturbances like fire, a decisive factor in the history of the Mediterranean landscape, as we explained in the first chapter. This dynamic, linked to the biological mechanisms that regulate the functioning of the ecosystem, escapes a large part of the public opinion, leading to the widespread belief that

[1] Italy accounts for only 0.21% of the world's forests.

the cutting of a forest, or the occurrence of fire, are synonymous with permanent deforestation.

This book also seeks to illustrate how the different ways of using the forest and interacting with local culture have contributed to the complexity of the landscape, which can be observed at different levels. The approach to the study of the landscape must take into consideration first of all the concepts of space and time, in fact every landscape has its own historical dynamics and is articulated in space in a peculiar way. Understanding the landscape value of a forest requires also the analysis of the landscape context in which it is located, while often today we tend to extrapolate the forest from the rest of the landscape. The variety of the Italian landscape has greatly reduced today: compact and homogeneous forests, mostly uncultivated, now occupy most of mountains and much of the hills, having now interrupted integration with agricultural and pastoral activities, except perhaps in Sardinia Island, one of the most important silvo-pastoral landscapes still existing in Europe. This has contributed to a polarisation of the landscape, now organized into three main large systems: mountain and high hill, predominantly forested and largely depopulated; low hills and part of the plains, with prevalent agriculture and low population density; extensive urban and suburban areas with high population density and industrial plants in the plains. Unfortunately, there are few areas where these systems interconnect, producing more complex landscape matrix. It is also important to remember that traditional Italian agriculture had a very high percentage of trees, typical of the so-called "agroforestry systems", often presenting more trees than some types of forests. A further diversification of the landscape is given by the forms of management, essentially linked to various types of coppice woods and high forests, but with a large variety of traditional practices at single tree, groups of trees and forest level. These two fundamental forms were historically articulated in a very wide range of intermediate structures, according to the further functions to which they could be destined.

Another fundamental aspect concerns density. Very dense forests, such as the high forest intended for wood production, or the shrublands of many coastal areas, were accompanied by a wide range of variants. There is also a "vertical" perspective with which to observe the forest landscape, which from patterns only a few metres high can pass to forests of tens of metres in height, within which we can often identify several horizontal planes. As some chapters document, even the composition of species has been modified by man over the centuries according to different needs, this has also affected the landscape.

The combination of all these variants has introduced into the landscape a diversity often greater than that of natural forests. This marks one of the most significant differences between the naturalistic approach and the landscape approach. The ideal stage of climax, that is, the state of maximum naturalness towards which the forest could theoretically tend in the absence of human activity, envisages a forest at full density, in which biodiversity is measured based on theoretical models linked to the potential natural vegetation expected for a certain area. Human intervention provides a variability that enriches not only the diversity at the landscape scale, but biodiversity itself, creating new habitats for fauna and flora. This is the biocultural diversity concept, recently defined by FAO, UNESCO and the United Nations Convention on Biological Diversity, precisely to identify landscapes that are the result of the historical interaction between man and nature.

In the last thirty years Italian foresters and forestry have been largely influenced by the ideas of "naturalness" and the "return to nature", mainly due to north European and north American scientific views, reducing the importance of silviculture and active forest management. The European Union has politically implemented this approach developing the habitat directive of 1992 and the network of natural protected areas called "NAURE 2000". Therefore, even Italy had to find and protect "natural habitats". Science has been an important driver of this process, but there is also a sociological aspect to consider in order to explain it. It happens quite often that Italians are influenced by ideas coming from other cultures rather than by the values of their own culture. Forestry is no exception. As a matter of fact, the concept of "natural ecosystem" and its "anthropic degradation", used to define nature protection, is poorly suited to the Italian situation. In this country, it is the phenomena that have occurred within socio-economic systems in combination with nature, over time, that have determined the landscape structure, building its main values, including biodiversity. It is therefore difficult to identify a reference point for the identification of truly natural habitats and it is often established a priori that, in the presence of certain characteristics, we can speak of natural or semi-natural forests. In reality, biodiversity itself is a dynamic phenomenon, but very little attention is paid to the fundamental biodiversification processes occurred in history. Reversing the current perspective, we could classify Italian forests as cultural or semi-cultural, according to the time elapsed since the suspension of anthropogenic practices. In the book "Atlas of Italian Forests" (Agnoletti 2020), several forests in all the regions have been identified to present some evidence of forests still presenting historical features, although submitted to multiple forms of strict nature protection. These forms of protections have often created the interesting result of degrading the object of protection because the human intervention needed to maintain it has been limited or forbidden. In fact, most of the traditional forest practices mentioned in this book today are forbidden by official forestry and nature protection regulations.

In the face of a few thousand years of history in which the forest has contributed significantly to social and civil progress, the last 70–80 years of history have swept away peasant civilization and much of the traditional knowledge associated to forestry and the related landscape and environmental values. The rural exodus has resulted in the creation of an urban society almost completely detached from agriculture and forestry, in this respect this book has also the ambition to document at least a part of the Italian forest heritage, before it will be lost. While for some millennia the socioeconomic conditions determined landscape change, the transformations that have occurred after the second world war have indeed paved the way for policies that now contribute to determining the history of the forest landscape. The fact that cultural values currently play a limited role in the EU forest strategy indicates the scant consideration given to the role of culture and history in the overall valuation of forests within the paradigm of sustainability developed in recent decades. Failure to effectively and coherently address culture and history may very well be an emerging weakness that needs be reconciled. This is both to give the public and local communities confidence in the protocols designed to recognize well-managed forests, and in moving towards the goals of sustainable management (Fig. 1.2). While we publish this volume in English, seven years after the first Italian edition, a new law establishing the protection of forests having cultural, historical and landscape value is going to enter action. This law, finally recognizes the need to preserve woodlands whose values are the results of the influence of man on the natural environment by completing the framework of forest protection in Italy, which now takes into consideration both natural and cultural forests.

Fig. 1.2 Map of Italy with the geographic areas described in this book

References

Agnoletti M (2000) The development of forest history research. In: Agnoletti M, Anderson S (eds) Methods and approaches in forest history. CAB International, Wallingford and New York

Agnoletti M (2012) Historical rural landscapes. Cultural values for the environment and rural development, Springer, Dordrecht and New York.

Agnoletti M (2017) Prefazione. In: Gemignani CA (ed) Emilio Sereni. L'origine dei paesaggi della Grande Liguria. Due inediti dei primi anni Cinquanta. Edizioni Istituto Alcide Cervi, Gattatico

Agnoletti M, Piras F, Venturi M, Santoro A (2022) Cultural values and forest dynamics: the Italian forests in the last 150 years. Forest Ecol Manage, Special Issue on Restoration of cultural values in forested landscapes, Guest Editors John Parrotta & Mauro Agnoletti 503:119655. ISSN 0378-1127. https://doi.org/10.1016/j.foreco.2021.119655

Gabbrielli A (1964) Divagazioni storiche e forestali sulla campagna senese. Annali Dell'accademia Italiana di Scienze Forestali 13:371–406

Moreno D, Piussi P, Rackham O (eds) (1982) Boschi: storia e archeologia. Quaderni storici 17(49):7–163

Piussi P (1966) Sulla storia delle utilizzazioni e del trattamento di alcune fustaie di conifere del Trentino. L'italia Forestale e Montana 21(6):253–266

Rackham O (1976) Trees and woodland in thè British landscape. In: Archaeology in thè field series. J.M. Dent & Sons, London

Sereni E (1981) Terra nuova e buoi rossi. E altri saggi per una storia dell'agricoltura europea. Einaudi, Torino

Tieskens KF, Schulp CJ, Levers C, Lieskovský J, Kuemmerle T, Plieninger T, Verburg PH (2017) Characterizing European cultural landscapes: accounting for structure, management intensity and value of agricultural and forest landscapes. Land Use Policy 62:29–39. https://doi.org/10.1016/j.landusepol.2016.12.001

Williams M (2002) Deforesting the earth. From prehistory to global crisis, The University of Chicago Press, Chicago

Chapter 2
The Domestication of the Natural Landscape

Abstract The transition from a society characterised by the prevalence of hunters and gatherers of fruits to temporary agricultural settlements and then to permanently cultivated areas, capable of providing stable harvests, was at the basis of the process of civilization in Italy. This process also includes the domestication of the woodlands which was not only reduced in size to create cities, pastures and fields, but mainly modified in its characteristics of density, structure and specific composition to provide a vast range of products necessary for life. Thinking about the landscape in protohistoric times, we must refrain from the idea of the prevalence of mountains covered by forests and of human settlements concentrated mainly in the plains. Numerous authors report the preference of primitive populations for the terraces and morainic amphitheatres of the mountainous areas. Fire plays a fundamental role in the construction of the landscape and the shaping of forest, particularly through the practice of slash-and-burn (debbio). This practice is widespread from the north to the south of the peninsula and is documented by numerous sources, particularly linguistic ones. As can be understood, the maquis scattered throughout our peninsula are one of the many historical landscapes that we could define as of "social" origin. Understanding the practice of debbio allows us to understand not only the processes of deforestation but also those at the base of the formation of the scrubland and other plant formations that derive from its use.

2.1 Human Colonisation

The history of the Italian landscape in ancient times is largely characterised by the transformation of the natural environment into agricultural areas, including the cultivation of the forest. The transition from a society characterised by the prevalence of hunters and gatherers of fruits to temporary agricultural settlements and then to permanently cultivated areas, capable of providing stable harvests, was at the basis of the process of civilisation. This process also includes the domestication of the woodlands which was not only reduced in size to create cities, pastures and fields, but

M. Agnoletti, *History of the Italian Forest Landscape*, Environmental History 19,
https://doi.org/10.1007/978-3-031-81631-4_2

mainly modified in its characteristics of density, structure and specific composition to provide a vast range of products necessary for life.

For a very long historical period that goes from the first development of agricultural activities until the early years after the second world war, it was unthinkable to do agriculture without having a sufficient supply of trees, both in the form of woods and scattered trees. Therefore, the reconstruction of a history of the wooded landscape, if it must necessarily also take into account the transformations of the extension of the woods, can only be based mainly on the uses of the wood and on its management methods. In this sense, the environmental characteristics of Italy in terms of tree species, and their geographical and altitudinal distribution, in relation to the soil and climate, certainly play an important role, but often secondary to the needs of the population. We mean by this that, while recognising for example the prevalence of certain tree species in some environmental contexts, such as for example the chestnut tree in the Alps and the Apennines, the need to cultivate this species over the centuries has led it to be present today from sea level up to 1500 m of Mount Etna (Sicily), in environments very far from its ideal ecological requirements.

Moreover, the substantial geographical unity of the peninsula which, as De Seta reminds us, united the peoples of Italy before even historical events (De Seta 1982). It also constituted a physical limit within which the circulation of cultures and animal and plant species was somehow forced to move in a limited and dense space. It is therefore not the intention of this work to focus particularly on the transformations of wooded surfaces over the centuries, or on the natural characteristics of forest vegetation, which are easily found in scientific literature, but rather to highlight the characteristics of the landscape determined by historical dynamics linked to the continuous relationship between human needs and natural features.

Thinking about the landscape in protohistoric times, we must first of all refrain from the idea of the prevalence of mountains covered by forests and of human settlements concentrated mainly in the plains. This does not mean that the valleys excluded inhabited areas: numerous authors however report the preference of primitive populations for the terraces and morainic amphitheatres of the mountainous areas, justifying it with the exposure conditions and the pedology of the slopes. The looseness of the soil fertilised by glacial muds and the absence of a dense wood cover undoubtedly constitute a decisive element for this location of primitive settlements, well documented also from an archaeological point of view (Godefroy 1940). This situation is in some way confirmed by historians like Strabo, who specified that the mountains and peaks overhanging the alpine foothills of the western Alps were well populated Bonaccioli (1562). The available information for the Bronze Age in the Alpine area speaks of semi-nomadic mountain agro-pastoral systems, where small cultivated areas are bound to the needs of livestock farming and the necessary movements in search of pastures (Guidetti 1998). The settlement of the mountain probably initially involved the most fertile debris cones or scree slopes, as a result of the removal of oak and beech woods. The most stable residential settlements were concentrated mainly in the altitudinal band of the mid-mountain, with gentler slopes and often more evolved soils, especially on the plateaus, where the wood gives way to the cultivation of the land (Borgioli 1946).

The intense frequentation of the Alpine mountain seems in some way corroborated also by recent discoveries linked to the retreat of the Alpine glaciers as in the case of the discovery of the mummy of Ötzi, otherwise known as the Similaun man, a glacier located in South Tyrol on the border with Austria where the body of this Copper Age shepherd was found. By analysing what was recovered in terms of tools and clothing on the man's body, who lived around 3200 BC, 17 tree species were identified, from pear to larch, from plum to juniper, from elm to willow, a sign of a considerable variety of uses of tree plants that suggest in-depth knowledge of their qualities in relation to the various uses ranging from clothing to the objects that the hunter had with him. Genetic research conducted on the corpse would confirm the news according to which Ötzi and his ancestors would have been part of agricultural populations belonging to a group called haplogroup G, typical of agricultural populations rooted in the Middle East. This would also confirm the Indo-European root of some terms related to tree species present in the Alpine arc, as well as the cultural flows that have influenced the Italian landscape from the sea and through the land routes that connect the Eurasian continental platform (Fleckinger and Steiner 2000).

The presence of the first settlements in mountain areas is certainly not a unique characteristic of the Italian landscape. Also in the near Middle East, archaeological finds and linguistic data show us how the elevations of the land were particularly sought after by primitive populations, from the Neolithic period. If the highest peaks of the hills and mountains were used as refuge and defence, the heights and minor elevations of the land were more widely and normally frequented, often being the first lands that primitive tribes chose for occupation. In territories like those of the Po valley emerged on Pleistocene alluvium, still subject to flooding and marshes, or otherwise occupied by post-glacial forest, this choice was probably almost obligatory and often the only alternative to pile-dwelling settlements. This also happened in the Rhone delta and along the coastal strip from the Berre pond to the Pyrenees, where every elevation of the land has marked, from the earliest times, a stage in the occupation of the soil and in human settlements.

The scarcity of pre-Roman linguistic and toponymic relics, which refer to the major valley depressions, therefore suggests that the large valleys did not play a significant role in the initial stages of man's occupation of space. It is only in a later period, in the Celtic, Roman, and often Romance phase that in the rich and cool valley bottoms, but precisely for this reason cluttered with a denser forest cover, clearings were developed and human frequencies multiplied. Valley settlements are more frequent where Naturally bare areas of tree vegetation offer more favourable conditions for a still relatively backward technique for deforestation. The first populations hardly face deforestation in valleys where the tree vegetation is dense and continuous and where the humidity of the soil makes the use of fire difficult, perhaps the main form of landscape management in the Mediterranean. The hypothesis of a first colonisation of the mountain areas would also find partial confirmation for the Apennine area, with the documented presence of numerous settlements from the Upper Palaeolithic, about 35,000 years ago, while there is much news about the regular seasonal exploitation of summer pastures at high altitudes for the Neolithic (Barker 1975). The hypothesis of mountain settlements of some importance also seems supported by the classifications of ceramics, whose basic types for our country begin with the phase of the "decorated and rough Apennine ceramics" of the Bronze Age, preceding the Villanovan phase. However, this information is accompanied by descriptions of settlements located at lower altitudes, where, compared to an economy based mainly on hunting, we find populations already dedicated to the breeding of domestic animals and the cultivation of cereals, as evidenced by the excavations in the Neolithic settlement of Pienza in Tuscany, a hilly area of the Val d'Orcia, but also by dozens of other sites present in Etruria and southern Italy, particularly in Puglia. Between the second and first millennium BC, the number of settlements seems to increase proportionally, in the plain and in the mountains, so much so that for the Italian Bronze Age the term "Apennine culture" has been coined (Potter 1985).

In the analysis of landscape dynamics, we cannot ignore that some important aspects related to forest vegetation and residential settlements have been conditioned by climatic evolution and orographic conditions. The shape of the peninsula projects it from the centre of the European continental platform to touch the African one with climates that vary from glacial to subtropical and semi-arid, according to the Köppen classification. This proposes a great variety of different environments for the development of forest vegetation to which are added the altitudinal variants, from sea level to the peaks of the Alpine chain, with a territory that is 35.2% mountainous and 41.6% hilly, but with only 23% of plains, thus giving it that connotation of a landscape mainly of vertical development, as Emilio Sereni wrote, so different from the landscapes of central and northern Europe (Sereni 1961). To this diversity of environments are added particular phenomena linked to long-term climatic variations. This is the case, for example, for the particular situation of the Insubric Alpine region, corresponding to the Lombard lakes area, characterised by vegetation typical of the Mediterranean area, little linked to the traditional Alpine and pre-Alpine environment, the result of the immigration of thermophilic and south-eastern elements in the post-glacial period. The subsequent climatic deterioration would have determined the "isolation" of these formations in the lake area, eliminating the penetration routes from the sea through the Po plain, except for the Euganean hills area that still testifies to it. The particular climatic conditions of the lake region (Insubria) have facilitated the development of human settlements and also a vegetation typical of Mediterranean areas located much further south: current examples are the cultivation of olives and citrus fruits in the agricultural field and of holm oak in the forest context. The remaining Alpine geographical regions, on the other hand, see the presence of vegetation more in tune with the characteristics of the continental climate, with important variations depending on the exposure of the slopes, which in those to the south have favoured human settlements, but significantly transformed by human activities, as we will see in the following chapters.

In terms of human activities, in the Iron Age the exploitation of mineral resources carried out by the Celts, especially in the pre-alpine areas, must have influenced the structure of the landscape, with the development of trade routes for the commerce of iron and salt. As reported by Sereni, it seems certain that, at the time of Roman colonisation, mountain forests, as well as those of the plains, were already involved in collection activities and timber production. Trees were already being felled with iron axes, tools that had gradually supplanted stone axes unsuitable for felling large trees, and which allowed trees up to eight feet in diameter to be felled on the Ligurian Apennines (Gemignani 2017).

In the times of Posidonius and Diodorus, in the first century BC, iron was already used for this purpose, allowing the production of planks and timber for shipbuilding, of which Genoa had become the emporium and which constituted the first item of Liguria's export trade in the times of Strabo (Bonaccioli 1562). The iron axe is an important technological innovation, which no longer forces us to open gaps in the forest only with fire, but allows selective felling of tree species for various uses of timber. An activity that is accompanied by other products that mountain people regularly trade with the plain such as resin, pitch, resinous branches for torches, cheese, wax, honey. However, it would be wrong to conceive of this transition from the world of stone to that of iron as abrupt and without mediation. In Diodorus' own account, in fact, the image of the stone re-emerges, with an insistence that is certainly not accidental, even when talking about the agricultural activities of the populations (Gianotti et al. 1986).

In pre-Roman times the condition of the country appears anyway different from that of other European territories. Central Europe was still covered by marshes and forests, populated by "aurochs, bison and elk" (Lachmann 1826), when in Italy from times preceding the foundation of Rome the effects of Middle Eastern and Asian civilization, the arts and commerce were already evident. According to Pliny, in the centuries preceding the Roman conquest the peninsula had as many as 1197 cities (Aragosti and Barchiesi 1984), was extensively cultivated, sold the surplus of cereals and the agricultural production of the *Campania Felix* and its other lands, while in the imperial era it would then resort to the granaries of Egypt and Africa to meet the needs of the population (Dati 1582; Adriani 1861; Machanei et al 1738). The Pontine marshes and those of the Venetian estuary were intensely cultivated territories, dotted with cities and villages. The Maremma were considered the granaries of Etruria (Tuscany), from which the populations drew food, livestock and timber for the construction of fleets.

However, it is with the arrival of the Romans in the Etruscan, Celtic and Ligurian areas, that the forests become centres of normal and fundamental productive activities, from hunting and gathering economy to a real forest economy, based on cutting and trading wood, as well as some important by-products. Timber for shipbuilding, wild honey and products of special forest industries, such as resin and pitch, played an important role. However, we should not think of a phenomenon exclusively linked to the Alpine area: also in Calabria the Bruzi population initiated significant trade exchanges with the Romans, based on timber and pitch from the Sila.

The fact that timber was the object of a real appropriation and commercial exploitation by the Alpine tribes is confirmed by the episode of Messala, recounted by Strabo (Bonaccioli 1562). Wintering near the territory of the Salassi, he was forced to pay not only for the wood used for fuel, but also for that used for the spears and military exercises of the troops. Even in later times, the industry and trade of coal were at the origin of the immense fortune of the family of Pertinax, the only emperor that Liguria gave to Rome: his father had already managed a significant taberna coctilicia, a factory with a coal depot (Lamboglia 1939), fuelled by the surrounding woods, which he expanded and developed when he was confined by Commodus in Liguria.

In protohistoric and historic times, therefore, the testimonies unanimously point out the importance that forestry activities assume alongside those of livestock farming, while, as we will see, the more properly agricultural activities occupy a position that is still relatively secondary. Those related to hunting and the economy of gathering, if they still have a significant complementary importance, no longer have that exclusive weight they had in more remote times. In the tall woods, a selection of species is already underway that changes the landscape, as documented for Etruria to the detriment of the white fir that reduces its surfaces, while also the coppice management extends. As described by Pliny the Elder and even more by Columella.

Bearing in mind what has been explained about the development of the first human settlements and the results of palaeobotanical research, it seems certain that in protohistoric times vast flat areas of the country, such as the Po plain, are still covered with extensive forests, interrupted by marshes and clearings, with a series of drier mounds and elevations (Martegani and Agnoletti 2003). These formations, attributable to the mesophilic forest vegetation of the oak-hornbeam, were dominated by the English oak (Quercus robur), to which species such as the black alder, poplars, hornbeam and others were associated. In the fourth and third century BC, the environmental situation already appears conditioned by the activity of populations present in this area that can be divided into three large ethnic areas, characterised by the Ligurians south of the Po, the Gauls between Garda and the Apennines and the Veneti to the east. According to what Sereni wrote about the "Great Liguria" in pre-Roman times, which included part of the Po plain, at this time the grazed forest had already assumed the connotation of a characteristic element of the landscape. The position of the Po plain from then on proposed it as a natural strategic area of connection between the north and the south of Europe, towards which the attention of Rome was directed, which in the first decades of the second century BC was engaged in the culminating phase of its conquest. By defeating the Boii, Insubres and Orobii Gauls, it indeed began the work of political and territorial organisation that was to incorporate the local ethnic groups into the Roman-Italic culture, which, however, left important traces of themselves not only in toponymy (such as Mediolanum, of Gallic matrix), but also in forestry terminology. From the forests to the origins of the Po, from which the "ragia" called in Gallic language pades, seems to come the name of Pado given to the river, but there is a large series of lexical relics related to species that due to their alpine location were little known to the Romans, and that the Latin borrowed from indigenous dialects. Among these we find "birch", of Gallic origin, or "sondar", of pre-Indo-European linguistic matrix, which indicate associations with rhododendron and dwarf pine, from which the names Sondrio and Sondalo derive; but the list is richer, and includes terms referring to pines, larches, firs and also broad-leaved trees.

In these centuries, the work of reclamation and cultivation of the plain began and the progressive contraction of the original large lowland forests. In the time of Strabo, many of the marshes that The domestication of the natural landscape 11 Hannibal had had difficulty crossing had already been reclaimed by Emilio Scauro in 109 BC, but both Polybius and Livy speak of pastures and abundant cultivated fields. Despite this, it seems that the structure of the oak forests, especially in the wide series of terraces of the middle diluvial that extend from the last mountain slopes to the Po, along whose banks prevailed poplars, alder and willow, was enslaved to the grazing of large herds of pigs. Cato notes the importance of the production of cured meats among the Insubres and the large size of the animals, which could only be transported on carts. Polybius, for the first centuries of Roman colonisation, describes herds of a thousand or more heads that, after having spent the night in rudimentary shelters, go out to graze in the oak forests. This is, therefore, a semi-wild breeding system, whose spread speaks of important modifications brought by man to the forest landscape, since pig grazing, eliminating natural renewal, favoured a sparse high forest devoid of undergrowth.

If the mountains are home to forestry, pastoral and gathering activities, in the plain the Roman colonists carry out a progressive expansion of agricultural activities and an organisation of the territory that gives the fields a geometric structure, the centuriation, partly preserved to this day. Also in this case, the Roman techniques confront with the pre-existing ones, which made wide use of indigenous forest species, such as the field maple and the elm. The "atinia", as reported by Pliny and Columella, was a widely spread elm used to tie the vines, while Varro speaks of trees called "opulus" and "rumpus", field elms or maples, used to marry the vines. The term "rumpotinetum" seems to already refer to the Padanian planting of vine trees, a technique that Columella reports to be in use in Lombardy (Carena 1997), but probably introduced by the Etruscans. In this work of colonisation and transformation of a plain often marshy and forested into one of the most fertile areas of Europe, the abundance of water played an important role. This was due mainly to the runoff from the Alpine and Apennine chain, with underground aquifers that emerge forming the band of springs near the Alpine slope and dividing the territory into a lower plain, more irrigated and fertile, and a higher plain, upstream of the same, but more arid. A band that man has progressively moved northwards to expand the irrigated areas and to which are added all the canalisations made in the plain, the real responsible for its current fertility, more than the climatic regime.

2.2 The Fire

Fire plays a fundamental role in the construction of the landscape and the shaping of forest areas over millennia, particularly through the practice of slash-and-burn (debbio). This practice is widespread from the north to the south of the peninsula and is documented by numerous sources, particularly linguistic ones. Slash-and-burn corresponds to the practice of burning tree vegetation and grassy bark for the purpose of cultivation and soil fertilisation; it could involve the definitive cultivation of the land but, more often, the periodic burning of vegetation to allow cultivations for a limited time. The spread of slash-and-burn continues almost to our days, as happens with the burning of fields after harvest in some regions of the south, although it is now almost everywhere an illegal practice. We believe it is useful to discuss it in this initial part of the text since its understanding provides a fundamental interpretative basis for much of the Italian landscape in historical and protohistoric times, whether it be wooded, pastoral or agricultural. The best way to succinctly understand its importance in the peninsula is probably to follow the distribution of the terms associated with it, thanks to the work of Sereni (Sereni 1981). According to this author, its roots lie in the Mediterranean culture of the pre-Latin peoples, as evidenced by the Piedmontese dialectal nomenclature of slash-and-burn (*motera and mota),* which does not have a Latin linguistic background but dates back to a time before the Roman conquest. A pre-European Mediterranean origin is also attributed to the term *narbone* in the Sardinian dialect, which also corresponds to a series of place names in the Mediterranean area. The term *debelis* is attested in the *Tavola alimentaria* of Veleia where it appears in the description of some rustic properties and is associated with the practice of slash-and-burn, further confirming the spread of this practice.

In Roman times, slash-and-burn often recurs in agronomic treaties and in the texts of jurists. Columella (Calzecchi Onesti 1948) associates slash-and-burn with the word *caesio* and with the *silva caesa* with the meaning of a plot of forest or scrubland subjected to slash-and-burn or in any case deforested. *Caedes* is used in the *Notti attiche* of Aulus Gellius (XIX, 12.7) not only with the value of "slaughter" but also with the meaning of "cutting of trees and branches". The Latin term *runcare* is frequently used by classical authors (Canali and Lelli 2000); in the fourth century AD, Palladio (Palladio *De re rustica*) lists, among the tools necessary for agriculture, the *runcones, quibus vepreta persequirum* where the notion of *runcare* is referred not only to the clearing of the land from weeds but also to the elimination of shrub vegetation. Only in toponymy does a trace remain of the use, as a technical term of the nomenclature of slash-and-burn, of another Latin voice, *bustum,* with the meaning of place of burning, pyre. In literary use, expressions such as *silvas cremarat in Lucretius (*Marchetti 1813*) or silvestrem flammis et ferro mitiget agrum (Horace Epistulae) in Horace.* In this Latin nomenclature of slash-and-burn, however, the most commonly used terms are those of the type *incendere,* that is to set fire, burn and, especially in popular use, those that result from expressions with a direct reference to *ignis* (and later to *focus*), "fire". As for *incendere,* the Roman agronomic and legal literature refers both to the practices of slash-and-burn, as reported by Virgil in the Georgics (I, 84), and to those of the burning of pastures of stubble, as well as those of slash-and-burn on scrub, or on forest, in the texts of the jurists. *Incendere* and *incendium,* used as synonyms for *runcare* and *runcus,* are found associated with the penalties inflicted on those who have caused fires as a result of the failure to observe the due precautions in the practice of slash-and-burn.

In the early Middle Ages the Latin term *caesa* appears in the forms of medieval Latin *cesina* (or *cisina),* as attested by the medieval sources of southern Italy from the eighth century in Apulia, Abruzzo and Campania, where it also appears in toponymy and in dialect as a synonym for slash-and-burn. In the regions of southern Italy occupied by Greeks and Byzantines, the Latin term *caesa* recurs in the form of *kesina* from 1149 as attested by documents from Byzantine Calabria, where the spread of the practice of slash-and-burn is observed mainly in the southern part of the region. The term *cesa* is the fundamental term of the nomenclature of slash-and-burn also in some regions of central Italy. Central (such as Lazio, Umbria, Marche, Tuscany) where it appears in feudalistic legal literature, in municipal statutes, in toponymy and in dialect. The term *cesa* also recurs in early medieval documents of northern Italy, as evidenced in Lombardy from the year 755. The semantic value of Latin *(silva) caesa* seems to have undergone, however, in northern Italy, a different evolution from that prevalent in southern Italy. In medieval Latin, as then in Lombard, Venetian, Ladin and Emilian dialects, *cesa* comes to mean live or dead hedge, and no longer a plot subjected to the practice of slash-and-burn, or in any case deforested; while to designate the slash-and-burn the term *tagliata,* from late medieval Latin *silva incisa.* From medieval Latin *silva fracta* derives the term *fratta* used from the thirteenth century in Tuscany to indicate slash-and-burn and the term *frata* used from the thirteenth century and still used today in Trentino, which indicated, in medieval documents, the right of populations to cultivate, with or without the use of burning techniques, a plot of forest subject to civic uses.

In the Latin agronomists of a later period and in medieval Latin, *extirpare* is the most used term to indicate the practice of debbio. Even more than what could be documented for Latin of the Roman age, in medieval Latin *extirpare* is widely present in the nomenclature of deforestation and ploughing in general, but also as a technical term of the nomenclature of debbio. From the sixteenth century, the nomenclature of debbio specialises by referring to particular burning techniques. An example is offered by some Ligurian Piedmontese statutes where the practice of burning with furnaces is mentioned: the so-called covered ronco. The furnaces for debbio, of a truncated cone shape, were built by overlapping grassy tiles one on top of the other, covering them with branches which were then set on fire; the ash obtained was then spread on the ground to fertilise the soil before sowing. Also during the 1500s, space will be dedicated to the technique of burning in agronomy treaties, as in the case of the *Treatise on burning meadows,* chapter of the work by Agostino Gallo entitled *The twenty days of agriculture and the pleasures of the villa.* Later, burning and scorching will be the terms that, in his memorial of 13 August 1768 addressed to the providers, the Venetian agronomist Pietro Arduino will use to indicate the debbio, and under the heading "abburning" he will deal with the techniques used in the *New universal and reasoned dictionary of agriculture* by Francesco Gera, published in Venice by the publisher Antonelli between 1834 and 1835.

Until the end of the 1800s and the beginning of the 1900s, the practice of debbio was still widely practiced in Calabria, particularly on the Sila, and in Liguria. Many place names, still widespread throughout the national territory, such as Monte Calvo, Monte Toso, Monte Caruso, Monte Nudo, Monte Biotto, derive from the Latin term *calvus* and indicated the deforestation and burnings that had reduced the forest cover on the mountains. In some cases, these lands are named with a more specific reference to the vegetation following the repeated practice of debbio and grazing, that is, the scrub. As for Sardinia, an island where fire has been the basis of what is still today considered one of the most important pastoral landscapes in Europe, Beccu reports that the term used to deforest and then burn the plants to fertilise the soil and grow cereals, usually wheat or barley, was "narbonare". To do "narboni" was such a practice that allowed to increase the yields per hectare of wheat by at least one and a half times the normal ones. Once the new fertility was exhausted, the surface was abandoned and people moved elsewhere: in this way, open spaces were created inside the forest and the wooded surface was replaced by the shrub one. The place names *narboni, narvones, narboneddu, nerboni, nalboni, narboneri, nalbunacci,* testify to a widespread practice conducted by expert staff, the narbonators, usually shepherds, the true custodians of such techniques (Beccu 2000).

Fire as a cultivation practice has been the subject of a progressive demonisation in recent decades, even though in reality the controlled burning has resumed in some more advanced countries.[1] In fact, the most recent historical-environmental investigations show how the combination of periodic fire practices, especially the surface fires typical of shepherds, combined with sheep grazing maintain an efficient and species-rich grass carpet. In other words, the high floristic complexity of the Apennine pastures has been fundamentally linked for centuries to the alternation of fire and grazing, just as in meadows repeated mowing can increase floristic richness (Cevasco 2007).

2.3 The Scrubland

Understanding the practice of debbio allows us to understand not only the processes of deforestation but also those at the base of the formation of the scrubland and other plant formations that derive from its use. Because of the historical dynamics previously described, the scrubland still constitutes one of the most extensive shrub populations in Italy.[2] Widespread in various parts of the country with terms such as *matta, tuppa, frutectum, fracta,* the Mediterranean scrubland is almost always a secondary plant formation, resulting from a process of transformation of the natural vegetation, due to fire and grazing (Bernetti 1995). For this reason, it appears widespread in almost all the coasts of the Mediterranean and its islands, constituting a landscape that, together with the vine, the olive trees and wheat, represents a fundamental part of the cultural identity of this area. The scrubland is a dense plant formation, made up of shrubs, or trees in the state of shrub, generally sclerophyllous, among which we find the holm oak, the strawberry tree, the mastic, the cysts, the heathers, the brooms, the tree euphorbias, the junipers, the oleander and other species. It constitutes a low, dense, often impenetrable vegetation cover, which may extend over surfaces of considerable extensions. The scrublands today are formally classified in forestry literature as primary scrublands if derived from factors such as soil poverty, and secondary scrublands if instead due directly to human action. In this case, they are almost always described as degradative phases of the theoretical natural vegetation that would have existed without human intervention. In the first case, it is mainly composed of species with an almost tree-like habit, with canopies reaching 4 m in height. In the low scrubland, the vegetation is instead mainly composed of species with a shrub-like habit, with canopies reaching a maximum of 2–3 m in height.

Scrublands usually settle as a result of the use of debbio as a cultivation practice. When other disturbance agents or practices such as stubble burning or hoeing did not intervene, already after the first year of cultivation on debbio, the forest species could predominate on the "cultivated steppe" of grasses, so much so as to severely reduce its production in the second year, and even lead to the abandonment of the land. In cases like these, the whole process could orient itself, after the abandonment of agricultural cultivation, towards the regeneration of the scrubland on which the practice of debbio could be repeated after a more or less long "rest" period. In our country, practices of this kind have been repeated for hundreds of years and have strongly conditioned the specific composition and structure of the forest stands. In cases like these, the whole process could orient itself, after the abandonment of agricultural cultivation, towards the regeneration of the scrubland on which the practice of debbio could be repeated after a more or less long "rest" period. In Italy, practices of this kind have been repeated for hundreds of years and have strongly conditioned the specific composition and structure of the forest stands (Fig. 2.1).

Among the most widespread tree species in the Mediterranean environment, the holm oak *(Quercus ilex),* other oak species such as the cork oak *(Quercus suber)* and various species of pine—like the maritime pine *(Pinus maritima),* the Aleppo pine *(Pinus halepensis)* and the stone pine *(Pinus pinea),* which play a decisive role in

Fig. 2.1 The scrubland of the Uccellina Mountains in Tuscany. This vegetation pattern results from the use of fire and it has been managed for centuries as coppice with rotation of 5–6 years to produce charcoal or even less to produce brooms. The watchtower was used to spot the ships of the Saracen pirates who threatened the Tuscan coast until the 16th century

the landscape of our coasts—are often the result of debbio, or vegetative successions associated with the maquis. The current predominance of the *Pinetum,* that is, the plant association characterised by pines along the coasts, cannot be considered a primary characteristic of the landscape since it was originally an association dominated by the holm oak *(Quercetum ilicis).* The spread of heliophilous and pyrophytic species like pines, species that resist fire and can even benefit from it, has occurred as a result of anthropic factors that have progressively reduced the coverage of the original Mediterranean forest not only and not so much by felling individual trees, but due to fires resulting from the fires of shepherds and more frequently those of debbio. These are the reasons why in the Mediterranean landscape the plant associations of the *Pinetum* have taken over as the prevailing associations from the primary one of the *Quercetum ilicis.* This process finds an important exception in some reforestations, particularly in the plantations of stone pine along the Tuscan and Lazio coasts that took place between the eighteenth and nineteenth centuries after the elimination of the natural vegetation with fire.

When the joint and repeated action of the fires of shepherds, those of debbio, cuts and grazing make it impossible to reconstitute a forest-type formation, the vegetational dynamics evolve towards the affirmation of a new plant formation which is that of the Mediterranean maquis. Most of the dominant species of the various types of maquis are indeed typically pyrophytic like the strawberry tree *(Arbutus unedo),* which is often the first to return to vegetation, just a few weeks. Precisely with the repetition of these fires, the holm oak often ends up transforming into a strawberry tree maquis, a plant association widely spread along the coasts, the hilly areas and the first mountain slopes of central-northern Italy. In the strawberry tree maquis, the *Arbutus unedo* is often associated with another characteristically pyrophytic species, the *Erica arborea,* or tree heath.

Among the other types of maquis widespread in Italy, also dominated by characteristically pyrophytic species, we mention the mastic tree maquis *(Pistacia lentiscus),* in which this species is often accompanied, especially in Sardinia, by the wild olive tree. The juniper maquis *(Juniperus oxycedrus macro-carpa* and *Juniperus phoenicea)* is particularly widespread in Sardinia, while the tree spurge maquis (*Euphorbia dendroides*) dominates the rocky slopes and the warmest exposures of Mount Circeo and the coasts of Calabria. The dwarf palm maquis *(Chamaerops humilis)* already In classical antiquity, it characterised the landscape of Sicily, Sardinia and the Tyrrhenian coast of the peninsula, from Tuscany downwards. All types of Mediterranean scrub, whether primary or secondary plant formations, have a specific composition that confirms the fact that a prominent part in their formation is due to the fires of shepherds and the fires of debtors. A huge influence has been and still is today on the formation and specific composition of the Mediterranean scrub, the practice of grazing. Grazing operates by favouring the development of some species to the point of conditioning the configuration of a type of scrub and then, possibly, the affirmation of other and different plant formations. Even among the pyrophyte species of the Mediterranean scrub that have resisted the repeated fires of shepherds and debtors, not all are equally appetising to livestock and not all offer the same resistance to grazing. Grazing livestock eat with particular eagerness, for example, the young shoots and fronds of the holm oak, the Phillyrea, the strawberry tree, a little less willingly those of the mastic tree; but the strawberry tree, together with the mastic tree, resists much more effectively and for a long time to the bite, thanks to its ability to defend itself and renew itself with the emission of vigorous suckers. In the Mediterranean scrub repeatedly subjected to the combined effect of grazing, fires and debtors, species such as the strawberry tree and the mastic tree often end up prevailing over those species that, like the holm oak or the Phillyrea, are less resistant to one or the other, or to both of these anthropogenic agents. Beyond a certain limit, even the species relatively more resistant to grazing must give way in the face of the competition of others that, to the characteristic of pyrophytism, accompany that of a full immunity from the bite of livestock. When grazing and fires are repeated with increasing frequency and intensity, the cysts, the helichrysum and the other species more resistant to the combined effect of these two factors still manage to renew themselves, but the scrub can no longer reconstitute itself, and therefore non-shrub, but bushy-shrub, or even herbaceous plant formations are established.[3] These, unlike what happened for the scrub, are open formations in which, that is, the plant cover no longer appears continuous and relatively uniform, but is characterised by discontinuity, with stretches of different vegetation or bare soil. These formations are called garrigues, steppes or Mediterranean prairies, and are more widely spread in most of the pastures that, in southern Italy and the islands, from the coasts and plains develop on the mountain slopes and on the more arid and sunny plateaus, at fairly high altitudes. Among the various types of scrub, the shrubs of *Erica scoparia* L. and *Erica arborea* L. are among the most interesting to explore to understand their role and their usefulness. These are shrubs that can reach between 3 and 6 m, widespread in the central and southern regions also beyond 1000 m of altitude, even if the areas of some importance for the uses and the quality of the products have been Sicily (Peloritani Mountains) both on the Tyrrhenian and Ionian side, Calabria (lower part of Aspromonte, Tyrrhenian side), Campania (Cilento), Tuscany (Maremma and Valdichiana) and Sardinia. Their extent has grown over the centuries only to significantly decrease in recent decades; this decrease is due to, in addition to the ceased utility in farms, the reduction of fire practices, the reduction of grazing and the reforestation carried out to improve these formations, considered degrading elements of the ideal forest vegetation. They are "evergreen" species that prefer acidic and siliceous soils; the heather is a xerophilous and "active pyrophyte" species, that is, a very flammable species, but at the same time fire favours its spread (Bernetti 1995). After a fire, numerous new root suckers develop from the stump, which quickly reach the original size of the bush. They are often found on very degraded and arid soils, and have the ability to quickly colonise areas otherwise unsuitable for tall vegetation, spontaneously covering bare hilly lands, protecting them from washing away and from the erosive action of water. In Tuscany, along with *Calluna vulgaris, Avenella flexuosa* and *Cistus salvifolius*, the strawberry tree forms mixed cenoses and grows from the coastal strip, where it represents one of the characteristic elements of the Mediterranean scrub, pushing up to the Apennines in the chestnut areas. From a forestry point of view, it is considered a pest species due to the competition it exerts on the forest, also in competition with the tree species of the overstory, of which it attenuates the growth and limits the settlement, and is indeed often considered a degrading form of the tall forest. In reality, the documentation coming from various Italian regions and in this case from the registers of Tuscan farms shows how cultivation practices, such as "heather cultivation", were widespread and took on the characteristics of a real small forestry industry, producing a particular landscape (Agnoletti 2002). To mention only the uses documented in Tuscany we have a differentiation that concerns both the above-ground and underground part of the plant. Vine poles were produced, averaging 2 m in length, produced with a cutting cycle of about 8–10 years, lasting (about 14–18 years) longer than that of the chestnut; branches to set up the "wood" in the breeding of the silkworm (hence the name of "wood heather" for the tree heather); branches for bundles, to feed bread ovens and brick kilns; charcoal, often used in cellars to accelerate the fermentation of wine; branches used in the construction of drains and sewers in agricultural arrangements, particularly for the "drainage" of vineyards, i.e. bundles placed at the bottom of the pits made to plant the vines; branches to set up the roofs of huts, a material particularly suitable in protecting from rain and external heat and also in preventing the nesting of mice and other parasites of the domestic dwellings; brooms for cleaning houses, roads, etc., an activity that began towards the end of the fourteenth century. From the underground area, the "block" for pipe drafts was obtained, from which were obtained, and are still obtained today, the "little ovens" or "braziers" for heather pipes. The scoparia heather also produces hard and compact root blocks that are used for the production of pipes, called "white heather pipes". The heather brooms for cleaning the streets, stables, squares and large open spaces of factories constituted around 1960–1970 about 65% of the production of brooms marketed in Italy and were mainly sold to municipalities that made extensive use of them for street cleaning. At the beginning of the year cities like Rome, Milan and Naples, each commissioned over 200,000 brooms from manufacturing companies. The prepared brooms were predominantly marketed in the central-northern regions, while in the south, especially in Sicily, purified bundles were delivered, which were subsequently transformed on site. Other consumption channels for brooms were large industries and stables in the Po Valley, while the waste was used to fuel the blast furnaces of the steel industries in Taranto, Bagnoli and Piombino (Marinelli 1972). The same waste, combined with other shrubs derived from the clearing of the undergrowth (broom, juniper, strawberry tree, etc.), was used in kilns for the cooking ("whitening") of lime. 35% of broom production was destined for the foreign market (France, Switzerland, Germany, Africa and the Middle East) where the product was highly appreciated for its qualities.

A landscape characterized by shrub vegetation, even if not directly linked to the Mediterranean climate, is that of the heath. The term "heath"—from "brucus", a term of Celtic origin, i.e. the Italian brugo, or "brug", a local name given to calluna, a shrub plant whose name also derives from a pre-Roman root and more often known as heather—indicates a type of landscape present in the Alpine and Apennine area, but historically widespread in the Po area, particularly in the high plain, especially in the province of Varese, Como and Milan. The heathland prevailed mainly on "ferretti", clayey soils, poor in humus, derived from deep alterations of moraine pebbles and

Pleistocene floods, but which were decisively influenced by agricultural practices. These uncultivated lands were essentially exploited for grazing and for the collection of calluna for use in stables, with a form of coppice management every 4–5 years, as was the case in many other parts of Italy for shrubs. In Lombardy between the eighteenth and nineteenth centuries, attempts were often proposed to modify the heaths by proposing reforestation with other species. The proposals to plant demanding forest species found little application, while reforestation with pines (especially Scots pine) and robinia, a species then widely suggested for the colonisation of poor lands, was somewhat successful and today characterises the landscape of the high plain between Ticino and Adda. However, the effort made by the state administration to convince owners to interrupt the traditional use of these formations is significant, both with the dissemination of educational publications, the establishment of prizes, and with criminal sanctions. However, it does not seem that the implementation of reforestation in the heaths has reached significant extensions. The abandonment of traditional practices, which still considered the heath as the "dowry" of the farm in agricultural contracts until 1910, probably had more important effects on the change of this landscape.

As can be understood, the maquis scattered throughout our peninsula are one of the many historical landscapes that we could define as of "social" origin, not favoured by the official indications of forestry sciences, or natural science, but the product of agricultural and pastoral activities. Compared to the heaths of the plain, those of the Apennines, especially the northern ones, are considered groups of substitution of mesophilic woods on more or less acidic soils. The dominant species are often the calluna, the broom and the heather and probably constitute phases of recolonisation of the prairies in relation to historical systems of summer pastures on poor substrates. In the nineteenth century, calluna was considered a good food to provide to sheep and goats; The Savoyard military topographers referred to areas covered by heaths and ferns with the term "gerbido", indicating plant populations with reduced productivity. Historical ecology research conducted on the Ligurian Apennines suggests a historical management within the framework of multiple systems linked to grazing and fire, confirming a practice that finds parallels at a European level along a north-south gradient from Norway to Italy (Cevasco 2007).

Notes

1. For almost thirty years, controlled fires have resumed in the south of France to preserve pastures, while in Italy we are still at the beginning.
2. Open plant formations, characterised by associations of typically xerothermic and xeromorphic species, with low shrubs and discontinuous bushes, which rarely exceed half a metre in height, and more often take on grey, almost dusty tones.
3. Formations that are also open, but characterized—unlike what happens for the garrigue—by herbaceous plant associations dominated by grasses adapted to dry, arid environments, which often accompany legume species.

References

Adriani M (ed) (1861) Plutarco, Le Vite parallele, vol III. Le Monnier, Firenze

Agnoletti M (2002) L'area di studio di Gargonza. In: Agnoletti M (ed) Il paesaggio agro-forestale toscano. Strumenti per l'analisi, la gestione e la conservazione. ARSIA-Regione Toscana, Firenze

Aragosti A, Barchiesi A (1984) Gaio Plinio Secondo. Storia Naturale. Calendario dei lavori agricoli. Torino, Einaudi

Barker GWW (1975) Prehistoric territories and economy in central Italy. In: Higgs ES (eds) Palaeoeconomy. Cambridge University Press, Cambridge

Beccu E (2000) Tra cronaca e storia le vicende del patrimonio boschivo della Sardegna. Carlo Delfino Editore, Sassari

Bernetti G (1995) Selvicoltura speciale. Utet, Torino

Bonaccioli MA (ed) (1562) La prima parte della Geografia di Strabone, di greco tradotta in volgare italiano. Francesco Senese, Venezia

Borgioli E (1946) Aspetti e problemi generali dell'alpicoltura e della industria zootecnia nella montagna alpina italiana. L'italia Forestale e Montana 1(4):138–150

Calzecchi Onesti R (ed) (1948) Columella, De re rustica, IV. Ramo editoriale degli Agricoltori, Roma

Canali L, Lelli E (eds) (2000) Marco Porcio Catone. De agri cultura, II. Mondadori, Milano

Carena C (1997) L'arte dell'agricoltura di Lucio Columella. Einaudi, Torino

Cevasco R (2007) Memoria verde. Nuovi spazi per la geografia. Diabasis, Reggio Emilia

Dati G (ed) (1582) Gli Annali di Cornelio Tacito, cavalier romano de' fatti, e guerre de' Romani, vol II. Bernardo Giunti e Fratelli, Venezia

De Seta C (1982) Presentazione. In: Romano R, Vivanti C (eds) Storia d'Italia. Il paesaggio, Annali 5. Einaudi, Torino

Fleckinger A, Steiner H (2000) L'uomo venuto dal ghiaccio. Museo Archeologico dell'Alto Adige, Folio Editore, Bolzano

Gemignani CA (ed) (2017) Emilio Sereni. L'origine dei paesaggi della Grande Liguria. Due inediti dei primi anni Cinquanta. Edizioni Istituto Alcide Cervi, Gattatico

Gianotti GF, Corcella A, Labriola I, Orsi DP (eds) (1986) Diodoro Siculo. Biblioteca Storica. Sellerio, Palermo

Godefroy R (1940) La nature alpine. Exposé de géographie physique. Revue de Géographie Alpine 28(3):459–461

Guidetti M (1998) Vicende politiche, forme dell'insediamento, organizzazione del paesaggio. In: Benetti D, Langé S (eds) Il paesaggio lombardo. Identità, conservazione e sviluppo. Cooperativa Editoriale Quaderni Valtellinesi, Sondrio

Lachmann K (ed) (1826) Der Nibelunge not mit der klage. Reimer, Berlino

Lamboglia N (1939) Liguria romana: studi storico-topografici, vol I. Istituto nazionale di Studi Romani, Roma

Machanei D, Vineti E, Schotti A, Gruteri J (1738) SextiAurelii Victoris Historia Romana, cum notis integris. Amstelodami, Janssonio-Waesbergios

Marchetti A (1813) Tito Lucrezio Caro. Della natura delle cose, V. Società tipografica de' classici italiani, Milano

Marinelli A (1972) L'economia dei cedui di erica scoparia nel Valdarno Superiore. L'Italia Forestale e Montana, Accademia Italiana di Scienze Forestali, Firenze

Martegani V, Agnoletti M (2003) Il paesaggio forestale lombardo. Immagini e storia. Cierre edizioni, Verona

Potter TW (1985) Storia del paesaggio dell'Etruria meridionale. La Nuova Italia Scientifica, Roma

Sereni E (1961) Storia del paesaggio agrario italiano. Laterza, Bari

Sereni E (1981) Terra nuova e buoi rossi. E altri saggi per una storia dell'agricoltura europea. Einaudi, Torino

Chapter 3
"Silvae", "Saltus" and "Lucus"

Abstract The relationship between human activity and the forest seemed to coincide, in Greek and Roman civilisation, mainly with the exploitation of its forage resources for livestock feeding, but we can say without mistake that the Italian forest landscape, from the Roman period and until the second post-war period, was largely dominated by wild grazing, as we will demonstrate in the following chapters. The terms pasture and the terms forest and pasture were indeed used synonymously, at least until the twelfth century, when referring to pasture, it referred to a forest granting grazing rights. The term wood, according to Di Bérenger, derives from the Greek equivalent βοσκω (wood) which also meant pasture (Di Bérenger in Archeologia forestale, ossia, Dell'antica storia e giurisprudenza forestale in Italia. Longo, Treviso, 1863); the terminology used in the Roman period referred instead to the words silva, nemus and lucus. As we will see, the term forest is a Germanic origin word (Forst), indicating a high forests, often enclosed, but also forests with different characteristics from those present in the Mediterranean area. For Italy, it is probably more correct to speak of a wood. In roman times the nemus referred probably to a high forest with a low density, suitable for grazing. The lucus was instead a sacred forest dedicated to a pagan god, and was entrusted to the custody of a priest. *Silva* was instead a term that indicated any wooded place, but was further distinguished into *caedua* (coppice) and *non caedua, silva palaris* (wood for producing poles) and *silva fructifera (wood for producing nuts),* the latter also called *silva pascuaria.* The Latin authors give us news of a few wooded areas present in the peninsula that were used both for the production of timber, were intended for religious purposes: they probably represent remnants of ancient forests still present in the country in Roman times. Some are described for their large extension, but also smaller woods are mentioned by the authors.

M. Agnoletti, *History of the Italian Forest Landscape*, Environmental History 19,
https://doi.org/10.1007/978-3-031-81631-4_3

3.1 The Cultivation of the Forest

Understanding the different forms that forests take due to human activity, and the same forms of management implemented by the Romans, requires a brief illustration of some fundamental characteristics related to plant physiology that have allowed humans to use the wood produced by trees without destroying the forests. Indeed, we would have no forest not only in Italy, but all over the world, if cutting down a tree meant the perpetual elimination of the forest.

In the case of felling a forest, the regeneration of the new forest can occur from plants born from seed, therefore by natural dissemination, or by artificial sowing, and in this case the management system of the forest is called "high forest", in which the plants grow and are then cut when they have reached the technical maturity for the different uses of timber. In the case where the regeneration occurs instead through the buds that naturally develop on the stump of the cut trees, that is the part of the trunk remaining near the ground after the cut, the management system is called coppice (low forests). From this it can be deduced that, in the case of forests managed as high forest, the renewal can be natural or artificial, meaning in the first case that it is entrusted to the seeds that the tree periodically releases and that the wind carries and deposits on the ground, while in the second new seedlings are planted to replace those cut. In the case of the coppice forest, however, the renewal always occurs naturally from the stump.

The coppice lends itself to multiple variants, essentially related to the "simple coppice", when all the plants of a forest are cut leaving only the stumps near the ground. We speak instead of a coppice with standards when, together with the cutting of the plants, in a given plot, some trees are released that will replace the old stumps, since over time the stumps lose the ability to regenerate new shoots. Therefore, while from a seed only one tree can develop, from a stump numerous shoots can develop. This explains how the coppice system has been the most widespread form of forest management in Italy, as it is capable of giving many different types of products (poles for vineyards, for power lines, brushwood for livestock, beams for roofs, wicker, common use objects, firewood and many other uses) compared to a high forest, generally used for the production of construction timber, carpentry, or for the production of acorns for the wild rearing of livestock. This ability to adapt the coppice forest to different needs also explains the many forms and names of the various ways of making cuts, as in the case of pollard system, to sgamollo (the cutting of the branches along the stem), to selective coppice, creating a great variety of landscapes linked to this type of management. Another fundamental aspect of coppicing concerns the fact that while seed renewal can be applied to all tree species, coppicing can only occur for some species of broadleaves as conifers, with the exception of larch in Italy, do not have the ability to emit shoots after a cut. This also explains the greater distribution of this type of forest in the Apennine mountains, at lower altitudes in the Alps and, more generally, where deciduous trees are more widespread.

From the above, it is understood that cutting a forest technically does not mean destroying it. For this to happen, it is necessary to uproot from the ground the stump left after cutting, including the roots, in the case of a coppice, or to prevent regeneration from seed in the case of a high forest. In this type of forest, in fact, the plants left after a cut, even in distant areas, fruiting and producing seeds that are carried by the wind, are able to generate a new forest. This is the mechanism by which a cultivated field or a pasture, even if abandoned, after a certain number of years are again colonised by the trees and shrubs, reproducing forests even where they had been eliminated. In this sense, the continuous cultivation of the forest produces over the centuries a landscape essentially of anthropic origin as the mechanisms of renewal are managed by humans (Figs. 3.1 and 3.2).

The rebirth of a wood on an abandoned field also generates forests that cannot be defined as "natural" or "primary", but secondary forests, since the new tree vegetation that develops on an area previously used for other purposes always suffers from the effects induced by anthropic activities carried out previously. The rules for coppicing and more generally for the cultivation of forests have remained almost unchanged for millennia, only some cutting tools have changed, such as the chainsaw compared to the hand saw or the axe: this has allowed to maintain or restore forest landscapes very similar to those present in ancient times.

3.2 The Woods in the Roman Period

The relationship between human activity and the forest seemed to coincide, in Greek and Roman civilisation, mainly with the exploitation of its forage resources for livestock feeding, but we can say without mistake that the Italian forest landscape, from the Roman period and until the second post-war period, was largely dominated by wild grazing, as we will demonstrate in the following chapters. The terms pasture and the terms forest and pasture were indeed used synonymously, at least until the twelfth century, when referring to pasture, it referred to a forest granting grazing rights. The term wood, according to Di Bérenger, derives from the Greek equivalent *βοσκω* (wood) which also meant pasture (Di Bérenger 1863); the terminology used in the Roman period referred instead to the words *silva, nemus* and *lucus.* As we will see, the term forest is a Germanic origin word (Forst), indicating a high forests, often enclosed, but also forests with different characteristics from those present in the Mediterranean area. For Italy, it is probably more correct to speak of a wood, taking the Greek matrix as a more suitable cultural reference to describe a variety of forms and types much more varied than those present in Northern Europe, even if nowadays common use employs the two terms practically as synonyms.

The *nemus,* the *alsos* for the Greeks, referred probably to a high forest with a low density, suitable for grazing (Gabbrielli 2002). The *lucus* was instead a sacred forest dedicated to a pagan god, and was entrusted to the custody of a priest, who, for the part that was not needed for religious ceremonies, could dispose of it by giving it in emphyteusis or renting it for five years. The death penalty was imposed

Fig. 3.1 Typical structure of a coppiced tree. From the stump resulting from the cutting of a high tree, numerous "suckers" branch off, which will be cut when they have reached the desired sizes for the various purposes for which their wood is intended. The cutting of suckers on the same stump can be repeated for a few decades as new suckers naturally develop from the cut stump (photo by Mauro Agnoletti)

for sacrilegious acts and for cutting trees located in the enclosures of public reason sanctuaries. A decisive blow to the sacred character of the forests was given by the Catholic religion, and in particular by Emperor Theodosius, who in the fourth century made Christianity the only and obligatory religion, but also imposed the cutting of sacred forests and banned religious ceremonies around trees (Di Bérenger 1863; Trifone 1957).

Silva was instead a term that indicated any wooded place, but was further distinguished into *caedua* (coppice) and *non caedua, silva palaris* (wood for producing

Fig. 3.2 The typical structure of a high forest, in this case it is a beech forest. The evenaged forest is characterised by plants of the same age (photo by Mauro Agnoletti)

poles) and *silva fructifera (wood for producing nuts),* the latter also called *silva pascuaria.* The use of these terms seems slightly different compared to the current meaning of this form of forest management. The distinction between *silva caedua* and *non caedua* was based on the different type of product that could be obtained and on the different cultural care. Some jurists, like Servius, considered only the forests that naturally reproduced, whatever the treatment used, therefore in the category *silva caedua* could also include the timber forests. The other definition, that is "non caedua", to which Di Bérenger also adheres, took into account exclusively the woody production that was obtained from the forest and that disregarded the mode of renewal. For this reason, in the caedua category could also include reforestation by sowing or, according to Columella's expression, *ope humana* or *manu nostra* (Trifone 1957).

In the management of a silva caedua, a rotation of cuts over time was used, suitable for the particular nature of the woody species present. and to the respective maturity of the forest in commercial terms (Di Bérenger 1863). The coppice forest from which firewood was obtained was considered a "natural" forest, that is, uncultivated, because, unlike the forest used for producing poles, it did not undergo any intervention before the expiry of the cutting period. The coppice was generally cut down every five or six years, and this gave rise to small-sized wood assortments (few cm in diameter) very useful for agricultural activities. The pollarded coppices generally had a biennial or triennial cutting cycle, those of willow quadrennial, the chestnut coppices seven-yearly, while a nine-year cycle was used for the oak groves. It is not clear that the Romans knew the selective coppice, that is, the one characterised by shoots of different ages present on the same stumps, which is first mentioned in the *Liber ruralium commodorum* by Pietro de' Crescenzi published in the year 1305 (Gabbrielli 2002). Among the species most frequently subjected to simple coppice treatment was the chestnut, if only firewood was obtained, but also the plane tree, the hazel and the alder, which was planted in wet places to protect the banks of rivers from the force of the waters. A particular importance was given to the willow coppices, used mainly for the production of wickers, not only for tying the vines, but also for the creation of wicker containers for the transport of various types of goods.

The pole forest was considered non-coppice because, unlike that, it was subjected to periodic cleanings that embraced two operations: one called uprooting, or clearing (*sti—patio*), the other called pruning (*supputatio*) (Di Bérenger 1863), these operations recurred every three or four years, until the expiry of the turn planned for the cut. The *stipatio* consisted in cleaning the ground from the shrubs of the undergrowth, the *supputatio* in the thinning of the shoots on the stump to improve and stimulate the growth of those released until the end of the turn. The pole forest, as can be seen from the Pandects of Justinian of the VI century regarding the right of usufruct, was governed by a *compound coppice* (Pothier 1836), the same laws applied to the grazed forests (Di Bérenger 1863) always considered non-coppice; in both cases the usufructuary could obtain assortments of various sizes (Pothier 1836). Regarding the relationship with agricultural activities, Columella observes that with one iugerum of chestnut coppice (2500 m^2, that is a quarter of a hectare) you could produce poles for at least 20 iugeri of vineyard (5 ha), while one iugerum of wicker was enough to make the ties for 25 iugeri of vineyard.

Among the various types of forest, a fundamental role was played by the *silva fructifera,* that is, a high forest, of beeches, of chestnuts, but especially of oaks, intended for the wild grazing of pigs that feed on acorns (Aragosti and Barchiesi 1984). This type of forest in the Pandects of Justinian is called *silva pascua, or silva pascuaria.* The term *silva circumsepta* instead indicated a forest enclosed by a dry stone wall, or a palisade, in which hunting animals were kept. The surrounded forest of the Romans was therefore a kind of hunting reserve, which the Greeks called Μουαγρία (Wagner 1787), the Germans forest, from *forestel,* from which the Italian word forest, which in Anglo-Saxon laws denotes bar, or barrier (Schmid 1858). To this corresponds also the *fachia* of the Celts, the *warenna,* or *garenna* (from the German *Waren,* that is to defend), of the Franks, the *merica* of the Slavs, the *gas, gaium, gazium* (Muratori 1739) (from the German *Gau,* countryside, domain), *or guiza, or viza* (from the Latin *vetitum)* of the Lombards, a term used in the Upper Piave Valley, called by the Tuscans defence, or ban, meaning anyway a forest whose use was subservient to specific functions (Cianciani et al. 1781). Given that the denomination *silva circumsepta* is generic, it probably had to include, in a broad sense, every other enclosed forest, such as the pomario (orchard, or fruit garden) and perhaps also the barbaric forest, populated by trees of mixed species, but especially fruit (Carena 1948) (Fig. 3.3).

As for the tall forests, the cuts were made after the moment when the plant was able to bear fruit, therefore in mature age as observed by Pliny. For beeches not before 60–80 years, towards 60 in firs, from 50 to 60 in pines, 40 for larches, between 30 and 40 for hornbeams, ashes, elms and other broadleaves. About the season of cutting, the ancient authors agree on the need to carry it out in autumn or in winter. Cato the Censor also indicated the best periods for some species: for the oak the winter solstice, for the elm and the willow the period of leaf loss. Varro, on the other hand, notes that for the cutting of the willow the best period was that between the spring equinox, 21 March, and the pleiades, 22 April. Regarding the lunar phases, it seems that the authors agree on the need to cut during the waning moon for construction timber and during the waxing moon for firewood in coppices. The tradition of performing cuts in relation to the lunar phases has been maintained almost until our days.

3.3 The Primeval Forests

The Latin authors give us news of a few wooded areas present in the peninsula that were used both for the production of timber, were intended for religious purposes: they probably represent remnants of ancient forests still present in the country in Roman times. Some are described for their large extension, but also smaller woods are mentioned by the authors. We report them without the possibility of comments given the scarcity of sources, considering such information useful for the compilation of a kind of register, naturally partial, of the woods mentioned in those centuries,

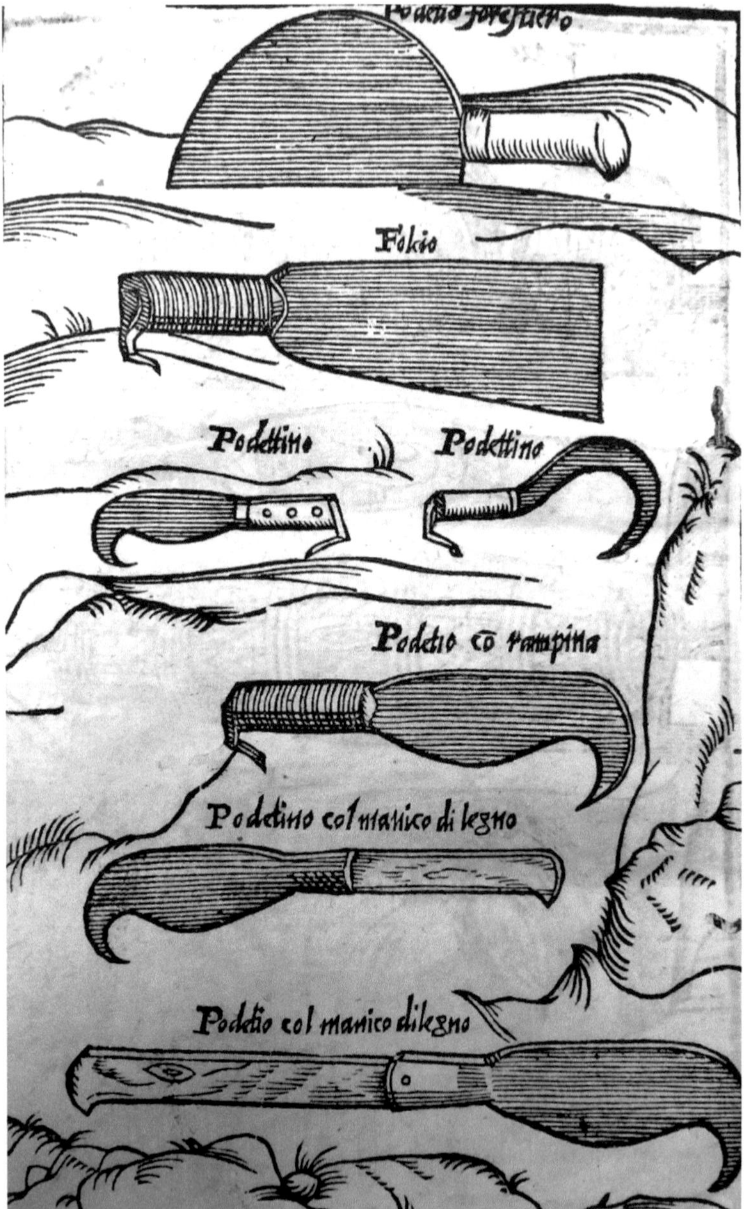

Fig. 3.3 Billhooks used for cutting tree branches and for pruning in the Roman period (Agostino Gallo, The twenty days of agriculture and the pleasures of the villa, Venice 1565)

often reaching up to our days. It can be observed that the scarcity of large forests definable as "virgin" of "pristine" in the first centuries after Christ is probably to be connected to the extensive agrarianisation of the territory of the peninsula, a hypothesis that emerges from the surveys carried out in various regions of Italy that would confirm the very high density of rural settlements during the Roman period. The river valleys of Puglia and Abruzzo, the coastal plains along the Tyrrhenian Sea and the hilly Etruscan countryside seem to have hosted a very large rural population that brought large extensions of marginal lands into cultivation (Potter 1985).

The Sila, around Cosenza, was 130 km long and covered the "Mount of Pitch" for an extension of about 90 by 100 km. It was the ancient territory of the Bruttii, who, in spontaneously submitting to the Romans, ceded half of it to them, from which they derived substantial rents every year, having leased it. It was a territory rich in woods suitable for civil and naval constructions, with an abundance of firs, poplars, pines—i.e. pines for the production of resin, hence the name Mount of Pitch-, beeches and ashes. The trees of the woods near the sea and the rivers were cut, transported to the nearest ports and sent to various parts of Italy; those located far from the sea and the rivers, on the other hand, were made into smaller pieces and often carried on men's shoulders. The bigger and resin-rich conifer trees were used to produce pitch: it was indeed known that the pitch called *bruzia* was the most sought after.

Strabo and Pliny the Elder remember the importance of the production of timber from the woods scattered throughout the peninsula that were floated down to Rome, particularly along the Anio (now Aniene)—which from the territory of Alba, a Latin city, passed through that of the Marsi to flow into the Tiber -, the Nera and the Tenea, rivers that through Umbria flowed into the Tiber (Bonaccioli 1562). The timber cut in the woods of the Apennines was transported, loose or tied on rafts on the Tinia (today Topino) and on the Chiana, rivers not suitable to float and even less to move large timber, therefore artificial floods were procured by river locks to increase the water level and facilitate transport (Aragosti and Barchiesi 1982). Other timbers, especially the larch cut in the Rezia Alps (the eastern part of the Alpine arc), already very precious at that time, were tied on rafts and were floated up to Ravenna (Gros 1997), then loaded onto ships and transported by sea to Ostia circumnavigating Italy and from there via the Tiber river to Rome.

To facilitate also the land transport from the woods to the river loaders and from the ports on the sea to the cities, roads were often built, partly paved (Bonaccioli 1562). Carts loaded with large firs and pines travelled in long convoys along the streets of Rome, making the buildings tremble as they passed (Manfrè 1702; Giordani 1839), so much so that to avoid disorder carters were forbidden from entering or crossing the city during the day (Bruns 1860). To recall the amount of timber brought to Rome, Pliny tells us that among that ordered by Emperor Tiberius for the construction of the Naumachiario bridge there was a larch beam brought from the Raetian Alps that was one hundred and twenty feet long (m 35.46), and a uniform thickness of two feet on each side (m 0.59) to which corresponded a weight of about nine tonnes. The beam remained on display to the public until it was used in the construction of Nero's amphitheatre. In antiquity, however, exceptional measurements of trees were not rare: the mainmast of the Egyptian ship on which Caligula had the famous obelisk brought to Rome, later moved to St. Peter's Square in 1586 by Sixtus V, was twenty-four feet in circumference. Another cedar tree from Demetrius's ship (third century BC), cut in the mountains of Syria, was one hundred and thirty feet long and had a circumference of about eighteen (Aragosti and Barchiesi 1984).

From these brief notes, it is understood that in these centuries few forests in Italy were relatively untouched by man. According to Latin authors, the forests that we would today call "virgin", meaning by this term woods not deeply modified by human activities, are already quite reduced in number around the first century AD. The Ciminia forest on Mount Fogliano, near Viterbo in the heart of Etruria, sacred to the assemblies of the twelve Etruscan peoples and admired for the monumentality of its holm oaks, was so wild that it was considered more impenetrable than the forests of Britain and Germany. When the consul Fabius proposed the passage of the Roman army through this forest, the Senate warned him not to face such danger as the place was famous as "miraculous" and dominated by mysterious and invisible powers (Gronovi and Sigoni 1740; Aragosti and Barchiesi 1984; Di Ligni and Bettoni 1823). The Litana forest near Bologna extended from the Apennines towards the Po up to the mouth of the Senio, "then called the forest of Lugo, or Padanea forest, from *pados,* which in Gallic means fir" (Di Bérenger 1863; Ginanni 1774) and perhaps passed through Faenza, where Lucius Postumus, falling into the ambush set by the Gauls, was defeated by Hannibal (Gronovi and Sigoni 1740). In the Gulf of Cumae the Gallinaria forest—whose trees served for the construction of the ships of Misenum, where the imperial fleet created by Augustus was stationed—was a refuge for bands of brigands and occupied the beach that ran from Cape Miseno to the Volturno, bordering the dense Aorna forest later cut down by order of Agrippa. In the *hiding places* of this forest, ancient mines dug by the Cimmerian peoples, lived the priests of the goddesses Hecate Cumana and Proserpina, famous because it was believed that they evoked the souls of the dead (Manutio 1608). The Anaizia forest, called in the nineteenth century the Albi forest, on Lake Fucino in Marsica, was composed of gigantic oaks and famous for the priests who were said to know how to tame poisonous snakes (Caro 1826). After the Sila, the Bruttii ceded to Rome also the willow groves of Minturno on the Garigliano (Zumpt 1861), the Scanzia forest in Campania (Zumpt 1861) and the Carmianensi forests in Calabria (Labbe 1651) which under Vespasian were surveyed, measured and confined (Blume et al 1848). This also happened due to the disputes that levellers and colonists moved to the State, and indeed they were subsequently placed under the protection of a procurator (Blume et al 1848).

The authors also remember the Diomedea forest on the Timavo river, especially for the beauty of its plane trees and for two nearby groves, one consecrated to Argive Juno, the other to Aetolian Diana (Bonaccioli 1562). The Lupanica forest was instead described as a refuge for packs of wolves, hence perhaps the name of lycophoros that the Veneti gave to their horses. It extended from the Isonzo to the Livenza rivers (Bonaccioli 1562; Filiasi 1781) bordering in this area with the Fetontea forest, the "Silva magna" (Magenta 1842; Filiasi 1781; Tentori 1792), rich in oaks and pines, thinned by the Etruscans and the Celts and perhaps destroyed by the Paphlagonians and the Euganei, when they arrived from the Adriatic Sea. Remnants of the Fetontea forest were the Torcellis, the Clocisca, the Lauretana, the Torunda, the Morgana, the Cavolana and the Montello forest, which covered the hills located between the ancient Postumia road (Azzoni 1840) and the Silis, or the Piave (Meneguzzi 1850), and would later become one of the most important forests for the naval constructions of Venice. In addition to the oak forests, distributed from the mouths of the Tagliamento to those of the Po, various pine forests extended in these areas that provided the timber for the construction of Venetian and Ravenna ships (Filiasi 1781; Mastrofini 1830; Casabuoi et al. 1671); a part of them, expanded thanks to the reclamation works, became today's Pineta of Ravenna (Buzio 1765; Ginanni 1774). The Lugana forest, rich in wild boars, which seems to have extended near Artelia, perhaps today's Peschiera, recalled the memory of the defeat of the barbarians who descended in 268 AD to conquer Veneto and was donated in 878 by Charlemagne to the monasteries of San Zenone and Desenzano (Rosa 1857). The marshy thicket of Modena composed of oaks, a residue of the Padana plain forest, was known for the defeat that the Roman colonies of Piacenza and Cremona had suffered against the Gauls before the arrival of Hannibal (Kohen 1824). In Etruria the Mesia wood was known because of Anco Marzio who took it from the Veienti (Gronovi and Sigoni 1740) and the Falena forest, near Civita Castellana, sacred to Juno Quirita (Oehler 1849; Henzen 1828; Orelli 1828; Mueller 1861). On the borders of Lazio, the Arsia and the Cere were both consecrated to the god Siivano. The first was very well known for the victory achieved by Valerio Publicola to the detriment of the Tuscans, who are said to have been frightened by the voice of the "woodland deity" (Kempfius 1854); the second, composed of large firs, was famous instead for having been the scene of the first festivals celebrated by the ancient people of the Pelasgians (Caro 1826).

In various parts of Italy, forests were then remembered dedicated to Feronia, an Italic goddess protector of forests and harvests especially venerated by the Sabines, the Umbrians and the Tusci, who the Greeks pared to their Persephone (Preller 1858). The most famous among these forests rose at the foot of Mount Soratte, near Capena near Rome, where there was a temple plundered by Hannibal, precisely in the place where the village of Sant'Oreste stood (Gronovi and Sigoni 1740; Bonaccioli 1562; Buzio 1765; Aragosti and Barchiesi 1984). A second forest *lucus Feroniae*, a remnant of the ancient Ciminia forest, He was in Petra sancta, today's Pietrasanta, in Tuscany (Blume et al 1848; Orelli 1828; Grutero 1707). Piccioli writes that in reality the forest extended from the territories of Volsinio (Bolsena) to the mountains of Viterbo, and in Livio's time it was renowned for its gigantic oaks and considered so large and wild as to be considered more inaccessible than the forests of Britain and Germany, an interesting parallel to understand what the differences between Italian woods and those of Northern Europe were according to Latin authors (Piccioli 1923). A third was located near Septempeda (San Severino in the Marche) (Blume et al 1848), a fourth on the Volsci coast near Terracina, bathed by a very clear spring (Bindi 1759; Aragosti and Barchiesi 1984), a fifth in Campania (Gronovi and Sigoni 1740), a sixth in Sabina, at the foot of Mount Albano, near Trebula Mutuesca (Monteleone Sabino), renowned as the place of the Latin diets, where the fairs and markets of Etruria and Lazio were also held (Mastrofini 1824; Gronovi and Sigoni 1740; Fabretti 1702). The woods in Amiterno, today's San Vittorino north of L'Aquila, in Aveia, a town of the

Fig. 3.4 The forest of the Sila

Vestini, near Fossa (L'Aquila), in Pisaura (Pesaro), in the district of Florence and in that of Verona (Mommsen 1850; Henzen 1828; Orelli 1828) were also dedicated to the goddess Feronia. According to Piccioli, who reports the writings of Theocritus, Sicily was rich in oak forests, among which those surrounding Enna were renowned, from which timber was extracted for the fleets but which were largely dedicated to sheep grazing (Fig. 3.4).

Many other woods known to the authors outlined the territorial boundaries or flanked the public roads, bordered the banks of rivers such as the Adda, the Mincio, the Pisauro (Isauro), the Tiber, the Tanagro, the Sele and all the other rivers of Italy protected by the hydraulic laws of the Romans (Sirmondi 1652; Biancoli 1768). The waters that sprang from the sacred spring of Clitunno, surrounded by a sacred grove of cypresses, flowed from Spoleto to Bevagna through a basin shaded by ash and poplar woods (Lemaire 1823), while at the sources of the Tiber there were forests consecrated to the god Tebro (The elegies of Propertius and Tibullus 1828). One of the oak forests along the Garigliano (the Clanis, Glanis or Laris) in Campania, near Minturno, was consecrated to the nymph Marica of Lazio (the Circe of the Greeks) and to Fatuello, or Faunus (Gronovi and Sigoni 1740). This is a woodland deity whose oracles were listened to in a valley of the Tiburtina forest (Caro 1826; Cartari 1745) through which the river Anio (Bonaccioli 1562), now Amene near Tivoli, not far from the house of Horace and from the Albuneo grove, consecrated to the tenth Sibyl (Fabricius 1786). Along the Appian Way were the sacred woods of the goddess Egeria and the Muses (Sermonti 2014) and the Aricina forest, which took its name from the rex nemorensis attributed to the head of a college of priests of the demon

Virbio, believed to be Hippolytus (Grutero 1707), and of Artemis Tauropolos. On the Salaria road stood the grove of the goddess Laverna (Leto or Letona), protector of thieves, a famous forest in which the Romans fled after the defeat suffered on the Allia (Mueller 1839; Janus 1852). This is a site just eleven miles from the city of Rome where the Allia river, descending from the Crustumini Mountains in a very deep gorge, flows into the Tiber just below the Salaria road (Perelli 1953). The Allia probably corresponds to the current "Fosso Maestro", a small left tributary of the Tiber. Towards the mouth of the Tiber, therefore not far from Ostia, there was the Stimula, or the grove of Semele, known for obscene bacchanals (Gronovi and Sigoni 1740), abolished by the decree "Senatus auctoritas de Bacchanalibus" of the year 186 BC and also mentioned by Cicero (Moser 1824). Along the Campana road, to the right of the Tiber and five miles from Rome, there was instead the grove of the goddess Dia and on the Nomentana road, outside the Canicula gate, there was the grove of the goddess Robiginia, in which the robigalia were celebrated (Mueller 1839), while on the Laurentina road, six miles from Rome, there was the grove of the god Terminus where the annual Terminalia festival was held on 23 February (Cartari 1745).

Other types of woods, distinguished by the variety of species present, surrounded the sanctuaries, tombs and springs of the rivers. On the Palatine Hill, around the year 201 BC, the first pine forest was planted to establish the cult of the goddess Cybele. The cypress was dedicated to Pluto and its spread in the Roman period is also linked to the art of topiary, the technique of pruning trees to give them a geometric shape different from the natural one for ornamental purposes. The laurel was dedicated to Apollo, the lime to Venus, while the yew was dedicated to the Furies. Among the woods known for the various species, we remember, in Calabria, the grove of Hera (Juno Lacinia) near Crotone, with pastures shaded by pines (Gronovi and Sigoni 1740; Bonaccioli 1562); in Bruttium, the grove of Temesa, composed of wild olive trees, sacred to Polites, companion of Ulysses (Bonaccioli 1562); in the Terra di Lavoro, on the slopes of Mount Tifata, the grove of Diana Tifatina (Riguez 1675), measured and surveyed by Roman surveyors by order of Emperor Vespasian (Orelli 1828), whose name seems to be due to the holm oaks (in fact tifata in the ancient language of the Osci means "holm oak"). We still remember the grove of Juno Noceria (Nocera dei Pagani) (Aragosti and Barchiesi 1984), in Sabina, and two forests of the goddess Vacuna, patroness of the agro reatino and goddess of Victory, one near Rieti on Lake Velino, centre of the Sabine diets and place of festivals and annual markets (Aragosti and Barchiesi 1984), the second on a hill in the Valle Digenzia (now Licenza), near a farm of Horace, with a temple rebuilt by Vespasian (Martin 1870).

The Maliziosa was instead a forest passed down in history for the first victory won by the Romans against the Sabines (Mastrofini 1824). In Lazio, the grove of Juno Sospita between Anzio and Lanuvio was also remembered, surrounded by houses inhabited by priests, who in addition to a temple mainly guarded a cave containing a prophetic snake, to which every year a virgin with blindfolded eyes had to sacrifice a cake, as it was popular belief that the year would be lucky if the snake had eaten it. The same grove, then, served as a common asylum for Roman and

Lanuvian citizens (Gronovi and Sigoni 1740). In these surroundings, that is, in the Roman countryside, were located the grove of Diana on the crossroads of the road to Anagna, capital of the Hernici (Gronovi and Sigoni 1740), the Laureto populated by wild boars, near Laurentum (Sermonti 2014), the Jupiter Indigetes' forest, believed by Aeneas, towards Lavinium on the bank of the Numicius (Aragosti and Barchiesi 1984). Then there was Ferentino, today's Marino scrubland, with its sacred spring, called caput Ferentinae, once dedicated to the assembly of the Latin diets (Gronovi and Sigoni 1740), at the foot of Mount Tuscolo with an ancient beech forest, named Come di Diana (Aragosti and Barchiesi 1984), and the forest of Mount Algido, also consecrated to Diana, out of devotion by the Equi, Hernici and Latins, later converted into a pasture by the will of the Roman pontiffs (Nobili 1784). Also well known was the forest called the Pomonal, between Ardea and Ostia, sacred to the goddess Pomona, the most graceful of the garden goddesses that Ovid says lived in the time of King Proca, cultivator of fruit trees but enemy of forest trees and rivers (Mueller 1839).

3.4 The Landscape of the Saltus

As already mentioned, the terms nemus and saltus express both forest and pasture, but the saltus is a form of landscape that has also played a significant role from the point of view of historical persistence. Until the nineteenth century in Sardinia, the mountains, the acorn forests, the shrubs and all the lands reserved for grazing were still defined as saltus-; a "salto" is still defined as an open wooded pasture, while a "tanca" is a pasture enclosed by a dry stone wall (Di Bérenger 1863). The custom of considering forests and pastures as two inseparable things, both in public and private economy, persists for many centuries up to the second post-war period, and represents a necessary point of view to understand the history of much of the Italian forest landscape. The saltus (Di Bérenger 1863) corresponded, according to the definition of Elio Gallo, to a shapeless landscape "ubi silvae et pastiones sunt", that is of forests and pastures, interrupted only by some small cultivated plot, for the use of shepherds (Sereni 1961). Varro describes it as a land that is not cultivated because it is wooded and suitable for grazing. The term saltus was often used to indicate an entire forest destined for grazing as indicated by Virgil (Biancoli 1768). Livy calls saltus Citheronis the Mount of Cithaeron, saltus Manlianus the Manlian forest, saltus Germanici the Ercinian forest and other forests of Germany (Gronovi and Sigoni 1740).

Since pre-Roman times, the lands that legally belonged to the villages were organised in such a way as to keep separate the agricultural areas and the grazing of domestic livestock—the first called "vidazzone", the second "paberile"—from those called saltus, which were located far from the inhabited centre and grouped the uncultivated lands and the forests reserved for the grazing of livestock and the collection of wood (Beccu 2000). The landscape of the saltus occupied large extensions of land in the Roman period, was linked to the importance of the pastoral economy and the accumulation of enormous wealth that contributed to creating extensive land estates. Pliny mentions a certain Claudio Isidoro who, on his death, left an estate of more than 3600 pairs of oxen and 250,000 heads of small livestock grazing on his properties, most likely a landscape of forests and pastures interrupted by some small cultivated plot that was protected from the bite of livestock, planting trees around it that served to define its boundaries (Di Bérenger 1863).

At the edge of every field, pasture and farm, poplars, oaks, and elms were often used; the latter had the advantage of also being used to marry the vines, for the collection of foliage for sheep and cattle, and for branches for the hearth. The trees placed at the edge of these enclosed fields demonstrated the need to defend against damage caused by livestock, especially goats, that grazed in the free territories. They were kept as tall trees but, being exposed to the danger of being mutilated by grazing animals, as soon as they were firmly planted, they were "pruned", that is, they were surrounded by thorns that were left around the plant until the so-called age at which it would have been immune from the bite of livestock (Di Bérenger 1863). They were then "scalped", that is, all the branches of the plant were cut except for the top, at different times of the year: those of holm oak and laurel in January, those of oak, poplar, ash, fig in autumn before they began to yellow, for livestock feed (Calzecchi Onesti 1948). Regarding the winter feeding of oxen, Columella advised feeding them from the beginning of July until the beginning of November, throughout the summer and autumn, with tree foliage that is not suitable unless it has matured under continuous rains and dews. According to the author, the best foliage is that of elm, then that of ash, that of poplar, and lastly comes the leaf of laurel, oak and holm oak, but fig leaves could also be used if there is an abundance and if it is convenient to defoliate this tree. The last foliage that was collected were "those for bedding or manure" for the animal stables (Di Bérenger 1863).

The saltus were divided into summer and winter, and their possession was considered legitimate even if they were abandoned for some time. As accessories of the saltus, the *saltuari* and the *topiari* were considered, that is, those who were assigned to the custody of the products and the borders, in addition to those who took care of the cultivation and ornamentation of the fund. The complexity of the saltus landscape is also evident in the management of grazing animals, in fact, those animals that remained both in summer and winter were considered relevant to the fund. However, a distinction was made between those that were kept in enclosures from those that roamed free, defining the former as belonging to the saltus, while the latter, being free, did not belong to it; therefore, it was not allowed that the purchase of a saltus automatically also implied the purchase of all the animals present. The servitude of pasture is in fact remembered in Roman law, but it does not seem to have found great application, given the "care to assign to each community and colony a portion of saltus for grazing" (Di Bérenger 1863). This ager publicus was rented annually, subject to the payment of a scriptura (Trifone 1957), that is, a tax for each head of animal that was to be introduced. Often the winter pasture was also exercised in private farms, but only with the consent of the owners and not as a servitude, as happened later in the Middle Ages. In addition, the shepherd had to pay the owner a toll or "fida", while the latter reserved the property of the wood and the pledge of the

Fig. 3.5 The landscape of the *saltus* is still partly present in Italy. Here we are in Salten, in the province of Bolzano, a historical landscape identified by the National Catalogue of Historical Rural Landscapes (Agnoletti 2010). Salten derives from the Latin *saltus,* indicating in this case a larch tree pasture (photo by Mauro Agnoletti)

animals as compensation for any damage caused by the livestock. However, goats were excluded from plantations, from woods that had been recently cut and from olive groves (Fig. 3.5).

For newly planted vineyards, Virgil advised to create hedges and keep animals away, especially when the leaves were young, explaining that more than cold winters and too strong sun, the greatest damage is done by wild bulls and goats, in fact, a young goat was usually sacrificed to Bacchus for this reason (Daniello et al 1734). On the contrary, it seems that pigs could be allowed into the vineyards "as long as they are prepared for it and even after the harvest, they will root out the weeds, and almost do as much good as digging" (De' Crescenzi 1852). After the winter, before May, the animals were gathered in the plains and taken to the mountain forests, generally state-owned or communal, kept as high forest. These were defined as natural if their treatment only involved interventions to preserve them, promote their growth and renewal. The artificial ones were not only those created by sowing or planting, but also those given particular care and attention such as the "fruit-bearing" ones, composed of plants that gave fruit or acorns to fatten pigs, such as oaks. Such was the importance of acorn collection in pasture oak forests that the first meaning of the word *lex* (law) indicated the discipline of acorn collection in these forests (Di Bérenger 1863). The value of this fruit, especially for fattening pigs, was such that it was called *appanagium* or *panagium*, literally "bread producer". Acorns were a food resource for many people, even in peacetime, in case of cereal crop shortage: from dried and ground acorns, in fact, a flour is obtained that is kneaded to make a sort of bread (Di Bérenger).

Acorns were nevertheless appreciated throughout the Mediterranean basin and some varieties are known and renowned even today.[1] Columella recalls that acorns could also be usefully employed for the feeding of other types of livestock; in the case of oxen "*in January... if we want to save grains, twenty bushels (172 kg) of dry leaf will suffice... to these, when the region allows it by producing them in abundance, add acorns; and they should not be given to satiety, because otherwise they cause scabies. In November and December during sowing, mostly a bushel (8.6 kg) of acorns per head and straw to satiety are enough*" (Calzecchi Onesti 1948). The great importance of the saltus is also linked to the fact that the Latin peoples particularly appreciated the meats obtained from the slaughter of pigs fed with acorns and beech nuts. It was said that the beech acorn made pigs thrive, made their meat easy to cook, light and healthy for the stomach; the holm oak instead made the pig lean, shiny and thin, while the oak made it gain weight and thickness, since the acorn is heavier and sweeter. The acorns of downy oak, turkey oak and cork oak instead seemed to make the meat spongy (Rozier 1832).

The *silvae fructiferae* were only cut when they began to bear fruit and this was repeated every five years. The density considered most suitable to facilitate fruit production is indicated by Pietro de' Crescenzi: "chestnut trees should, on each side, be spread apart from each other, at least forty feet (about 12 m), so that the branches can spread and produce a lot of fruit, without hindrance, because they naturally spread out, on the sides a lot. And it is necessary, that for the oaks they have no less distance. The oaks, and the turkey oaks, for twenty feet, or less, must have distance. And these three trees produce acorns, which are excellent food for pigs" (De' Crescenzi 1852). To promote the natural renewal of the woods, pigs were introduced into the clearings to break the surface layers of the soil and stir it up, thus creating better conditions for seed germination, but the same animals were also used for this purpose in the fields after sowing (Di Bérenger 1863).

With the transition from the Republican Rome to the Imperial Rome, the meaning of the term *saltus* seems to change, becoming in practice synonymous with large noble or imperial property (Sereni 1961). The small farming property disappears, whose existence is still testified today in Tuscany by numerous place names of Latin origin, especially those called predial, that is, which have taken the name of the *praedium*, of the owner of the agricultural fund, such as for example: Marciano, Rignano, Volognano, Nipozzano (Gabbrielli 1997). In its place comes the villa, the result of land consolidations cultivated conquered with wars and purchased by the rich Roman patricians. In fact, Roman law required dethroned sovereigns and subdued cities to surrender their own public lands, and often also private ones, to the treasury which resold them to its citizens. Instead, the part made up of woods and pastures was generally not alienated but went to increase the *ager publicus*. In the organisation of these agricultural businesses cultivated with the labour of slaves, livestock farming remains of considerable importance, so much so that Cato in his work *De agri cultura*, written in 53 BC, in order to ensure that the business obtained the maximum profit, advised to reserve a portion to the *glandaria silva,* a high forest of oaks, but also beech and chestnut, suitable for the livestock of the company (Gabbrielli 2002).

With the expansion of Roman domination, a progressive organisation of rural space is realised, in which the order of crops contrasts with the disorder of the "natural" landscape. Thus, a new landscape layout is realised, which represents one of the many characteristics of Roman civilisation, which by expanding agricultural crops simultaneously reduces forest spaces, so much so that the expansion of crops in the early imperial age appears doubled compared to the past in Cisalpine Gaul (northern Italy). In this way, initially only functional needs were met, which later also became aesthetic, contributing to increase the value of the agricultural property, So much so that Horace complained that some villas favoured ornamental plants, such as the plane tree, over the "humble but very useful elm". The order of the agricultural landscape is therefore contrasted by the extension of the *saltus* which occupies significant spaces, in homage to a pastoral economy of fundamental importance that seems to increase in the late empire, not only for the beginning of a slow process of disintegration of the landscape, but also for its further extension to the detriment of the natural landscape. As can be seen from these brief notes, in Roman times the boundary between forest and "non-forest" is already diluted into a more complex mosaic, in which the trees of the pastures, single or in groups, give way to tree plantations, with a careful but widespread use of forest species.

Notes

1. The ballota (from *belutah,* the name of the holm oak in Arabic), which the Greeks called βαλανοξ and the Spanish *ballota* (acorn), *ballotera,* in Tunis was called "bread tree". The famous Spanish ham "bellota", still today produced from pigs raised in broad oak pastures and which feed on acorns, takes its name from this acorn.

References

Agnoletti M (ed) (2010) Historical rural landscapes. For a national catalogue, Laterza, Rome-Bari

Aragosti A, Barchiesi A (1982) Gaio Plinio Secondo. Storia Naturale. vol I, Libro terzo: Europa. Torino, Einaudi

Aragosti A, Barchiesi A (1984) Gaio Plinio Secondo. Storia Naturale. vol III, Libro sedicesimo: Alberi selvatici. Torino, Einaudi

Azzoni A (1840) Considerazioni sopra le prime notizie di Trivigi. Tip. Andreola, Treviso

Beccu E (2000) Tra cronaca e storia le vicende del patrimonio boschivo della Sardegna. Carlo Delfino Editore, Sassari

Biancoli A (ed) (1768) La georgica di P. Virgilio Marone tradotta in verso toscano. Stamperia Amatina, Pesaro

Bindi A (1759) Le satire di Orazio Flacco tradotte in verso italiano da Oresbio Agieo pastore arcade, con varie annotazioni secondo l'edizione bentleyana, Libro I. Siena

Blume F, Lachmann K, Rudorff A (1848) Die Schriften der Römischen Feldmesser. Erster Band, G. Reimer, Berlin

Bonaccioli MA (ed) (1562) La prima parte della Geografia di Strabone, di greco tradotta in volgare italiano, vol V. Francesco Senese, Venezia

Bruns G (1860) Fontes iuris romani antiqui quos in usum praelectionum. Libraria Lauppiana, Tubinga

Buzio M (1765) Raccolta di tutti gli antichi poeti latini colla loro versione nell'italiana favella. Tomo XXXV. Contiene altri sette libri di Cajo Silio Italico della seconda guerra cartaginese, Libro VIII. Stampatore Regio Camerale, Milano

Calzecchi Onesti R (ed) (1948) Columella, De re rustica, VI. Ramo editoriale degli Agricoltori, Roma,

Carena C (ed) (1948) Columella, L'arte dell'agricoltura e Libro sugli alberi. Einaudi, Torino

Caro A (1826) Eneide di Virgilio, vol II, Libro VII. Società tipografica de' classici italiani, Milano

Cartari V (1745) Raccolta di tutti gli antichi poeti latini alla loro versione nell'italiana favella, vol XXIII, Libro II. Regio Ducal Palazzo, Milano

Cianciani P, Coletium S, Pitterium F (1781) Barbarorum Leges Antiquae cum notis et glossariis, vol II. Statutorum civitatis Belluni, Venezia

Daniello B, Lori A, Ibichense P (trans) (1734) Raccolta di tutti gli antichi poeti latini tradotti in versi italiani, tomo VII. Regio Ducal Palazzo, Milano

De' Crescenzi P (1852) Trattato della Agricoltura. Vicentini e Franchini, Verona

Di Bérenger A (1863) Archeologia forestale, ossia, Dell'antica storia e giurisprudenza forestale in Italia. Longo, Treviso

Di Ligni C, Bettoni N (1823) La storia romana di Lucio Anneo Floro. Milano

Fabretti R (1702) Inscriptionum antiquarum. Ex Officina Dominici Antonii Herculis, Roma

Fabricius JA (1786) L. Coelii sive Caecilii Lactantii Firmiani opera omnia, (De divinae institutionibus I, 6. Ex Typographia Societatis, Biponti

Filiasi G (1781) Saggio sopra i Veneti primi, vol I. Pietro Savioni, Venezia

Gabbrielli A (1997) Le trasformazioni del paesaggio forestale in Toscana: un tentativo di sintesi storica. Accademia italiana di scienze forestali, Firenze

Gabbrielli A (2002) Il bosco ceduo nell'azienda agraria dell'antica Roma. In: Ciancio O, Nocentini S (eds) Il bosco ceduo in Italia. Accademia italiana di scienze forestali, Firenze

Ginanni F (1774) Istoria civile e naturale delle pinete ravennati. Generoso Salomoni, Roma

Giordani G (1839) Le satire di Giovenale, vol I, III. Tip. Antonelli, Venezia

Gronovi JF, Sigoni C (1740) Titi Livii patavini historiarum quod exstat. Thurnisios fratres, Basilea

Gros P (ed) (1997) Vitruvio. De Architectura, Libro II. Einaudi, Torino

Grutero I (1707) Corpus Inscriptionum ex recensione et cum adnotationibus Ioannis Georgii Graevii, vol I. Franciscus Halina, Amstelodami

Henzen G (1828) Inscriptionum latinarum selectarum. Amplissima collectio ad illustrandam Romanae Antiquitatis. Typis Orelli, Fuesslini et sociorum, Turici

Janus L (ed) (1852) Macrobii Ambrosii Theodosii. Operae quae supersunt, I. Quedlinburg und Leipzig

Kempfius C (1854) Valerli Maximi Factotum etdictorum memorabilium. G. Reimer, Berolini

Kohen I (1824) Le storie di Polibio da Megalopoli volgarizzate sul testo greco dello Schweighauser e corredate di note, Libro III. Tip De' Fratelli Sonzogno e P. Andrea Molina, Milano

Labbe P (1651) Notitia dignitatum imperii romani. Cum pluribus altis opusculis indicibus ac notis, Sezione XLIII. Typographia regia, Paris

Le elegie di Properzio e di Tibullo recate in verso italiano, II (1828) Bettoni, Milano

Lemaire NE (1823) C. Plinii Caecilii Secundi Epistolarum libri decem et Panegyricus. Pars altera, vol I, Libro VIII. Editionis Schaeferianae, Paris

Magenta P (1842) Gli epigrammi di Marco Valerio Marziale, Libro IV, epigramma XXV. Tip. Antonelli, Venezia

Manfrè G (1702) L. Annaei Senecae. Tomus secundus in quo Epistolae et Quaestiones naturales sunt. Ex Typographia Seminarii. Apud Joannem Manfrè, Padova

Manutio A (1608) Le epistole famigliari di Cicerone, IX, epistola 23. Girolamo Polo, Venezia

Martin T (1870) The odes, epodes and satires of Horace, Epode II, 59. Oxford University, Oxford

Mastrofini M (1824) Le antichità romane di Dionigi di Alicarnasso, vol III, Libro XX. Tipografia dei fratelli Sonzogno, Milano
Meneguzzi G (1850) Del corso antico del Piave. Tip. Alvisopoli, Venezia
Mommsen T (1850) Inscriptiones regni neapolitani latinae, Inscr. 5753, 5983. Georg Wigand's Verlag, Leipzig
Moser GH (1824) M. Tullii Ciceronis De legibus libri tres, Libro II. Typographeo Branneriano, Francofurti ad Moenum
Mueller CO (ed) (1839) Sexti Pompei Festi: De verborum significatione quae supersunt. Libraria Weidmanniana, Leipzig
Mueller L (1861) P. Ovidii Nasonis Carmina amatoria, Amores III, 13. Sumptibus R. Gaertneri, Berolini
Muratori LA (1739) Antiquitates Italicae medii aevii sive dissertationes, vol II. Mediolani, ex typographia Societatis palatinae
Nobili GO (1784) Le odi di Q. Orazio Flacco tradotte in versi toscani di vario metro, I, 21. Gio. Vinc. Falorni, Livorno
Oehler F (1849) Q. Septimii Florentis Tertulliani Apologeticum et Ad nationes libri duo. Halae Saxonum, Anton
Orelli CI (1828) Inscriptionum latinarum selectarum. Amplissima collectio ad illustrandam Romanae Antiquitatis. Fuesslini et sociorum, Turici
Perelli L (1953) Livio, La prima deca. Utet, Torino
Piccioli L (1923) Selvicoltura. Seconda edizione riveduta e ampliata. Utet, Torino
Pothier RG (1836) Le pandette di Giustiniano, vol I. Tip. di Antonio Bazzarini, Venezia
Potter TW (1985) Storia del paesaggio dell'Etruria meridionale. Archeologia e trasformazioni del territorio. Carocci, Roma
Preller L (1858) Römische Mythologie. Weidmannsche Buchhandlung, Berlin
Riguez R (1675) C. Velleii Paterculii Historiae romanae, II. Fredericum Léonard Typographum Regii, Paris
Rosa G (1857) I feudi e i comuni della Lombardia. Pagnoncelli, Bergamo
Rozier F (1832) Nuovo corso completo di agricoltura, XXIV. Tip. della Minerva, Napoli
Schmid R (1858) Die Gesetze derAngelsachsen. F.M. Broackhaus, Leipzig
Sereni E (1961) Storia del paesaggio agrario italiano. Laterza, Bari
Sermonti V (2014) Le metamorfosi di Ovidio. Rizzoli, Milano
Sirmondi J (1652) C. Solii Apollinaris Sidonii arvernorum episcopi opera, Libro I, epist. 5. Sebastiani Cramoisy, Paris
Tentori C (1792) Della legislazione veneziana sulla preservazione della laguna. Giuseppe Rosa, Venezia
Trifone R (1957) Storia del diritto forestale in Italia. Accademia italiana di scienze forestali, Tip. Coppini, Firenze
Wagner IA (1787) Alciphronis Rhetoris epistolae, epist. I. Officinae Muellerianae, Lipsia
Zumpt AW (1861) M. Tullii Ciceronis orationes tres De lege agraria, Libro II. Ferdinandum Duemmlerum, Berolini M. Agnoletti [ed], *Historical Rural Landscapes. For a National Catalogue,* Laterza, Rome-Bari 2010

Chapter 4
Wooded Pastures and Pastured Woods

Abstract With the crisis of the late empire and the first incursions of the Germans the agents of disintegration of the Italian agricultural landscape accelerate significantly, a process accompanied by the progressive decay of urban centres. With the arrival of the Lombards in 568 AD and their stable settlement in Italy, a process of disorganisation of the landscape is consolidated, with a reduction of agricultural spaces and a regrowth of woods and pastures, linked to an unstoppable demographic decline favoured by famines and pestilences, which reduced the population from about eight million in the imperial period to the four million estimated for the sixth–seventh century. The change in the landscape is also reflected in the redefinition of social roles: in Lombard society, the pig keeper certainly has more importance than the small farmer. We are, therefore, in a regressive phase that brings the landscape of the plain back to forms pre-existing the Roman colonisation and that affects also the mountain, in which the change is less dramatic, but where the woods are growing again, giving new impetus to the beech forests, probably favoured by a cooler climate. In Tuscany, with the birth of the farm economy and the diffusion of share crop system, the management form adopted in the woods was generally the coppice with standards, but also pollard coppice was very common, while shredding occurred mostly at single tree level. Pastured wood and wooded pastures were quite common in all Italy. An interesting case to explain both the wooded pasture, but especially the management system in which such cultivation was inserted, is the practice of alder cultivation, widespread in the Val d'Aveto on the Ligurian Apennines between the XVIII and XIX centuries. The case study of Moscheta, a site in the Appennine Mountains north of Florence, shows that in the early nineteenth century the complexity of the landscape mosaic could be higher than the one that can be found on the plains. Even the Sila woods, in the Calabria region, appear as largely subject to grazing, in the form of wooded pastures, giving us an interesting cross-section of the situation in the south of Italy. A configuration of the forest sometimes described by the land registries, quite widespread not only in the Apennines mountains but also in the rest of Italy, are the forest managed for hunting, in particular for the birds.

M. Agnoletti, *History of the Italian Forest Landscape*, Environmental History 19,
https://doi.org/10.1007/978-3-031-81631-4_4

4.1 Woods in the Early Middle Ages

With the crisis of the late empire and the first incursions of the Germans, but especially from the early years of the fifth century AD, with the invasion of the Goths and the sack of Rome (410 AD) perpetrated by Alaric, the agents of disintegration of the Italian agricultural landscape accelerate significantly, a process accompanied by the progressive decay of urban centres. With the arrival of the Lombards in 568 AD and their stable settlement in Italy, a process of disorganisation of the landscape is consolidated, with a reduction of agricultural spaces and a regrowth of woods and pastures, linked to an unstoppable demographic decline favoured by famines and pestilences, which reduced the population from about eight million in the imperial period to the four million estimated for the sixth–seventh century. This is, in reality, a great crisis also of the Greek-Latin "cultural codes", which only the clericalisation of knowledge attempted by the church tries to preserve, with a parallel crisis of the collective memory of the populations that is also reflected in the structure of the landscape (Renucci 1974). The regression of the cultivated landscape is probably accompanied by an unfavourable climatic phase (Le Roy 2009), and it is in this context that the natural dynamics of the landscape regain strength, with the extension of the forest in the mountains and in the plains.

Along with the extension of the forest, however, there is also a worrying spread of marshes in many other areas of the peninsula once described as places with flourishing agricultural activities in Roman times (Jones 1974). The process of castle-building that is recorded in the seventh century on the Tuscan coasts of the Upper Maremma places the centres of territorial control in dominant positions on the hills, beneath which wooded and marshy plains cover the remains of the Roman villas that will then be rediscovered only in the first archaeological excavations of the mid-nineteenth century (Agnoletti 2009). On both sides of the Po, a strip of woods and marshes returns come to cover a large part of the Mantovano and Reggiano areas. In the middle Po plain, the patches of forest are more discontinuous, while in the high plain and in the subalpine strip of low hills, outside the area of the heaths also used for grazing, a more or less numerous peasant population persists, with cultivated spaces but also with vast areas of uncultivated land. The change in the landscape is also reflected in the redefinition of social roles: in Lombard society, the pig keeper certainly has more importance than the small farmer. We are, therefore, in a regressive phase that brings the landscape of the plain back to forms pre-existing the Roman colonisation and that affects also the mountain, in which the change is less dramatic, but where the woods are growing again, giving new impetus to the beech forests, probably favoured by a cooler climate. The return of the woods is reported both in the plain, as shown by the situation described for the Piacentino, and on the Apennines (Galletti 1988; Salbitano 1988).

As a new centre for organising the life of rural populations, the so-called "curtes or domuscultae or massae", as they are variously called in different parts of Italy (Sereni 1961), were born, where the new conquerors created a different form of territorial control. The new rulers, in occupying the peninsula, superimposed their rights on those of the conquered populations, considering as their own property the state forests previously intended for the exclusive use of the state. They jealously preserved these forests not so much for the products they provided, but for the game that was abundant there. The passion for hunting, which had already greatly developed in the last days of the Roman Empire, grew even more under the rule of the barbarians, who came from northern regions where hunting was still one of the main means of subsistence. Hunting was elevated to the honour of being one of the main occupations of the wealthy feudal lords and was the main concern of the state. Therefore, the state properties passed from the Romans to the new lords were jealously guarded and defended by the latter for the sole purpose of serving as hunting reserves.

In this way, the hunting bans, i.e. forests declared exclusive for the use of the lord, were greatly extended: the word "forest", in fact, probably meant not a wood but simply a reserved place, derived from the Latin words "foris stare", meaning that those who had no right had to stay out. This right was opposed to the civil liberties of using the wood, even though it contributed greatly to the conservation of the wood itself, since with the development of feudalism the woods were then ceded in fief to the nobles, as Charlemagne did in 800. In fact, Lombard laws like others of the time, for example the Salic law, established both the privileges of possession and those of passive servitude (e.g. collecting wood and grazing), not recognised by Roman laws. This is the origin of almost all the ancient forest servitudes legally and abusively existing in Italy, largely based on barbarian laws.

Later, as hunting became more and more a privilege of the nobility, the right of ban, or reserved hunting, even on those public goods on which the populations had until then exercised the uncontested right to collect wood and graze livestock, took shape. Initially, the right to ban was exclusive to the king, but later it was given as a gift to the larger feudal lords and from them gradually to the smaller ones, until anyone who had jurisdiction over a given region also had the right to ban. The ban initially consisted only of the right to hunt, but later it extended to preventing all those operations that damaged the game, as the felling of trees, the collection of grass, acorns, etc. Worse still, it was accompanied by the right of *sequela venatoria,* by which the owner of the ban could pursue the wounded animal on other people's land for 24 h. Hence it often happened that the farmer, who was about to collect his own crops, suddenly saw a horde of horsemen pouring onto his own field who, pursuing some wounded prey, quickly destroyed the fruit of so much labour. The penalties for violating the ban were very severe and sometimes cruel. To ban a forest, it was enough for the banishers, equipped with trumpets, to walk around its circuit, proclaiming to the four winds that it was under protection; then all around were placed placards with the inscription "foris stare". Thus, it is remembered that in 1130, Ruggero II surrounded the vast Ficuzza forest near Palermo with walls, introduced the bread tree and the white mulberry, recreated it with underground canals, and populated it with deer, roe deer and other abundant game.

These bans, which eventually became a real usurpation of property, became numerous over time, so much so that they hindered the development of agriculture. The king Emanuele Filiberto alone dissolved as many as 4465 in Piedmont in 1560. They certainly contributed to the conservation of the forests—in fact, our peninsula is depicted by historians as very wooded during that period—, but this conservation of the forests was to the detriment of the attention that the population had for the forests. Almost as if they were the expression of the arrogance of the feudal lords, the bans with all their restrictive laws were indeed hated by the agricultural populations.

The system that prevailed in the early Middle Ages was the so-called "curtense" where the agricultural estate was divided into *mansi*, which in medieval law corresponded to the extent of land that could be cultivated with a single pair of oxen or a single plough, and entrusted to a peasant family, while a part was managed by the lord, the *dominus* (Duby 1970). The *mansi* were granted to settlers who, in most cases, were former small landowners who, driven by the search for security and protection, or by poverty, were incorporated into the *curtes* of the lords and subjected to their power, with the obligation to pay rents and services (the corvée). The *mansus* also included forest and pasture where silvo-pastoral activities could be practised, and these elements took on a progressively greater importance in the landscape that was being built in the centuries following the fall of the Roman Empire, becoming a mode of management closely linked to the curtense economy and rural society, as it concerned the vast majority of wooded lands. The part managed by the lord, called *dominus,* made use of servants owned by the *dominus,* disposing of them at his pleasure, and of the corvée imposed on the tenants of the farms. However, there were also free peasants gathered in independent communities, especially villages, who could avail themselves, together with the tenants tied to the lord, of the so-called civic uses, later guaranteed by municipal statutes. These allowed the use, in part and at certain times, free or onerous, valid for all types of livestock or only for some, of woods and pastures both private and communal. The *ius pascendi* allowed grazing animals on someone else's land after the owner had harvested. The *ius lignandi* allowed access to communal woods to collect a little firewood. The ruspo was exercised on chestnut groves after the owner had finished harvesting, and ensured a little chestnuts for all the needy. The *ruspo* of the poor was followed by the *rumo*, that is, the opening of the chestnut grove to pig grazing (Bellandi 2000).

Around the mansi, large expanses of forests, uncultivated lands, marshy lands where all forms of cultivation had been abandoned, extended. In this context, pastoral activities were privileged over agricultural ones, given the semi-nomadic habits of the Conquering populations, as is evident in the Edict of Rothari of 643 AD, the first collection of Lombard laws. These populations, essentially made up of warriors and breeders, knew almost only marginal and precarious crops: in the 388 articles of the edict, the landscape of the forests seems to dominate, which is mentioned 44 times, hunting, mentioned 42 times, falconry, with 9 times, wood, with 20 times, and pig grazing, 23 times (Sereni 1972). Some place names present in Tuscany still recall the past Lombard domination: the terms of gaggio, gaio, caggiolo, gaiole (Trifone 1957) are of Lombard origin and concern places that were grazed forests and that did not designate a specific point, as today, but a rather extensive wooded region (Casabianca 1941). Moreover, in Tuscany today at least 570 place names referring to the forest can be found, with the names "bosco" to 200, as many as "selva", about 100 as "macchia" (Cassi 2000). In the forests of the less rugged territories, an important pastoral activity was born, which focused particularly on the large wild breeding of pigs, in order to ensure the local populations that food reserve of fat that replaced the cultivation of the olive tree, declined with the fall of Rome. This breeding was often practiced within wooded and uncultivated areas owned by the *dominus* who left them open to common use, in return for a tax: for example, the glandaticum (the acorn tax), consisting of a part of the collected acorns, or the decima porcorum, that is the tenth part of the grazed pigs (Baruzzi and Montanari 1981).

The rest of the territory was left completely wild, often groups of bandits took refuge there and there was an abundance of wild fauna such as the bear, the wild boar and the wolf, which in those centuries constituted serious threats from which it was necessary to defend oneself since for such predators man is just one of the many possible prey. Such was the importance assumed by the wild breeding of pigs that the forest was not estimated in terms of surface, but in terms of pigs that it was able to feed (Galletti 1988), a practice that is maintained in Tuscany until the twentieth century. In the ninth century the extension of the monastic-owned oak forests in the northern Apennines was valued in the accounting writings in number of pigs that could be fed there, whose measure ranged from one hectare to one and a half hectares per animal (Gabbrielli 2006). The herds of pigs, raised in the woods, were gathered in groups of a few dozen up to hundreds of heads led by the swineherd and the head of the flock, the major boar, which the Lombards called *sonorpair*, when he commanded a flock of at least 30 heads and a sow, the *ducaria,* equipped with a bell (Baruzzi and Montanari 1981). The wild grazing happened for almost the entire year, only in winter the pigs were brought back to the pigsties for a brief period of stabling. The breeds of pigs used, as shown in the iconography of the time, were dark in colour, red or blackish, and more similar to wild boars: *"the head was larger and longer; the snout pointed, not corked; the ears short and erect; the bristles upright on the back and the tusks prominently protruding from the snout"* (Baruzzi and Montanari 1981). Their lifespan was longer compared to current times, they were rarely slaughtered within the first year, but more frequently between the second and third year. The killing of the pig took place at the end of autumn, after grazing in the woods, when the season most suitable for preserving its meat began. In the illustrated medieval calendars, this activity was given great importance in the depiction of the months of November and December. The practice of grazing in the woods did not lead to the destruction of the forest, but rather to a modification of its density and structure that envisaged a smaller number of plants compared to a natural forest. However, large extensions of woods were necessary to raise both pigs and other types of livestock, and if the number of heads per hectare was excessive, this put the natural regeneration in crisis, which, starting from the soil, was often devoured by grazing animals; a problem similar to what happens today in Italy with the excessive presence of herbivorous wildlife species, such as roe deer or deer.

However, we are close to a new period of change in landscape dynamics. After the Saracen invasions between the ninth and tenth centuries and the Hungarian raids, the disintegrative processes have now reached their peak and a reversal of the trend is beginning to be seen. In Lombardy, in Milan, the establishment of the monastery of Sant'Ambrogio marks the first step of a new growth in the importance of the church, once again privileged during the struggles for the crown of Italy in the ninth century and which has as its strength its possessions on Lake Maggiore and Lake Como, where agricultural and pastoral activities are very flourishing. Possessions that increase through continuous donations of forests and pastures, but also with the acquisition of state-owned goods, and which are structured according to the more general patterns of the curtense system, as well documented for the constitution of the patrimony of San Salvatore in Brescia (Toubert 1983). At the beginning of the year 1000, a process of landscape organisation is therefore underway that takes place in the hill and in the mountain, on lands again preferred to those of the marshy low plain. Areas such as the Bergamo hill are now deeply influenced by the cultivation of olive, chestnut and vine: wine will in fact become one of the main products of this area, while the oil productions and agricultural exports from the lake region are already well structured (Andreis 1993). The fruit chestnut grove in this period appears to replace, especially on the south-facing slopes, the oak groves of downy and English oak, but also cenoses typical of cooler areas characterised by lime, maple and ash. This is also highlighted by the medieval statutes of the communities where the obligation to plant this species often appears, which on the higher slopes replaces the beech. In the highest part of the mountains, the need for pastures resulting from the new zootechnical development leads to the progressive removal of the forest vegetation, but the forage must often have been insufficient since in the eleventh century, to meet the great demand from Bergamo, carts of hay were sent from Milan and Novara.

4.2 The Woods of the Sharecropping System

The coppice forest historically represents the most flexible and widely distributed form of forest management in Italy, especially at the medium–low altitudes of the Alpine arc and much more extensively on the Apennines and throughout the hilly belt including the plain. In addition to timber production and especially charcoal production, the coppice forest also allowed grazing, and its distribution follows the historical phases linked to the agricultural colonisation of the peninsula, taking root in areas that assume particular importance for its widespread presence. In Sardinia, all coppice forests were extensively grazed, even those characterised by shrub species, by cattle, pigs, goats and sheep, continuously and without any limitation; in other Italian regions such as Tuscany, historical sources allow us to detail its characteristics that will lead it to represent the most relevant form of forest governance in the region still today. Throughout the peninsula from the 11th to the thirteenth century, there was a significant demographic recovery and a new phase of land clearing, to obtain new lands to cultivate and meet the increased food needs at the expense of the forest. The new colonising push has been interpreted by some as a real aggression towards the environment, exacerbated by the fear of wilderness. In the early Middle Ages, the fear caused by natural environments indeed seems to have been less than that which spread from the communal era, when wild nature began to be considered as something alien, a place populated by ghosts and dangers (Andreolli et al 1988).

We do not know if this interpretation represents the prevalent idea of the population in the communal period; certainly, as evidenced by the fresco on the *Wise Management of the Territory* made by Ambrogio Lorenzetti in 1338 in the municipal palace of Siena, and by that on the Bad Governance of the Territory by the same author present in the same building, the ideas of the time regarding the directions of good land management appear consolidated. Just as the first fresco shows a countryside and forests entirely and harmoniously cultivated, the second identifies in war, in the destruction of cities and in the return of wilderness an idea that opposes the very concept of civilisation. Man's dominion over a wild and uncontrolled nature, essentially hostile to human life, was considered at that time a sign of civilisation. Not coincidentally, the recovery of agricultural knowledge from the Roman period and of agriculture are a fundamental part of the rebirth of Italy in the medieval period as evidenced by the book on agriculture by Pietro de' Crescenzi. The agricultural landscape will also be object of admiration by all travellers who from the fifteenth century onwards will recognise in a landscape shaped by man some of the essential values associated with the Italian landscape. The fear of natural and wild environments, if it was actually representative of the period, is therefore not strange, but testifies to the lesser familiarity with a survival model of previous centuries that required a greater ability to relate to natural phenomena and the need to procure food by hunting, rather than cultivating the land, an activity to which wildlife and extreme natural events have always caused damage.

Due to the Black Plague of 1348, demographic growth suffered a drastic halt, with the depopulation of cities and countryside. This is a population loss of 70% compared to that existing before the catastrophe; in the Maremma of Siena alone, it led to the death of 80% of the population, which resumed its growth only after several decades. Starting from the second half of the fourteenth century, a new demographic growth involved the Tuscan territory and favoured a revival of economic activities linked mainly to trade and crafts. In this way, the emerging bourgeoisie and the city merchants became richer. The power of the communal civilisation expanded to rural areas at the expense of feudal lords, against whom bitter battles were undertaken. These were later forcibly removed from their castles and their lands and, according to Sereni, it was the bourgeois owners themselves who, having freed the serfs for the purpose of their struggle against the feudal lords, felt the need to keep them tied to the land (Sereni 1961). Thus, the system moved from the curtense system—which was based on cultivation only around the *curtes*, with large portions of woods and uncultivated lands where wild breeding was practiced—to one based on the farm conducted in sharecropping, which became the main system of agricultural management at least until the second post-war period in the centre and north of the country. The farm became the new peasant company, in which production differentiated to meet the new food needs of the cities which, in addition to livestock, needed more easily transportable and better preserved goods. Thus, the woods were removed to make way for old and new cereal and tree crops, while livestock was pushed towards other areas (Cianferoni and Fattori 1992).

According to the sharecropping system the cultivation of the farm and its constant maintenance was taken care of by the farming family, to whom the landowner immediately provided the house and livestock, which was mainly used for agricultural work, for sale and for the subsistence of the family, provided that the family settled permanently on the land. With this management system, the bourgeois owner could follow his own business in the city without having to worry about the organisation of work, which was entirely entrusted to the tenants. If anything, he built country villas for himself as master residences adjacent to one or more farms, complete with a garden, or *meadow*, and park, or *wild* woodland, where he spent his holidays (Rombai 2002). The production of the farm was divided equally between the owner and the sharecropper, who survived thanks only to his work and what he managed to produce. As Pazzagli writes, in the sharecropping system the principle of wanting everything for everything and of making the farm a real Noah's Ark, as autonomous and self-sufficient as possible, capable of providing the necessary products not only for the needs of the farming family and the owner. The ultimate goal of the cultivating family and the owner is to achieve the maximum product, but also the most varied possible, in homage to the logic of food and productive self-sufficiency, and to guarantee against the events of the climate and the unfavourable course of the seasons, against which the farmer tries to protect himself by focusing simultaneously on crops with different vegetative cycles (Pazzagli 1989). This philosophy is at the basis of the construction of a landscape mosaic made up of a multitude of small patches, i.e. fields cultivated with different crops, which has characterised all the territories affected by sharecropping, with a diversification of spaces that adds to the morphological variability, creating that diversity of landscape that still partly distinguishes central Italy, compared to landscapes linked to the large estate heir of the feudal system.

In the sharecropping system, the owner's part of the product was often put on the market, thus supplying the cities with foodstuffs to support demographic growth: in the 1400s Florence had about 100,000 inhabitants and it is conceivable that it would not have been possible to reach this size without the contribution of the countryside (Cianferoni and Fattori 1992). From the fourteenth century, the system of scattered farmer's houses was organised around new settlements called *farms*, which became the new centre of economic, administrative and commercial organisation for estates, even those not contiguous territorially (Pazzagli 1989). The farm became the depot for products and the place where the mill and the oil press were located, to which the farmers were forced to resort for the processing of their products, mainly grain and oil, under the control of the farm manager and the owner (Riccioni 1994). The birth of the first farms was recorded from the end of the fourteenth century, on the landholdings of the Medici family in the Mugello valley and in the plain west of Florence, while in the sixteenth century the system expanded to the assets of hospital, knightly and ecclesiastical bodies and of large city families, also located in other Tuscan areas (Rombai 2002). In this historical period, the civic uses of timber (Bicchierai 1995), acorn, sowing and leaf collection still survived in the woods, in favour of the local populations, but the form of management that increasingly asserted itself in association with sharecropping is the coppice forest. Generally, the previous medieval statutes limited grazing in certain periods of the year, especially for the collection of acorns and chestnuts for human consumption (Zagnoni 1997). The right to graze in communal woods was, in part, granted in return for a rent in the so-called 'bandite' to foreigners, in part to local livestock. One of these 'bandite', that of the Tombolo di Grosseto, appeared around the second half of the eighteenth century as a coppice with many standards and made up of ash, wild olive, lilac, cork, oak, field maple, holm oak and hornbeam. For the larger plants, branching was prescribed with the release of a guide (Gabbrielli 1985) (Fig. 4.1).

With the birth of the farm economy, more rational forms of treatment were applied to the woods attached to the estates (Biffi Tolomei 1833), documented more specifically from the seventeenth century, which combined forage production with timber production, useful to the company or for sale, as firewood, faggots, charcoal, splitting wood and work wood in the case of oak forests, vineyard poles and splitting and work wood, such as beams and smaller assortments from chestnut coppices. Grazing in the forest was limited to preserve timber production; it was often forbidden to allow grazing to livestock not belonging to the farms, while for the latter rules and restrictions were established such as not being able to approach the wood in the three years following the cut, respect the load of livestock established to be introduced into the forest and absolutely ban that of goats. According to Biffi Tolomei, the Coppice woods in Tuscany were among the lands left to the landlord, that is, not handed over to the sharecroppers, even if the owner allowed grazing in them, essentially because they did not require any annual work, but only cutting every 10 or 20 years. The management form adopted in the woods attached to the farms was generally the

Fig. 4.1 Typical landscape of a sharecropping farm in Tuscany. At the top, you can see the fruit chestnut grove, below on the left the coppice forest, on the right the oak forest for acorn breeding of wild pigs and below a tree pasture. In the centre are the mixed terraced crops, often vine with olive and maple. From the thirteenth century to the second post-war period, almost no sharecropping company could exist without the forest (Agnoletti 2002)

coppice with standards, that is, a forest in which the plants were cut at the base, but in which some tall trees were also released, both in the oak groves and in the chestnut groves, unlike the simple coppice in which no tall trees were released. The coppice component was subjected to clear cuts with pre-established rotations, reducing the number of shoots on the stumps, preceded by the elimination of the shrubs of the undergrowth. The thinning of the chestnut shoots was more complex than that of the oaks, since for the former the pruning of those released was carried out to contain the development of the crown, while for the oaks only the young oaks that gave hope "to become bigger and longer" (Piussi 1982) were maintained. The purpose of this practice was to collect small-sized wood, otherwise destined to dry up and rot in the forest, to promote the herbaceous vegetation of the undergrowth, which together with the shoots of the stumps represented a food for grazing livestock, and finally to allow a better growth of the remaining shoots. Sometimes the coppice was kept only as a pollard, but often both clear-cut stumps and pollards coexisted, as happened, for example, in the coppices of Turkey oak and holm oak (English oak, sporadically) of some farms in the middle Valdarno valley at the end of the 1500s. From the documentation of the time, it was found that the oak groves attached to these farms were managed as coppice with standards (Piussi 1980). The tall tree component of the coppice with standards was made up of oak seed trees, that is, the seed plants to renew the stumps—which at the end of the shift the farmer or the forest ranger marked, surrounding the trunk with a willow, or with a thin shoot of oak or chestnut -, chosen among the most beautiful plants, so as to exclude them from the cut and obtain the future timber for work and acorns for feeding the grazing livestock (Piussi 1982).

The landscape of the mixed coppice thus created was therefore characterised by tall trees scattered among the stumps of the coppiced plants, from which new shoots would be born: it is therefore a wood arranged on two levels linked to the different ages of the two types of forest present in the same space. In the Cerbaie estate (Pisa), at the end of the 1600s, the holm oak groves were managed as coppice with standards with the release, after each cut, of 80 seed trees per hectare, distributed by age classes (Piussi 1982). All fruit-bearing species (holm oak, Turkey oak, cork oak, English oak, chestnut, apple) were particularly protected, whose cutting was forbidden by the tenants, and the company had set up a nursery for the production of seedlings to be used in the renewal of these acorn woods. The coppice with standards system was also planned for pure chestnut coppices, with a numerically lower release of seed trees, the so-called "porrine", compared to the oak ones. In order to provide acorns and chestnuts for animal feed, tall trees were kept isolated or in small groups near the crops. Such was their importance that in sharecropping contracts the tenant was forbidden to cut them, to take care of the natural renewal and not to beat the tops to accelerate the harvest of the fruits (Piussi 1980).

In the 1700s, in Tuscany, there was a further expansion of coppicing, thanks also to the grand ducal laws that eliminated cutting restrictions between 1768 and 1781, and to the liberalisation of the timber trade carried out in 1770, in the framework of several economic reforms carried out by the Lorena Family. This family, belonging to the Royal Augsburg Family, ruled the grand duchy of Tuscany after the end of the

Medici family rule, from 1737 to 1801 and from 1814 to 1860, contributing to an important economic and social development of the region which went far beyond the duration of their dominion. These reforms were accompanied by the privatization of part of the agricultural and forest property, obtained through alienations of part of the grand ducal ownership, that of the institutes and ecclesiastical orders, leading to the gradual disappearance of common goods and civil uses that had survived from the barbaric period (Agnoletti 2002). For some grand ducal-owned farms, some forest management plans were proposed to improve the timber yields of the woods and at the same time reconcile the needs of grazing livestock (Fig. 4.2).

In 1796, for example, the farm factor Mr. Bicchi proposed for the farm of Cascine dell'isola (in the municipality of Campi Bisenzio, Florence) a wide thinning of oaks, elms and poplars was done in such a way that the remaining ones were 10 m apart from each other, and the cutting of the low scrub and the coppice timber for stacking, warning to leave intact all the acorn-bearing plants to raise them partly as high trunk, partly as pollard and partly as ground-level stump (Gabbrielli 1985). The factor Lorenzo Carniani, on 3 August 1768, in a lecture given at the Academy of Georgofili in Florence, stated that, to ensure the maintenance of pasture in the woods, the cutting of junipers, brooms and myrtle should be avoided because "these plants, especially if they are in quantity, break the fury of the waters, defend what is under and around

Fig. 4.2 Beech stump at the twenty-fifth year of age; starting from the fifteenth year, the shoots have managed to escape the bite of the livestock thanks to other shoots that have protected them (Piccioli 1923, p. 405)

their branches and wonderfully protect the herbs that are near and that vigorously vegetate in that flower of earth that is retained by these plants, courted by their green shoots" (Imberciadori 1953). For the woods attached to the grand ducal farms of Artimino, Ginestre and Calappiano, in 1775 a plan of intervention was proposed that envisaged in a mixed wood of oaks and chestnuts with undergrowth of broom, myrtle and juniper as the main species, the cutting of the chestnut trees now dry or decaying so as to be able to put new shoots from the stumps, useful as poles to the farms. Access to livestock was to be prevented for the next three or four years after cutting, to avoid the bite of livestock on young shoots and to do cleanings, and the thinning in the palisades because they came at the same time straighter and larger. In addition, the tenants of the farms were to be allowed to let the animals graze the leaf without breaking the branches (Gabbrielli 1985).

At the beginning of the 1800s in Tuscany, an unstoppable process of population growth began, which would go from 1,303,044 inhabitants to 2,317,004 in 1889 and a simultaneous increase in the number of sharecropping farms which only between 1830 and 1854 will go from 12,000 to 15,000, causing a further extension of coppicing. However, we must not think that sharecropping had extended to the whole Tuscan territory: in fact, the vast territory of Maremma, in the south of Tuscany, would see the arrival of sharecropping much later, well into the 1800s. This process was implemented on pre-existing large estates, often still in the hands of Lombard nobility, who reluctantly submitted to the reforms imposed by the Grand Duke aimed at severely limiting political power of the feudal lords, such as the law on fiefs of 1749, wanted to create a new class of agricultural owners. In this area of Tuscany, the landscape presents historical dynamics and characteristics that are quite different and autonomous from other parts of the region, well represented also in the descriptions of Tuscan agriculture of the second half of the 1800s (Agnoletti 2009).

4.3 The Customs of the Paschi of Siena

After the great plague of the mid-1300s, the coastal and inland area of the Tuscan Maremma (Bellucci 1984) became the subject of intense grazing, sheep and cattle, of transhumant flocks coming from the Apennine areas of Tuscany, where the winter grazing surfaces were covered with snow and lacked adequate forage resources to feed the resident livestock. The forms of organisation of grazing and management of the woods are one of the most interesting examples to understand the structure of the landscape of this vast part of Tuscany, but also the economic activity that represented the pivot of the Sienese economy in these centuries. The municipality of Siena organised the exploitation of the communal Maremma lands by elaborating in 1419 the Sienese Statute of the Paschi, which remained in force until the 1700s, which regulated the grazing of transhumant livestock within the so-called Customs of the Paschi di Siena from 1st September to 1st May. In 1472 the wealth produced by these activities led to the creation of the Monte de Paschi di Siena, which is considered the oldest bank still active from the time of its foundation at world level.

The Customs extended for much of the Maremma in the valleys of the rivers Bruna, Albegna and Ombrone, from the municipalities of Sassofratino to Paganico, from Samprugnaio to Capalbio (Barsanti 1987). The transhumant flocks came from the valleys of Casentino and Mugello in the Apennines, from Lucca and other cities, even non-Tuscan ones, such as Bologna and Perugia. The pastures, open to livestock, varied according to the period both in extension and for the duration of grazing, based on their productive capacities. The transhumant livestock was subject to the payment of a fida, or tax per head, including the price of the grass, and the balance of the fida took place by May, before undertaking the return journey. Article 28 of the Statute specified that foreign shepherds "cannot cut down cork oaks, oaks, beeches or any other tree that bears acorns or makes from the foot, excluding the branches that they can branch off to make fire and for their livestock" (Barsanti 1987).

In addition to the Customs, there were reserved forests, which were often "trusted", i.e. given in use to foreigners in return for the payment of the fida. Half of the income obtained had to be paid to the municipality of Siena, the quality of the pastures of the Customs was different and they had various names depending on the different quality of pasture. "Local" was defined as the pasture of any type taken directly by the livestock on the ground and on the trees without human intervention. "Separated" was that collected from the ground and preserved dry for food like hay, straw acorns and chestnuts. "Cetina" (Sereni 1981) was called the grass of the land cleaned by the Trees and from the stipa (Sereni 1981) and then burned to reduce to cultivation. "Cavallino" was the pasture of a plot located in the middle of the forest and dominated by the sun. The grass that the land covered by tall oaks and oaks produced was then called "macchia serena" pasture; "macchia bassa" was that covered by thick patches of broom, albatross or strawberry tree, holm oak, mastic and other evergreen plants (Barsanti 1987) (Fig. 4.3).

As you can understand, these are various types of landscape characterised by a great diversity of environments and structures all functional to the exercise of grazing, but which did not exclude other uses of the forest. Pastures were open to thousands of heads of livestock of all kinds, but precise data on the extent of the livestock load are only available for some customs and bans in the second half of the eighteenth century. It was estimated that 36,000 moggia of land (about 108,000 ha) were subject to customs in the Maremma; "in 1761 with 958 trusted… there was a load of 3043 cows, 1085 horses, 10,046 pigs, 73,246 sheep and goats… [about 0.8 heads/ha] in the bans grazed 5206 cows, 2199 horses, 3578 pigs, 134,929 sheep and goats" (Gabbrielli 1980). The shepherds knew well the dietary needs of their livestock and fed their flocks with different types of pasture depending on the time of year. Sheep were led to stubble, olive groves and vineyards, on uncultivated land, on maggesi and in cetine, avoiding if possible that in tall forests which was considered of mediocre quality. Goats were fed with leaves of wild olive, oak and acorns, they were led to graze in the forteti and in areas covered by poor and bushy pastures managing to eat everything, including thorny shrubs, tree bark, holm oak shoots, oak, etc. Cattle, in addition to pastures, were also led into the macchia serena, a predominantly oak

Fig. 4.3 Ancient pollard oak (Oak of the Witches, Capannori—Tuscany). The abandonment, the prohibitions of pollarding and of the activities of the Shepherds have led to a significant reduction in these monumental plants that today are protected. The little knowledge of their origin often leads to them being classified as "natural monuments", while they are the product of traditional practices associated with grazing

forest, especially from October to May and fed with tree branches from November to April. Pigs spent the autumn and winter in the woods and forteti feeding on holm oak and cork oak acorns, while in summer they were led to the fields to feed on the ears of cereals left after the harvest. After two or three months from his arrival, the transhumant shepherd was forced to buy the bans, since the pastures of Dogana were now unproductive due to the high livestock loads they were forced to bear. The Statute provided also provided for the so-called *uses*, pastures near the villages for the exclusive benefit of the poorest local populations, or for those communities that not having their own pastures or insufficient for their needs could enjoy those of neighbouring communities.

4.4 The Selective Coppice

The need to draw a frequent periodic product, even if more limited in quantitative terms, together with the need to have a multifunctional forest management, both in terms of products that can be obtained and in relation to the presence of activities such as grazing, led to the development of various forms of coppicing. The selective coppice is a management systems also known as "cutting of the ant" (Figs. 4.4 and 4.5).

Fig. 4.4 Pollard coppice of Willow. Pollard forests are no longer present in Italy, although in Tuscany they were the third type of coppice forest until the early 1900s (Piccioli, Silviculture cit.)

The "ant's cut" was widespread in various regions of Italy and had the characteristic of maintaining constant ground cover, as the shoots on the same stump were cut at different ages and therefore the stump never appeared bare. It was very common in Tuscany, especially in the mountains of Pistoia, where it seems that in the eleventh century some hospitals run by Benedictines monks for pilgrims and merchants crossing the Apennines, such as the Badia della Fontana Taona, the Spedaletto farm and the Hospice of Pratum Episcopi, managed coppice forests. The Benedictine culture, as we will see in the chapter on monastic silviculture, used to cultivate the forests and it seems that the Benedictines were the first to develop this form of treatment which they then spread to other Italian regions (Menicacci 2002). The practice allowed, for example, to exploit a coppice ensuring the owners a modest but constant and continuous income over the years. The technique involved coppicing every 6–8 years, leaving on each stump shoots of different age classes, 1–2 shoots

Fig. 4.5 Beech managed as selective coppice, also known as "ant's cut". The cut did not eliminate all the shoots on the stump but only some of them over time to have a regular wood production (Piccioli, Silviculture cit.)

of the third class (18–24 years), several shoots of the second (12–16) and many of the first (6–8), derived from the regrowth of shoots after the last cut. The succession of cuts meant that the whole operation was the natural consequence of the previous one and the premise for future ones, and from this type of cut usually charcoal and firewood were obtained. The practice continues until the advent of the chainsaw, in the 50s–60s of the last century, which no longer allows to easily cut by selecting the shoots present on a stump as was done with the axe.

4.5 Pollard Coppice Woods

The forests of Maremma underwent significant utilizations over time, especially from the sixteenth century, when the Medici laws created a monopoly regime on their cutting in favour of government finances. According to this ordinance, the forests within a strip up to 5 miles from the sea, about 8 km, even if privately owned, were used "to extract timber, wood and charcoal to trade "overseas" outside of Tuscany" (Gabbrielli 1997). The substantial profits obtained from this trade with foreign countries, especially with Genoa, were considered fundamental for the public treasury, so much so that these forests were called "forests for cutting for Genoa". Often private owners, whose forests fell within these 5 miles, tried by all means to remove them from this constraint by turning them to agricultural areas, or at least to wooded pasture, given that the State was directly interested in the cutting of wood (Gabbrielli 1980). These cuts continued for over a century and from the second half of the eighteenth century, to increase this "overseas" trade, almost all the Maremma woods became "woods for cutting for Genoa", such as those of the communities of Capalbio, Marsiliana, Montauto and Scansano. The cuts made were aimed at obtaining mainly wood and charcoal (Rozier 1832). For this reason, the oak, turkey oak, holm oak, and cork oak high forests were converted into coppices, often treated with pollarding, with a generally 12-year rotation and with the release of a certain number of mother trees capable of fruiting and providing timber for work. The cut in the pollarded coppice consisted of cutting the plant at a certain height from the ground (1–3 m) and was therefore called "aerial coppice": in fact, being the coppicing far from the ground, this prevented the bite of grazing animals from damaging the new shoots that regenerate from the stump (Orlandini 1833).

Traditionally, softwoods were "pollarded" around 10–20 years and hardwoods between 15 and 30 years, with a smooth cut and slightly inclined plane so that the pollards had a long-life avoiding water stagnation (Piccioli 1923). The emission of shoots occurs around the cutting section which, as the coppicing continues, thickens and takes on a round shape. According to Di Bérenger, "it is called a pollard or gabba or scamollo the tree pruned in such a way that, having all its branches cut, it represents with the top of the trunk almost a head….if forked they are pollarded to three or more heads, leaving he main branches one meter long, which is said to cut to crown" (Di Bérenger 1887). The cutting of the shoots that sprouted from the pollard was carried out in spring. If the pollard was young, i.e. with an average trunk diameter of 8 to 16 cm, it was cut flush with the point of insertion of the shoots on the trunk, while for the adult ones the cut was always made "long in the new wood, pruning the shoots at a distance of 6 to 9 cm from their point of insertion" (Di Bérenger 1887). Finally, for those that had exceeded 40 cm in diameter with their trunk, the cut was made even longer, sometimes leaving a couple of shoots (Di Bérenger 1887).

Precisely to favour this pastoral use of the forest, the vast wooded area between Capalbio and Marsiliana dell'Albegna was, in the seventeenth century, subjected to a cutting plan devised by a charcoal burner from Pistoia, Felice Antonini (Gabbrielli 1980). The area was divided into plots, that is, into parts of equal surface area, so that each The coppice was expected to be large enough to provide 3–4000 loads of charcoal (about 3000 quintals of wood); the proposed cutting cycle was every ten years because it was said that at that age the best quality of charcoal for trade was obtained. Thinning of the larger trees was planned to "allow the subordinate ones to progress and the other nearby ones to expand to produce acorns, while the old and twisted trees were to be reduced to pollards, leaving them the lateral guide and not the main one, if they do not put anything else but at the top, to leave good pasture". In those coppices already cut and returning to cut, "10 percent of the trees will be left to become construction trees, pollarding the twisted and old ones, while the trees suitable for construction, but not yet reached perfection, must have their branches removed and leave the main guide so that the plant can raise and thicken the trunk". Pollarding was usually practiced in Maremma and was preferred over cutting the plant "at the foot", because with the latter the production of acorns, leaves and grass was lost. Cutting at the foot cluttered the ground with lateral shoots, which regenerated even more than a meter away, reducing the pasture. It was therefore considered better to cut high to pollard (Gabbrielli 1980) and this explains how pollarded woods were still the third type of coppice forest in terms of extension in Tuscany at the beginning of the 1900s. It constituted a very particular forest landscape, in which at a certain height from the ground from the trunk of the plant not large and varied branches were divided to compose the typical architecture of a mature tree, but many small shoots, even several hundred, which formed a dense and minute crown characterized by a low woody mass but a very rich leaf mass.

4.6 "Scalvo" Woods, or "Meadows on Trees"

The collection of leaves and twigs used to feed livestock was a very widespread practice. It is already mentioned, for example, for the community of Premilcuore in the "Statute of 1379 and in the additions of 1515… The farmer-shepherds went to the community or farm woods to cut the new shoots of the plants with fresh leaves, dry them and collect them…" (Brusi 1987). Even in the Florentine and Mugello valley territory of the late 1700s Tolomei reports verbatim: "the farms around Florence and up to 10 miles can be said to have no meadows. Once the fresh grass is finished, they feed the animals with cut straw, mixed with tree foliage" (Imberciadori 1953). The technique adopted was the so-called "scalvatura", which consisted in cutting all the branches of the trees, leaving the trunk and the top intact, on plants that had reached about ten years, thus favoring the regrowth of young shoots with many leaves; "to have good shoots it is not convenient to prune before 5 or 10 years and not after the plants have exceeded the maximum height increase, beyond 10 years the plants began to suffer damage" (Piccioli 1923; Giumelli 1932). As Pazzagli writes, thanks to this traditional technique, the farmer manages to multiply the cultivated surface available, creating a sort of second layer on which the crops extend, raised above the trees, on which, as the proverb says in Tuscany, hay also grows (Pazzagli 1989).

It is a technique demonstrating the great variety of landscapes resulting from the management of forests combined with agricultural activities (Fig. 4.6).

The species subjected to scalvatura were oaks, elms, limes, hornbeams, maples, hazels, sometimes even chestnuts, beeches, alders (Di Bérenger 1887). The cuts were repeated every 2–4 years, up to a maximum of 10 (for example for elms); that of the Turkey oak was practiced every year: the period of time considered minimum for a decent reproduction of the leaf (Salvi 1982). For isolated trees, the cycles were shortened, given the faster sprouting. The scalvatura began between the end of August and the beginning of September and it was tried to finish it before the leaves began to yellow.

To speed up the operations, the branches were cut, "cutting the thigh of the branch at a distance of 5–10 cm from the respective armpit" (Di Bérenger 1887), starting from the base of the plant and continuing along the stem upwards, "up to 1 or 2 m from the top, without leaving suckers" (Piccioli 1923). The operation was still dangerous,

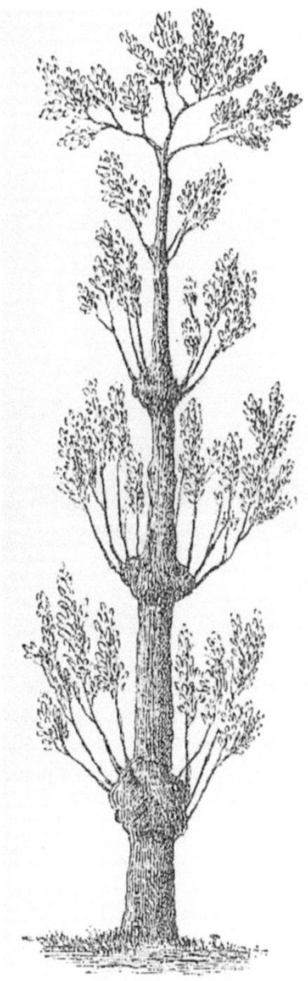

Fig. 4.6 Black poplar managed applying the "scalvo" technique (Piccioli, *Forestry,* cit.)

since the crowns were subjected to strong oscillations, and therefore had to be carried out with a certain speed and caution. Even though this technique generally involved the removal of all branches except the apical ones, sometimes in the case of already adult plants, topping was practiced, or rather a particular topping consisting in cutting the top when the plant had reached 10 m (Salvi 1982). Its purpose was to stop the growth in height in favour of the expansion of the crown and the growth of the stem. The branches were collected in bundles and the leaves, left to dry, were given in winter to feed sheep and goats. The twigs served as weaving material and bundles to burn, while from the stem was obtained construction timber, work and firewood (Piccioli 1923). The collection of leaves also took place in autumn. The litter, especially of chestnut, was removed from the forest and used in the stables as a bed for the animals (Fig. 4.7).

Fig. 4.7 The Landscape of the Monti Lucretili—Lazio region (Jacob Philipp Hackert 1780). The meadows with trees and groves were a major feature of the Apennine mountains. Many travelers of the "Grand Tour" visiting Italy not only wrote travel reports but also represented the Italian landscape with drawings or paintings

4.7 Wooded Pastures

The historical origin of wooded pastures has been documented for Molise, Abruzzo and the Tuscan-Emilian Apennines where traces of a system—which seems to have been established in the Lombard age (Cevasco 2007)—aimed at the production of hay, meadow and leaf fodder, in the sixteenth century maps concerning the "alps", i.e. the summer pastures (Moreno and Poggi 1996). These are structures characterized by the prevalence of meadows with scattered trees or small groups, both of the tree and shrub type, which must be grazed continuously to not evolve into a forest and which were therefore linked to an economy based on livestock farming. These traditional practices contributed significantly to diversification of the forest landscape, creating, along with pastured woods, a diversity determined mainly by the number, the species and the arrangement of trees in the meadows. Documentary evidence from the early decades of the eighteenth century suggests that livestock farming was also a key part of the economy in the Kingdom of Naples. The most common types of use are always more or less connected to grazing: pollarding for the production of forage branches, wooded pastures of oaks or beeches for the production of seed essential for animal feed. In the more inland areas (Abruzzo, Molise, Calabria) wooded pastures, hedges and pollards for the production of branches are described (Zanzi Sulli and Di Pasquale 1993).

Currently, the remaining wooded pastures in Italy can be found along the Alpine arc where, at higher altitudes, the larch constitutes the tree cover while the beech replaces it at lower altitudes. The Ligurian and Maritime Alps preserve traces of ancient wooded pastures of larch and white fir, as found in the Monregalesi Valleys and for the Gerbonte alp (Cevasco 2007). Along the Apennine chains of the central-southern regions, beeches and oaks form the tree canopy, while on the islands and in the coastal areas, holm oak and cork oak become prevalent (Del Favero 2008). According to the 1929 land register, the definition of wooded pasture is divided into: 1. Bushy pasture: grazing land, covered by scattered bushes that limit the production of grass to restricted areas. 2. Wooded pasture (and meadow): pasture (or meadow) on which there are aligned or scattered woody plants, the product of which significantly increases the land income. The degree of coverage of the projection on the ground of the crowns must be between 5 and 50%. These are, therefore, very articulated forms of landscape in which the tree component with various types of bearing, trees and shrubs and species, constitutes a strong element of diversification compared to more monotonous landscapes made up of compact and homogeneous forest covers.

The origin of the wooded pastures can be traced back to both the use of fire and the return of sporadic tree vegetation, and to the selective cuts practiced without silvicultural rules, and to the grazing practiced for centuries within the forests as happened in the oak or beech forests of the Apennines where large areas of pastures are wooded with beech trees of considerable age. This occurred especially in the central-southern and southern regions of Italy, where the woods were often state or communal property. The various historical and political vicissitudes, the frequent famines, the political and social changes such as the strong demographic growth that occurred from the seventeenth century, combined with the difficult geomorphological and hydrological conditions of the territory, had led to a progressive and exasperated search for fertile lands at the expense of the forest. A picture of the conditions of the woods of Molise in the early decades of the nineteenth century was offered by Pepe, an economist of the time, who observed how the irregularity of the uses and the absence of silvicultural rules had led to the creation of what the author defines as "wooded pastures". The practices commonly applied in the woods subject to grazing were the pollarding to obtain leaves and branches for the winter feeding of the livestock. To provide the flocks with food for the winter, the shepherds indeed cut a large number of branches. of oak in September and October. In other cases, such as the wooded pastures in Sardinia (the "salti"), the more extensive holm oak and downy oak forests, long grazed and subject to civic uses, had remained almost intact for centuries. In the years immediately following the unification of Italy, they were first alienated and subjected to intense cuts for thousands of hectares, radically transforming their original physiognomies. According to Alias, the main cause of the degradation of the salti in Sardinia is linked to the lack of respect for the ancient knowledge of the shepherds and the changed habits of the breeders, who from transhumant shepherds have become sedentary, with serious consequences linked to animal overloading such as the triggering of erosive processes and the lack of regneration of the forest, except in isolated patches covered by thorny bushes. As we have seen, with the term "defence" were indicated those wooded pastures or pasture woods of municipal property and burdened with rights of civic use such as the woodcutting, the acorn gathering and the grazing in favour of the local populations. Even today the term "defence" frequently recurs as a toponym of places in Southern Italy and, as the very meaning of the word indicates, the defences were particularly protected by the local populations who built, where possible, on the perimeter of the interested properties, dry stone walls that today represent important aspects of the historical identity of the landscape (Fig. 4.8).

Regarding the rules that regulated grazing and cutting within each defence, Bortolotti argued that in Abruzzo, like those of Pescasseroli, Opi and Pescocostanzo, the only animals to which the right of grazing was granted were cattle and horses, while there was an absolute prohibition of access to goats and sheep as well as absolutely prohibited were cuts of any kind except the pollarding which, evidently, were carried out when the coverage of the tree canopies became too dense and continuous making the grassy crust disappear (Bortolotti 1985). The situation was however different depending on the locality since, according to the testimony of De Crescenzi, in Molise, for the communal woods, a rotation of cuts had been established for sections where civic uses were exercised even if the population did not respect such rules (De Crescenzi 1865). These were constraints and prescriptions imposed by one of the first forestry laws promulgated in pre-unification Italy, the forestry law of 1826, issued by Francesco I king of the Two Sicilies, which imposed for the state, municipal and other entities the subdivision of the woods, especially if located in conditions of steep slope, in sections in "such number as to ensure (art. 72) their conservation, in general, and spontaneous renewal, in particular, for the purposes of their *perpetual income*" (Gualdi and Tartarino 2006). In the stands, it

Fig. 4.8 Beech trees shaped by grazing in the Aspromonte Mountain (Calabria region). Both the small bushes on the right and the tall tree on the left are beech trees (Fagus sylvatica). The first have been adapted to the browsing of the grazing animals

was made obligatory to carry out clear cuts with the release of at least 15 trees called "hope trees" per moggio (1/3 of a hectare), in order to guarantee their renewal and to put the cuts in defence, that is to prohibit the access of livestock for grazing (Del Favero 2008). It was also forbidden (art. 37) the execution of those by choice or by jump unless they had a phytosanitary or thinning character. It was also specified which species should be kept as high trunk both for the production of timber for work such as the beech, oaks (Turkey oak, downy oak, English oak, holm oak, the Trojan oak) and the field elm, for the production of fruit appetised by livestock. Compared to the wooded pastures of the South, it seems that in the Alpine arc, especially when they belonged to a private owner, they were subject to particular care especially in relation to the renewal of young seedlings which were protected with stakes and with branches, until they had reached a development in height such that they would not be damaged (Gori Montanelli 1939) (Fig. 4.9).

Fig. 4.9 A century old, pastured wood of holm oak (Quercus ilex) in Sardinia. The similar distance from the ground of the lower part of the crown of the trees is determined by the height reached by the grazing animals, mostly cows. This is a widespread feature of wooded pastures and pastured woods in the Mediterranean landscape

4.8 The Cycle of Alder Cultivation in Liguria

An interesting case to explain both the wooded pasture, but especially the management system in which such cultivation was inserted, is the practice of alder cultivation, widespread in the Val d'Aveto on the Ligurian Apennines between the XVIII and XIX century, at least according to the documents found during investigations carried out by Diego Moreno and Roberta Cevasco. In particular, the source that attests, even if implicitly, its existence was *The delivery of the woods and forests* (Moreno 1990). This act was produced by the forestry administration of the Kingdom of Sardinia in 1822. Each owner was obliged to describe, on the basis of a statistical procedure, their own woods in front of a judge ("delivery") according to the categories proposed by the legal terminology ("coppice wood", "high forest", "mixed forest"). The owners did not accept the imposed terminology but adopted the local one. The classifications reported are, for the most part, coinciding with grazed woods and wooded pastures depending on the stage of development and cultivation carried out within the forest stands. In particular, the particles to white alder *(Alnus incana)*, pure or mixed with beech or Turkey oak, and, to a lesser extent, those pure to beech were the object of a traditional system that included the practices of temporary cereal cultivation ("arable land", "nursery", rye clearings and harvest) (Moreno 1990). These agricultural cultivations were included in a cultivation cycle called alder cultivation (Fig. 4.10).

The cycle lasted about 12 years and was made up of successive stages that took different names depending on the development level of the forest. The cycle began by cutting down the existing alders with rotations of 3–12 years, depending on the fertility conditions of the plot. The other species present (beech and Turkey oak) were either pollarded (beech) or lopped (scalvatura) every 1–3 years (Turkey oak). The term "ground with trees" it was used to the indicate the overgrowth that was created

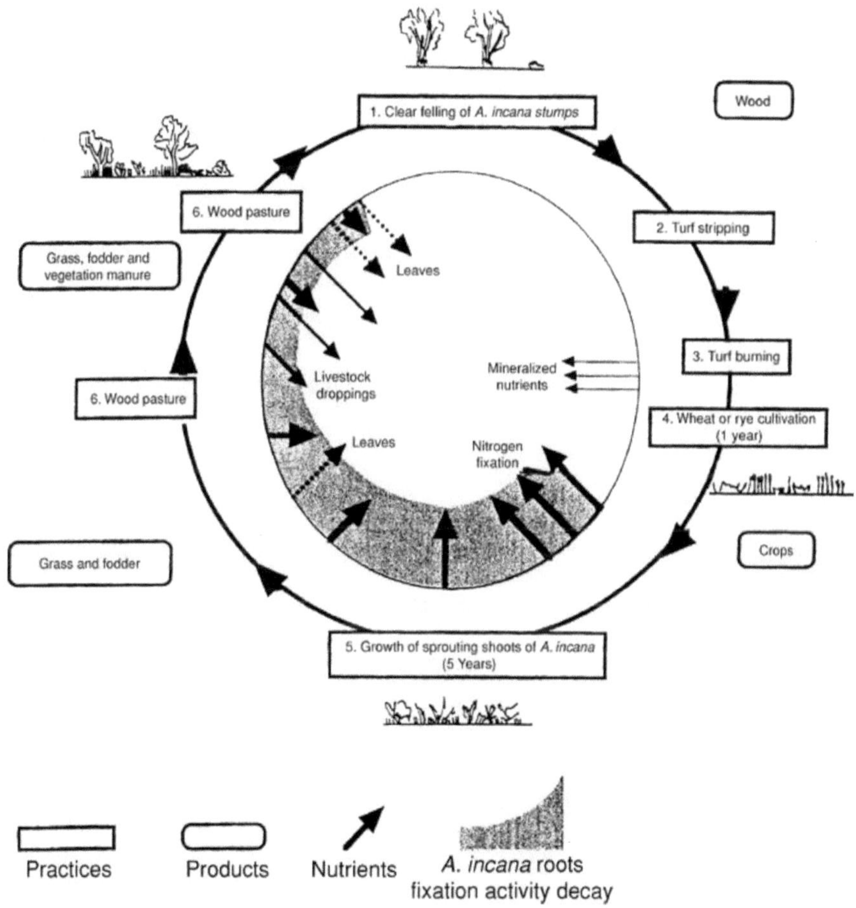

Fig. 4.10 The "alnoculture" cycle on the Ligurian-Emilian Apennines in the early nineteenth century is a typical example of the multiple systems of traditional agriculture, capable of "activating" local environmental resources and enhance biodiversity. The peculiarity of "alnoculture" was its utilization of the fertilizing properties of alder, which contains nitrogen-fixing bacteria in its root nodules (Bertolotto and Cevasco 2000, pp. 190–202, p. 198)

as a result of the coppicing of the alder, while other species were kept pollarded or as high forests. The "ground with trees" were formations that seem to correspond to the wooded pastures, even though the lands were liable not only to mowing but also to temporary crops. Beyond the strictly economic value of cereal production, which could be obtained from the cultivation of the woods in the forest properties of the Genoese Apennines, we know that this was also carried out with the function of defending the tree vegetation in the stages of renewal of pastures (Cevasco 2007). In order to cultivate the lands, the "ronco" was practiced, using the wood obtained from the cutting of the alder coppice and from that of the thorny shrubs (such as brambles and hawthorns). They proceeded in winter and they "made strips of branches and wood as long as the ronco, then they loosened the clods and covered them all like a charcoal kiln, then, when the wood was dry, they set fire; when everything was burned, they hoed around them and spread the ashes. In autumn they sowed wheat, and if the soil was acidic and poor, oats. Around the ronchi they blocked with branches and thorns to prevent the animals from going there" (Cevasco 2007). The practice of using fire confined in small furnaces (covered ronco) is likely to have helped preserve fruit trees such as cherry and other forage trees such as oak and ash; large trees of these species are still present in the historical sites of alder cultivation. The burning with covered fire is mandatory in grassy lands where the vegetative carpet, which burns poorly in summer, regenerates with increased vigour if the burning with current fire is applied. However, it has fewer drawbacks for the overgrowth. This system, which can be practiced from the beginning of spring, does not pose dangers for the stumps or for the present trunks and gives less risk of fire to the whole forest (Bertolotto and Cevasco 2000). The ronco allowed "the fattening of the soil" with the ashes and eliminated the thorny shrub species (Moreno 1990). While some thorny shrubs, especially hawthorns, were kept as hedges for the protection of the temporary seedbed (rye or oats) from grazing. The seedbed or sowing land lasted 2–5 years, generally the cereals sown were rye or oats (Cevasco 2007). The next phase was called "bosco alberato di costi" and "bosco alberato"; both indicated two evolutionary phases following the coppicing of the alder. The first described a young overgrowth where the alder stumps began to sprout again. The grazing of livestock, generally sheep and goats, to a lesser extent cows, was carried out after the last harvest, when this did not constitute an obstacle in the growth of the young shoots (Bertolotto and Cevasco 2000), and the grazing lasted until the "bosco alberato" returned to be mature for the next coppicing. The peculiarity of this agro-silvo-pastoral practice lies in the knowledge of the fertilising properties of the alder: "they usually make ronchi in the same territories and also collect the leaf for the manure". For the same purpose, the ronco was practiced, in this way "the soil was enriched by burning wood" (Bertolotto and Cevasco 2000) and livestock was introduced to the pasture.

This empirical knowledge of the fertilising properties of the alder extended throughout the eastern arc of the Ligurian Apennines: "the alder cultivation system of the Val d'Aveto must be compared to a second agroforestry system in which the genus Alnus, in this case of the glutinous species (black alder), was distributed in co-planting in the terraced fruit chestnut groves". Traces of it are found in Val di Vara and

Val Fontebuona. In the latter valley, the planting of alder in the fruit chestnut grove has been attributed to a parish priest in the early decades of the nineteenth century. However, there is interesting business documentation that indicates the existence in the nearby valley, where alder and chestnut appear to be grown together in plots from the mid-eighteenth century (Cevasco 2007). The practice of alder cultivation ceased during the last 150 years (Bertolotto and Cevasco 2000).

4.9 Woods, Pastures and Agriculture Between the Eighteenth and Nineteenth Centuries

Between the seventeenth and eighteenth centuries, the decline of Venice's trade, somewhat delayed compared to that of Florence, also caused a significant reflux of capital towards the countryside in Veneto. However, only in the second half of the eighteenth century will the activity of a new middle-bourgeois class of the mainland become visible, as an alternative to a feudal land aristocracy that was nevertheless the architect of a renewed interest in land ownership. The many 18th-century villas built in the Venetian countryside are an important fact, given that—more than in Tuscany or Lazio—they are the centre of a real noble agricultural business. Agricultural investments were not only resolved in the construction of buildings but also in the works of transformation and agricultural colonisation, the clearing of uncultivated lands and forests. This Venetian trend finds wide confirmation in other sectors of the peninsula: Piedmont, Lombardy, Liguria, Sicily, where the second half of the eighteenth century marks the flowering of noble villas that assume importance not only as leisure but as centres of capitalist investment. In Tuscany this had already happened in previous centuries, but now the development of the farm as a centre of a complex organisation of the large noble estate, annexed to a large manor villa, is beginning to be outlined. These are new capitalist production relationships intertwined with the old feudal relationships; the aristocracy is certainly on the way to its bourgeoisification, with attitudes tending to change the old power relations especially in relation to ecclesiastical abuses, but it will have to change again taking into account the effects of the French revolution and the "Risorgimento", the historical period during which Italy achieved national unity. Throughout the eighteenth century, the predominant form of management in Tuscany, as in the Po Valley and other regions of central and southern Italy, was sharecropping, based on plot sizes that, as in the Carolingian manor, had an extent corresponding to the working capacity of a peasant family. However, the conditions of the sharecroppers were becoming increasingly impoverished everywhere, with a progressive modification of the sharecropping system that would lead, especially in the north, to the formation of a proletariat that created the conditions for a new form of capitalist management.

In the second half of the eighteenth century, the communal uses inherited from the barbaric period, such as the right of locals to use the land after the harvest to exercise their rights, began to be increasingly questioned. These limitations were added to the rights of landowners to enclose their lands, which together with the liberalisation of grain trade and the claim of mortmain became the welding point between the landed aristocracy and the new middle agricultural class. Indeed, if the reduced investments typical of open field regimes could allow such a model, now tree and shrub plantations, systems to favour cereal cultivation, dry and irrigated artificial meadows, with the mass of related investments could no longer allow the tradition of communal uses. This is contributed to by a favourable evolution of market prices for cereal cultivation, so that throughout Italy measures are taken to limit grazing, facilitate the closure of lands and remove impediments to the freedom of cultivation. In Tuscany, the measures of Pietro Leopoldo are remembered, which restored the freedom of cultivation in 1775 and 1880 and the freedom of deforestation and clearing in 1776 and 1780, those that abolished the servitudes of grazing in the Pisa province and in the Pistoia mountain (1776), and those that ensured the redemption of grazing uses in the Sienese Maremma and imposed the closure of redeemed lands in 1778. Already in 1765, moreover, Venice had proceeded to limit grazing uses; in 1786 this limitation is then aggravated, and grazing uses are abolished, while the closure of lands is favoured. More tenacious was the resistance of the old regime in the south and in the State of the Church. In the Neapolitan area, only in 1792 was the redemption of grazing uses and the closure of lands allowed, but the measure remained without even a principle of implementation and it was necessary to wait for the Napoleonic laws subverting feudalism before closed fields could find a legal regime of implementation.

This progressive liberalisation of markets and property favoured a general increase in deforestation, a phenomenon that historically seems to reach its maximum intensity precisely in these years. Indeed, the problem of forest restoration of the mountain slopes degraded by deforestation is already urgently imposed at the end of the eighteenth century. But perhaps the spirit of the times responded more to the freedom of deforestation granted by Pietro Leopoldo, than to the protective laws promulgated in the Neapolitan area in 1749, 1756, 1762. The increase in deforestation is however greater than the extension of the cultivated lands and this is probably explained by the need for greater use of forest resources for commercial purposes implemented in various areas of Italy from Veneto to Calabria. In the nineteenth century, the need to cultivate new mountainous or hilly slopes becomes more consistent, also because the increase in population, which went from 11 million at the beginning of the eighteenth century to 20 million at the end of the century, does not find correspondence in an increase in the manufacturing industry—as in England or France—that could absorb the excess labour. The new generations seek subsistence in agricultural activities, and this would be impossible without new management forms of the hilly areas, a phenomenon widespread throughout Italy, albeit with different aspects: the arrangement into terraces or steps, with dry stone walls is adopted not only for the most valuable plantations, but also for fruit chestnut groves. With land reclamation, hillside arrangements, the construction of new farm buildings and the densification of tree plantations in mixed farming, large sectors of the Tuscan landscape begin to take on an aspect not very different from the current one. From the 16th to the eighteenth century, progress, albeit slow, of science and agronomic techniques had been essentially linked to the progressive penetration of commercial and capitalist relations in the economy of the manorial estate. The driving centres of such progress, the centres of diffusion of the new literature and of the new agronomic techniques were Venice and Florence. But from the end of the eighteenth century, the north undoubtedly surpasses Florence: the capitalist development of agriculture in the Po Valley gives a more decisive impetus towards innovation and development than in Tuscany. In the Po Valley, the farmstead replaces the single farm units, it benefits of constant capitalist investments and employs wage labour that replaces the peasant family.

In Tuscany, the phenomenon of the expansion of large capitalist property is recorded in two aspects. On one hand, we have the increase of country houses for the families of the sharecroppers, that from 1830 to 1854, only considering large farm estates, pass from 12,000 to 15,000. On the other hand, we have their consolidation: in fact, out of 50,000 farmhouses existing in Tuscany, at least 12,000 were combined into 1000 large farms. This was the direct result of the Lorena reforms and the Napoleonic era. A further influence on the forest was however the technical development of agriculture that occurred towards the middle of the nineteenth century, when the development of new forage crops favours the continuous rotation as an agricultural system, leading to the slow abandonment of pastures and to their cultivation. In the hills, the terracing arrangements extend and throughout central Italy (Umbria and Marche) the tree-lined system continues to prevail, which is perfected and extended, with a further extension of the hillside land arrangements.

In Southern Italy, things go differently. The evolution of agricultural systems had always been conditioned by the widespread diffusion of civic uses of grazing and sowing, on feudal and communal lands, in a regime of open fields. In fact, the vast majority of lands appear divided, on the eve of the overthrow from feudalism, into feudal, communal or ecclesiastical properties. Where the use of wage labourers has not been established, the land is offered to people who work the lands with their own means and pay the lord a share of the products. There are therefore no sharecroppers and stable peasant business (except for some exceptions), but the great mass of the rural populations is made up of "land tenants" who year by year cultivate a multitude of small plots of land variously located on immense expanses of large estates. These techniques, which exclude forage crops for climatic reasons, are based on regimes of open fields, which with the uses of grazing also allow to maintain some livestock. However, the feudal lords tend to exclude from civic uses their best lands, until the middle of the eighteenth century (Fig. 4.11).

Between 1806 and 1835, legislation tends to abolish feudalism. In the Neapolitan area, there is a mass distribution, that is, a part of the lands is attributed to the communities, but the farmer's right to stability, on the land on which he had he has passed on his rights of logging, grazing and sowing, is not recognized. This is not offset by the assignment to the community of lands in proportion to the civic uses exercised in the past. In general, however, the lands are reassigned to large owners and many of the most conspicuous estates are formed in this way. The peasants, unable to enjoy the rights of civic use, cannot even feed the livestock with free grazing. All of this leads to a massive extension of deforestation due to the free hand that

Fig. 4.11 As we see in this detailed fresco of Tuscany the agricultural landscape of the region was characterized by trees and groves scattered in the countryside with genial care and attention for the most minute details (Benozzo Gozzoli 1459—Palazzo Medici Ricciardi—Florence)

is thus left to entrepreneurial activities, without this involving a reorganisation in a productive sense of the agricultural landscape that remains of open fields. Some areas of the country, such as the Tavoliere (in the Puglia regiona), the Agro Romano, the Maremme, deserve a separate discussion since here the grass fields and wild breeding, often transhumant, prevail. In Maremma, the reclamation undertaken by Leopold II of Lorain in 1828 is completed on the eve of Unity of Italy on just under 10,000 ha against the 50,000 that remained to be drained. In the province of Grosseto, between 1828 and 1843, thousands of hectares of uncultivated fields are now cultivated and no less than 453 new rural buildings built. The landscape of the forest, the scrub, the pastures, typical of a grass field agricultural system, although still dominant, now begins to shrink compared to the less formless one marked by the regular alternation of fallow.

4.10 The Greek Sila

The name "Greek Sila" refers to the Greek settlements established in these area starting from the 8th century, communities speaking a dialect derived from the Greek language still exist today. Even the Sila woods, in the Calabria region, appear as largely subject to grazing, in particular in the form of wooded pastures, giving us an interesting cross-section of the situation in the south of Italy (Gallo and Jovino 2001). During the French domination over the Kingdom of Naples (1806–1814) legislative provisions are issued that deeply affect the organisation of the State and local administration. Already in 1806 feudalism was suppressed; a similar measure undergoes the ecclesiastical mortmain with the consequence that all goods, and in particular the properties of the clergy, are expropriated and aggregated to the Royal Domain; the overthrow involves for the nobles the conservation of the goods acquired with regular deed and the loss of goods of feudal origin. A feudal Commission was also established in 1807 with the aim of dealing with and deciding in a short time and definitively the disputes between the civic administrations and the feudal lords: this Commission in the years 1809–1810 issues sentences that settle the disputes between the municipalities of Bocchigliero, Calopezzati, Caloveto and Crosia and the prince of Campana; between the municipality of Campana and the former baron Sambiase; between the municipalities of Corigliano and Rossano and the duke former baron of Corigliano; between the municipalities of Cropalati, Longobucco, Paludi and Rossano and Prince Borghese; between the municipality of Scala Coeli and the former baron Sollazzo and the former baron Abenante; and finally between the municipality of Umbriatico and the prince of Pallagorio. In the procedural acts and in the sentences themselves, there is information on the state of the forests, on the use of the soil and, above all, indications on the exercise of civic uses in the Sila territories where for centuries the populations have logged, grazed their flock in certain periods of the year, when it did not damage the vegetation, they have collected fruits such as chestnuts and acorns at the bar and have sown. From the intendant ordinances on the definition of the causes between the civic administrations and the former barons and on the dissolution of promiscuity between two or more municipalities, it is possible extract useful elements to define the state of agricultural agreements, the exercise and quality of civic uses, the methods in cultivation of fields and in the practice of grazing. The lands with a more pronounced agricultural vocation such as those located in the territory of Bocchigliero, albeit very limited in extension, are subject to sowing every other year; in the sowing year the state imposes a tax on the land and the citizens "graze the spike"; "in the empty year the municipality sold the grass for a period of six months, remaining for the other six months for the use of the citizens". As far as concerns civic uses, it is noted that they are limited to "watering, overnighting and dry wooding". But in the same municipality of Bocchigliero other funds are sown every five years and "in the four subsequent years that were destined for rest, the grass remained for the free use of the animals".

In the acts relating to the municipality of Calopezzati, news is found on forested territories: "in the wooded places the animals can overnight, and citizens use the timber for farm use and other civil expenses. In the sowing year the citizens use the grazing of the spike and the grass, until the first Sunday in October. In the following two years they exercise grazing from the 1st of May until the first Sunday in October". In the woods, citizens have the right to cut wood indiscriminately and only in a few cases is it specified that the firewood right is limited to the acquisition of timber intended only "for uses of rural tools". In the best lands, such as Mirto di Crosia, olives, mulberries, oaks and fruit trees are planted. Sometimes it is possible to obtain data on the different quality of the lands on which the superintendent issues an ordinance: this is the case of the fiefs Dragonara and Malena located in the territory of Cropalati: "Dragonara: 350 tomolate, of which 250 cults sown with oaks and 100 of rocky and wooded lands; Malena: 3000 tomolate of which 200 are sown and the rest all uncultivated, wooded with oaks and wooded".

In these territories, the feudal lord Prince Borghese derives the revenues from the acorns and has right of grazing from September 8th to December 24th; subsequently the municipality sells the grazing and the acorns from December 25th to September 7th. In the sowing year (the fifth after four years of grazing) the municipality "sells the spicaggio. Citizens can graze there for the entire year, even fruit hanging from the acorns, three plough oxen and two cows for each of them; they can also cut wood for fire use, for farm tools and for buildings". A situation emerges overall uncertain in land ownership and in the exploitation of agricultural and forestry resources: consider the effects of free grazing and the state of open woods not only for the collection of dry wood but also for cutting for the construction of rural tools such as the plough or the cart, and for the construction of houses. The rotation of sowing has a generally triennial trend, which means that wheat is produced for bread making only every three years; this practice, however, is inherited from the '700: in the land register of Campana of 1742 it is said that the Tacavallicata fund of 100 tomolate is sown in favour of the clergy "two years every six years and for the other 4 must remain for the grazing".

The "sowing" lands, usually wooded with oaks and fruit trees, are mainly exploited for grazing. The new French government inherits a difficult economic situation and establishes a system of relationships based on the balance between the interests of the state (which receives the income from the sown land for a semester), the civic administrations (which rent the grazing land for another semester) and the citizens (who exercise their uses for two semesters). In this situation, the uses and utilisation of the funds are foreseen: but it is clear that no one has a direct and exclusive interest in land improvements. The protection of the forest is not seen as an investment for the future but only to ensure a more lasting exploitation. The forests, which constitute the most substantial part of the territory of the municipalities located in the inner mountainous area, are used for the supply of timber and grazing, but to a lesser extent due to the distances and the lack of roads and means of transport. The interests of the citizens are instead directed towards the hill and the funds located near the urban centres.

The government, after having issued legislative measures to protect the forests of the Kingdom, intervenes specifically in 1810 for the safeguarding of the Sila forests considered a "national importance asset" for the timber obtained from it and destined for the shipyards; the trunks of the centuries-old Sila pine trees are indispensable to the Navy of the Kingdom. From the examination of legislative measures and local documentation, a policy of forest conservation in a global sense does not emerge; ultimately, it only seeks to limit the destruction of the forest to prolong its exploitation over time. The French only manage to establish an economic policy capable of bringing about changes in the social structure: the quotation of ex-feudal and ex-ecclesiastical lands with the assignment to peasants and particularly to the most needy, if fully implemented, would have given rise to a new class of small owners interested in supporting the new regime and potentially committed to improving rural lands and protecting the forest. With the final fall of Napoleon, the Bourbons return to the Kingdom of Naples: this then determines the return to the clergy of much of the properties expropriated by the previous government; the nobles also regain power and influence, but do not resume the feudal prerogatives that enlightenment has swept away.

4.11 The Diversification of the Apennine Landscape: The Case of Moscheta

The analysis of a portion of the northern Apennines straddling Tuscany and Emilia-Romagna regione, the so-called Tuscan Romagna, allows us to better understand the structure of the Apennine mountain landscape and the importance of wooded pastures in the process of diversifying the Apennine landscape which, as we will see, had nothing to envy in terms of complexity to those of the hills, or the plains (Agnoletti 2007). This territory is located around the Abbey of Moscheta, in the municipality of Firenzuola, in the province of Florence. The Abbey of Moscheta was founded by Saint Giovanni Gualberto, the father of the Vallombrosan order, a minor order of the Benedictines monks, in the first half of the eleventh century. It is said that, stimulated by an old monk from the Florentine Badia, he went to the Apennines in search of suitable places to build a monastery and visited the areas of Ronta, Razzuolo, Biforco on the Lamone river and perhaps also Moscheta. Having learned of these searches, Count Anselm of Pietramala, of the powerful Ubaldini consortium, to which the important Archbishop Ruggieri also belonged, decided to donate to the Saint a property he had in Moscheta with a mile of forests around to build, as the hagiographic narration says, an abbey at his own expense. According to the sources cited, it was a territory of about 800 ha with beeches, chestnuts and oaks.

Characteristic of Giovanni Gualberto was the intention to improve the living conditions of the population through fieldwork, setting an example with his monks. Thus, the lands donated to the abbeys were tilled to obtain arable land or made suitable for livestock grazing. Later, thanks to bequests, the abbey became increasingly wealthy, owning both arable land and pastures even in distant areas, so much so that in the thirteenth century it appeared as one of the most prosperous that the Vallombrosan order had throughout Tuscany. Despite in the fifteenth century, 1400 staia of wheat and 100 of grain were harvested, the people of Moscheta consisted only of four families of workers from the lands adjacent to the abbey. Despite the scant news on the subject, we can reasonably think that the monks extended the cultivation of chestnuts, teaching their breeding, pruning and grafting. It is also likely that the fruit chestnut grove was sometimes the site of small temporary agricultural crops, especially near the monastery, in whose garden and small field, in addition to vegetables, medicinal herbs were also grown for the management of the hospital (Fig. 4.12).

In an undated drawing, but presumably from the eighteenth century, deposited at the State Archive of Florence, the abbey of Moscheta is depicted surrounded by chestnuts and beeches and also by a certain number of firs. The surrounding beech and oak forests must have been, then, mostly high forest in which sheep and cattle grazed, for whose winter feeding some leaves were collected cutting the most leafy branches in some perhaps coppice wood, before adopting transhumance in Maremma. The exclusion of goat grazing is certain, at least in the chestnut groves, as evidenced by the complaints of the population according to which the wandering of these animals made it impossible to raise graft shoots in the chestnut grove, so much so as to fear the loss of the chestnut grove itself which then, as in other parts of the Tuscan mountain, had to be kept very dense, with slender plants and low production.

Fig. 4.12 In this undated drawing, presumably from the eighteenth century, deposited at the State Archive of Florence, the abbey of Moscheta is depicted surrounded by chestnuts and beeches and also by firs

On the local woods, a visit made in these places in May 1626 informs us. "As for the Alps of Firenzuola—the visitors report—they are all facing north and the waters rain in the ecclesiastical Romagna and in them is included the Abbey of Moscheta of Cardinal Capponi "... and we found that the thickets and woods of these Alps are well preserved except in two places where several cultivated plots have been found, both old and new and 500 staia of wheat are sown there every year". Not so in the Alps of Mugello that are found "ruined and damaged by iron and fire, in particular those of the municipality of Ronta and neighbouring municipalities, deteriorated for the charcoal that is made there and that all serves for the city of Florence" made and sold by that poor people, who with such industry tried to live as best they could. From the sixteenth-seventeenth century begins the period of decline of Moscheta as a religious centre which was first commended and then suppressed during the reforms of Pietro Leopoldo (1748), had its assets sold at auction. The new owners were the Martini family of Florence who reorganised it into a farm, creating several estates according to the typical sharecropping system, where the main income came from livestock farming and timber production and with a significant production of chestnut flour. The crops, sometimes used in rotation, were mainly wheat, vetch, potatoes, barley; chestnuts and marrons were a very important resource; the livestock heritage was significant and mainly composed of sheep and goats, although cattle and donkeys were crucial for the management of the estates; the woods were used for brushwood, firewood and grazing.

By the 1800s, the territory of Moscheta can be studied in detail through the general Tuscan land registry, a work that began in the French era and led to the creation of a precise mapping of the territory of almost all of Tuscany, with an analytical description of land use and owners. At the time of the land registry, around the second-third decade of the 1800s, the territory of Moscheta appears organised as a farm with a series of estates, each of which housed a tenant farming family. The available archive documents, although allowing a continuous analysis for a period after 1865, show the presence of a well-structured company, with eight estates whose lands do not always appear consolidated around the rural building, but often split into several lots distant from each other. The breeding of wild cattle is the main economic activity, which justifies the dominant role of grazing spaces in the local landscape, as generally highlighted for other parts of the Tuscan mountain. The timber assortments listed in the balance sheet from 1866 onwards list sixteen types: poles, staves (barrels, vats, barrels), logs, poles, mother vines, sorb tops, chestnut halves, walnut footboards, planes, currents, to which are added stacks, bundles, embers and charcoal. Among the timber assortments, walnut footboards (sold for 25 lire each) and beams (8 lire each) stand out in terms of economic value. The wealth and variety of types confirm the presence of traditional productions, which however see the chestnut as an important protagonist also from a quantitative point of view (especially staves, poles, halves, currents) in addition to the planes, and the particular importance of relatively little present species, such as the walnut, but highly appreciated for their resistance qualities. In reality, the staves were also produced with beech, but preferably from coppice woods of variable age from 12 to 20 years. The pasture is mainly located on Monte Acuto and on the slopes of Monte Fellone, occupying both the high areas and those located lower down. In most cases, these are wooded pastures, in fact, bare pasture is only present in 45 ha, partly located on the top of Monte Acuto. The pasture with beech trees represents the most extensive type, with a good 82 ha, but the beech characterises in total nine different types of pastures in combination with hornbeam, chestnut, broom and turkey oak. The landscape is dominated by wooded pasture and then by the forest, followed by bare pasture and fruit chestnut grove, but the land registry records as many as 89 different categories of land use. The wooded pasture is divided into as many as 36 types, which differ for the presence of different tree species, which constituted as many pieces of a complex landscape mosaic and as many "habitats" expression of a great biodiversity linked to human activity and a form of landscape widespread from the north to the south of Europe. In practice, except for one type reported by the cadastre, all the pastures of Moscheta are wooded pastures. It is the element that dominates the local landscape, an expression of that typical structure of the Apennine pastures where the presence of trees in the pastures responds to specific needs, such as the mitigation of summer temperatures to allow the rest of the livestock, the production of fruits, leaves and timber, the lowering of the soil temperature, as documented in many Mediterranean countries, according to the logic of the multicultural systems (Table 4.1).

The categories reported by the cadastre also include meadows and wooded meadows with different types of tree species. This is another significant element of the rural system and landscape, which brings the component assigned to meadows and pastures to about 55% of the total of land uses. The emphasis placed in the cadastral definition also indicates the main function of the identified land use, so that "meadow and forest" and "meadow and forest of beeches and chestnuts" suggest to include these surfaces among those pertaining to the meadow. The meadows are divided into as many as 15 different types, among which stand out those with beech (which appears to be the most important in terms of extension), chestnut, oaks and fruit trees. The two categories "natural meadow" and "artificial meadow", the first of which concerns 31% of the total, indicate a different origin of the meadows, which the surveyor must somehow have identified. It is known how the practice of mowing can increase the specific biodiversity of meadows, qualifying as a real practice of "activation of resources" necessary for the increase of the natural capital available for the life of populations. The detailed statistics relating to the extension of individual land uses show the strong prevalence of "pasture with beeches", equal to 18.6% of the total, then follow the "pasture and beech forest" with 15%, the "pasture" with 11%, the "wooded pasture" with 10% and the "pasture with oaks" with 9.5%. The predominance of the beech is a constant element also in the cadastral category "forest", as we will see, and is mainly present in the southern area of the study area which is predominantly characterised by north exposure. Overall, the pastures that see the presence of the beech, in various percentages, extend for 191 ha, equal to 43.5% of the total pastures and 21% of the entire surface of the study area. This confirms the predominance of this species both in the form of forest and of isolated trees, constituting in the case of pastures a landscape typology of historical character of regional importance.

Table 4.1 The list of existing land uses in Moscheta according to the Tuscan nineteenth-century land registry

Cadastre classes—1832 (tot. 59)
Pasture, pasture and untilled land, pasture with rocks, pastures with caves
Pasture with trees, pasture with green oak, young oak, chestnut, green oak and horn-beam, fruits, horn-beam and beech-tree, beech-tree, beech tree, broom and horn-beam, chestnut green oak and beech tree, hornbeam and young oak, rocks and beech tree, sucker and chestnut, rocks and green oak, with oak
Natural meadow Meadow with trees, e.g. with green oak, young oak, green oak and oak, fruits, beech-tree, chestnut
Green oak wood, green oak and chestnut, beech-tree, horn-beam and chestnut, young oak, acorn wood Orchard, orchard with walnut-tree, orchard with mulberry-tree
Ploughed land
Ploughed land with trees, with chestnut, oak, walnut-tree and mulberry tree, sucker and fruit, green oak
Chestnut-grove Pasture with suckers
Meadow with suckers Meadow and pasture
Wood and pasture, with green oak, green oak and oak, green oak, beech tree, young oak and chestnut
Ploughed land and pasture
Ploughed land, wood and meadow
Chestnut-grove and pasture
Meadow and wood, meadow and wood of beech tree and chestnut

There are 59 land uses, testifying to the fact that the Apennine landscape could be more complex and diversified than many plain landscapes. In these landscapes, wooded pastures and pasture woods are very numerous, while dense and extensive woods are not dominant as in today's landscape. In these landscapes, overall biodiversity was also higher

The forest in total only concerns 39% of the territory, and is therefore minority compared to the pasture. The most extensive type of forest is that of beech, while the oak grove only concerns 2.8%. It was used not only for wood and charcoal but also for the production of brushwood, regularly used and sold for Livestock feeding. 20% of the woods is made up of fruit chestnut trees, an important symbol of the cultural identity of this territory, especially for its ties with the life of the populations, while conifers, more representative of monastic culture, are not mentioned in the land registry, even though they appear in a eighteenth-century map where it is said they occupied an area of about 21 ha. It is possible that the chestnut grove present in 1832 overlapped with the fir grove in the previous century and that this was then destroyed. It would indeed be unlikely that the grand ducal surveyors would not indicate the presence of a forest of such importance, given that at Moscheta areas smaller than a hectare and of lesser economic value than a white fir forest are reported. In reality, the various forest species appear distributed in a large number of land uses and it is not possible to identify the total areas occupied by each of them. The presence of the beech forest and pasture, as well as the turkey oak and oaks, testifies to the importance of grazed woods for livestock farming, another fundamental type of historical landscape also present in the plains and well described by Latin authors, as we have seen. At least five categories of land use concern the chestnut grove: "marroneta", "marroneta with turkey oaks and pasture", "marroneta and beeches", "marroneta and pasture", "marroneta and tree-lined pasture". About 60% of these surfaces concern "marroneta and pasture"; however, it should be noted that chestnut trees also appear in other types of work and pastures, testifying to the wide use of a species of which everything was used: leaves, fruits, bark, wood, down to the soil resulting from the decomposition of the inside of the trunk in the plants. The land registry data show that in terms of complexity of structure the landscape of the Apennine mountain does not appear more simplified than that of the plain, but on the contrary it shows a more complex landscape mosaic: comparisons with similar studies done in the hilly area of Tuscany or in the plains indeed show a total number of land uses often lower. As for the woods understood as dense and homogeneous formations, they are minority in the landscape, here as in general in the rest of the Apennines mountains. In the following century, the abandonment of the territory of Moscheta leads to a profound change in the landscape. Today it is characterized by just 29 land uses, pastures have been reduced by 271% and wooded pastures in particular are only 12% of those present in the 1800s, while the regrowth of uncultivated forest is 90% (Fig. 4.13).

4.12 The Woods for Hunting

A configuration of the forest sometimes described by the land registries, quite widespread not only in the Apennines (as in the area of Moscheta previously treated) but also in the rest of Italy, are the forest structures created for hunting, in particular for the birds. The woods for hunting would deserve a long chapter, since they were applied management rules aimed at shaping the plants for the capture of different animal species and also to create an alternation between open and closed spaces, fundamental for hunting needs, but also for the landscape diversity that, especially in this last century, has been disappearing due to abandonment (Casanova and Sorbetti Guerri 2003). It is therefore a form of management that shaped even very large, wooded areas, significantly characterising the landscape. Descriptions and representations related to hunting have been present since the Etruscan period, as well as the great hunts of the imperial Roman period are well described. In our case, however, we are talking about management models, mainly inherited from Germanic culture, which have effectively influenced the forest landscape, modifying its natural characteristics. In Germany, until the beginning of the 1800s, the responsibility for managing

Fig. 4.13 Monumental chestnut trees in Moscheta

the forests of large feudal properties was assigned to the Jaeger, the body of hunters, through the creation of schools, the Meisterschulen, to teach such practices to young people. It is only later that, with the growing importance of timber, management is entrusted to the cameralists, administrators who mainly dealt with forest production. Unable to devote adequate space to this topic, we will therefore limit ourselves to hinting at their existence to understand their role in the forest landscape (Fig. 4.14).

In Italy, examples of forests managed for hunting are not very widespread, nor are there treaties where it is taught how to manage a forest for hunting, while there is a vast literature in both French and German. In the Middle Ages, hunting was an activity reserved mainly for the nobility: in 1549 Cosimo I de' Medici issued a decree that drastically restricted the rights of citizens and peasants, even though founded on centuries-old customs, to hunt, but similar measures can be found in various parts of Italy even in earlier times (Galloni 2000). An example of a hunting management model can be observed in an eighteenth-century land register that represents the grand ducal estate of San Rossore, near Pisa, divided into various "hunts", each of which managed for a specific game. The Grand Duke tackled the forest problem by resorting to the advice of a Flemish technician, Enrico Van Buggenhoudt, who in 1762 visited the estate according to a trend common of the period that saw the Gran Duke of Lorraine often use foreign technicians. A forest management plan was drawn up, oriented mainly to the integration of cutting, grazing and hunting activities within the estate, with fences to protect the seedlings following uses from the bite of

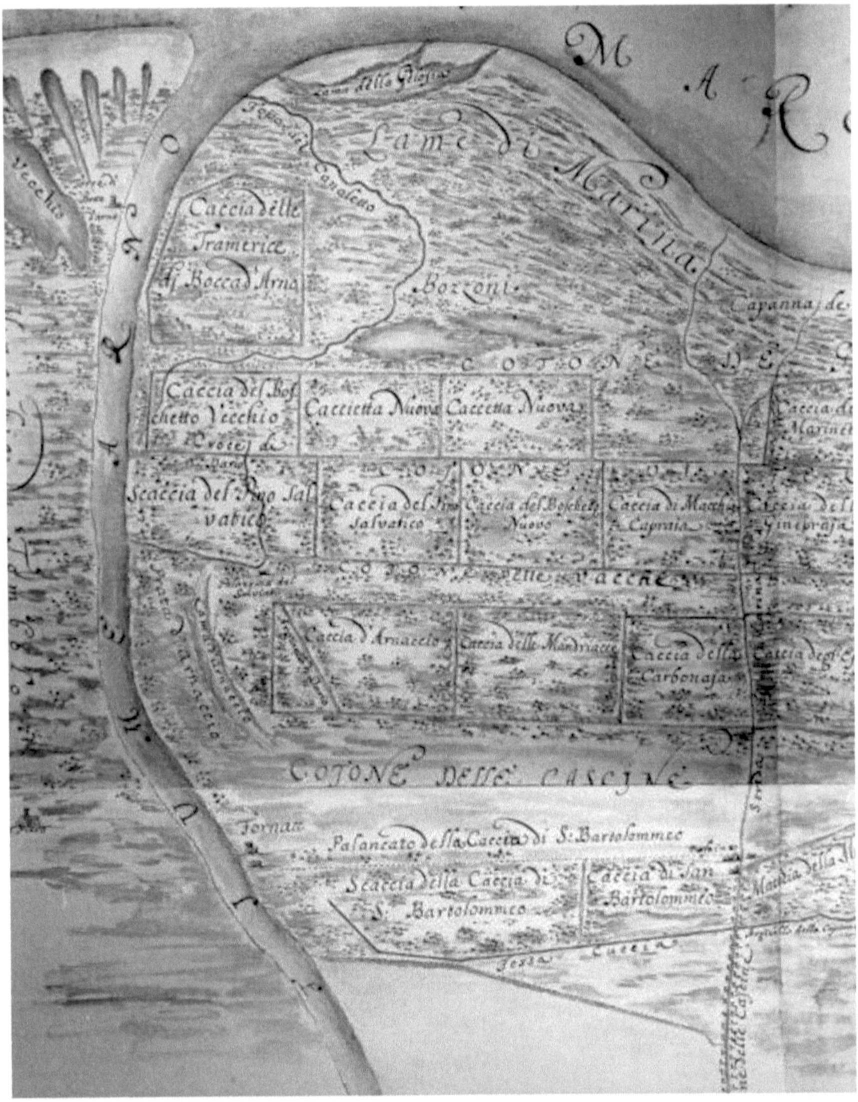

Fig. 4.14 1736 land register in which the forest of the grand ducal estate of San Rossore (Tuscany) is divided into "hunts" equivalent to different plots of forest managed to hunt a different type of game (State Archive of Florence)

livestock. Following a model diverging from that implemented in the "Medici farms", with the advent of the Lorraine the estates were not neither fractionated nor sold, but instead gradually adopted an organisation oriented towards large hunting reserves. Consequently to the hunting activity, the forest interventions at San Rossore were reduced both in number and in intensity, and were anyway aimed at implementing the wildlife heritage through the creation of numerous refuge areas for deer, wild boars and hares, of which the area was already particularly rich (Agnoletti 2005).

In Sicily, the Ficuzza forest, near Palermo, is remembered, whose historical origins date back to the end of the eighteenth century, when Ferdinand I of Bourbon, having occupied the Kingdom of Naples by the French, took refuge in Sicily and there built the Royal Casina where he could reside to exercise his passion for hunting, transforming this place into a hunting reserve. The same village of Ficuzza, home of the royal Bourbon palace and heart of the homonymous forest, was born and grew in function of the hunting reserve, to which a farm was attached: "beautiful and well-kept woods, numerous herds of all kinds, cultures and excellent pastures, abundant game were the pride of the King". Once he inherited the reserve, Ferdinand II abolished the hunting reserve and an agroforestry system based on the utilization of wood, the production of charcoal and pastoral activities was established in the forest (Barbera et al 2010) (Fig. 4.15).

Structures often described in land registers but widespread not only in the Apennines, but also in the rest of Italy, are those created for hunting, particularly for bird

Fig. 4.15 "Uccellare". Circular structure made of beech trees to catch birds, usually placed across mountain passes (Moscheta—Tuscany)

catching. These are plant structures with various configurations, relatively limited in extent, identified by various names such as "bird traps", "roccoli", "bresciane", "paretai", "ragnaie", to name just a few, and which were very widespread in the Middle Ages. They are often dedicated chapters in nineteenth-century treatises, being very complex creations, characterised by configurations in which various tree species were used, given particular and very varied forms, which were placed in different places depending on the types of birds to be caught. The "ragnaie" were often created near the gardens of the villas, about 15 m long, sometimes with internal paths for ornamental purposes and with trees planted very close together to create a kind of cage for capture.

The bird traps were placed in the birds' passage places, such as the ridges of the mountains and the passes, and had to be placed in open spaces, therefore without large trees in their vicinity, which in case were eliminated; they are sometimes still found today in the Apennines, although no longer used, often for the capture of thrushes. The "paretaio" was instead located lower down than the passes, requiring for this the construction of a dry wall, to give space to the capture nets supported on plant structures. Both the bird traps and the paretai provided for a hut of branches for the bird catcher, where the nets and calls were manoeuvred. As an example, it is worth mentioning the description in the National Catalogue of Historic Rural Landscapes of the Ministry of Agriculture, concerning Lombardy region (Cazzani 2010). Their total extent is still about 1550 ha in Lombardy, and in many cases they are reported on the cartography of the Military Geographic Institute. The Lombard area where many bird traps are still preserved extends between various municipalities of the provinces of Bergamo (particularly in the Hills of Bergamo, in Val Seriana, Val Brembana, Val Gandino and Val Cavallina) and Brescia (in the Upper Garda Bresciano, in Val Sabbia and Val Trompia). Many of these structures are currently located in the regional parks of Colli di Bergamo, Campo dei Fiori, Alto Garda Bresciano, Adda Nord, Groane, Valle del Lambro, Pineta di Appiano Gentile and Tradate and in the SCI (Site of Community Importance) Canto Alto and Valle del Giongo. The distribution of these bird traps, concentrated in the pre-Alpine area, is usually related to bird migration currents, at obligatory passage points such as ridges, hills or passes and connected to wooded areas or agricultural areas that attract birds. These structures were once widely spread in the national territory, made with tree individuals modelled for the purpose of facilitating the capture of various species of birds (Fig. 4.16).

The "roccolo" has been documented since at least the fifteenth century in Lombardy and Veneto, and a 1931 census reports that, of the 1890 identified in Italy, 880 were in Lombardy and 587 in Veneto. Its design is characterised by a circular or horseshoe-shaped plan, 50–60 m in diameter, surrounded by trees planted and pruned to create windowed plant galleries—often of hornbeam (*Carpinus betulus* L.)—in which vertical nets for bird capture are hidden. Attached to the pergola, called "tondo", made up of a wooden structure completely covered by regularly pruned vegetation, there is a three-storey tower building (the "casello") or a hut, hidden by thick vegetation, where the hunter would hide. The inner part of the roccolo is occupied by the tallest tree of all, surrounded by a grove of progressively decreasing height. In cultivated lowland or low-hill areas, the layout can also be

Fig. 4.16 "Roccolo del Sauch". This roccolo is a structure made of fir trees shaped to catch birds (Val di Cembra—Trentino)

rectangular and in this case it is called bresciana: it is distinguished from the roccolo by the absence of tall vegetation inside the pergola, left instead to grass or with pasture shrubs. The roccolo is therefore a complex "vegetal architecture" that requires continuous and planned maintenance that was once carried out by the "roccolista" (or "roccolante"). The roccolo was often associated with other roccoli, forming a complex landscape system in which the utilitarian-hunting value was connected to the aesthetic-decorative one, highly characterising the territory. Today, many roccoli still have characteristics of integrity. Some of them are kept active with traditional techniques, for the supply of calls of huntable species in Lombardy, where about forty roccoli are authorised to capture over 40,000 birds, or for scientific ringing. Some roccoli preserve the tree layout and are used as ornithological observatories, as public or private green spaces; most, however, have disappeared, or been reduced to ruins, integrated with the surrounding woods. Specific censuses recently carried out in the territory of the Val Gandino and the regional parks affected by the presence of the roccoli have identified over 200 roccoli at different levels of conservation. Of the roccoli surveyed, a minority (about 10%) is currently used for hunting purposes, while the majority are wild or destined for private greenery.

References

Agnoletti M (2002) Il problema scientifico e la metodologia della ricerca. In: Agnoletti M (ed) Il paesaggio agro-forestale toscano, strumenti per l'analisi la gestione e la conservazione. ARSIA, research report, Firenze

Agnoletti M (2005) L'evoluzione del paesaggio nella tenuta di Migliarino fra XIX e XX secolo. Regione Toscana, Firenze

Agnoletti M (2007) Il parco del paesaggio rurale appenninico di Moscheta. Pacini, Pisa

Agnoletti M (2009) Il paesaggio come risorsa. Castagneto negli ultimi due secoli. Edizioni ETS, Pisa

Andreis C (1993) Il manto vegetale nell'ultimo millennio. Storia economica e sociale di Bergamo, vol I. I caratteri originari della Bergamasca, Fondazione per la storia economica e sociale di Bergamo, Bergamo

Andreolli B, Fumagalli V, Montanari M (1988) Man and environment in the middle age. In: Salbitano F (ed) Human influence on forest ecosystems development in Europe, Pitagora editrice, Bologna

Barbera G, Cullotta S, Marino E (2010) In: Agnoletti M (ed) Paesaggi Rurali Storici. Per un catalogo nazionale. Laterza, Bari

Barsanti D (1987) Allevamento e transumanza in Toscana: pastori, bestiami e pascoli nei secoli XV–XIX. Edizioni Medicea, Firenze

Baruzzi M, Montanari M (1981) Porci e porcari nel Medioevo: paesaggio, economia, alimentazione. Clueb, Bologna

Bellandi F (2000) L'uomo, la terra, il bosco. Evoluzione e vicende del patrimonio boschivo del Mugello, della Val di Sieve e della Romagna Toscana dalla preistoria ad oggi. In: Guarducci A, Rombai L (eds) I problemi del bosco nel territorio fiorentino e toscano: realtà e prospettive. Centro editoriale toscano, Firenze

Bellucci P (1984) I Lorena in Toscana. Gli uomini e le opere. Edizioni Medicea, Firenze

Bertolotto S, Cevasco R (2000) The 'Alnocolture' system in the Ligurian Eastern Apennines: archive evidence. In: Agnoletti M, Anderson S (eds) Methods and approaches in forest history, vol 3. Wallingford, CABI Publishing

Bicchierai M (ed) (1995) Beni comuni e usi civici nella Toscana tardomedievale. Marsilio, Venezia

Biffi Tolomei M (1833) Saggio d'agricoltura pratica toscana e specialmente del contado fiorentino del senatore Matteo Biffi Tolomei. Giuseppe Tofani, Firenze

Bortolotti L (1985) Problemi di conservazione dei pascoli arborati («difese») d'Abruzzo. L'italia Forestale e Montana 45(2):65–72

Brusi G (1987) Premilcuore: vita e territorio di una comunità della Romagna toscana. MDM, Forlì

Casabianca A (1941) Notizie storiche sui principali luoghi del Chianti: Castellina, Radda, Gaiole, Brolio. Multigrafica editrice, Roma

Casanova P, Sorbetti Guerri F (2003) La caccia in Toscana negli ultimi sessant'anni. Evoluzione sociale, dell'ambiente e della caccia, vol I. Edizioni Polistampa, Firenze

Cassi L (2000) I toponimi. Regione Toscana, La Toscana dei boschi. Edizioni Vallombrosa

Cazzani A (2010) Uccellande lombarde. In: Agnoletti M (ed) Paesaggi Rurali Storici. Per un catalogo nazionale. Laterza, Bari

Cevasco R (2007) Memoria verde. Nuovi spazi per la geografia. Diabasis, Reggio Emilia

Cianferoni R, Fattori M (1992) La collina delle città e del sistema di mezzadria: profilo storico. In: Cianferoni R, Mancini F (eds) La collina nell'economia e nel paesaggio della Toscana. Atti dell'Accademia dei Georgofili, serie VII, XXXVIII, Firenze

De Crescenzi E (1865) Sullo scadimento dei boschi nell'Abruzzo Ulteriore II e sui mezzi di arrestarlo. L'Aquila

Del Favero R (2008) I boschi delle regioni meridionali e insulari d'Italia. Tipologia, funzionamento, selvicoltura. Cleup, Padova

Di Bérenger A (1887) Selvicoltura: trattato scritto per uso degli agenti forestali, ingegneri e possidenti di boschi. R. Marghieri, Napoli

Duby G (1970) L'economia rurale nell'Europa medievale. Francia, Inghilterra, Impero (secoli IX–XV). Laterza, Bari

Gabbrielli A (1980) Selvicoltura toscana nel '700. Parte prima. Annali Dell'accademia Italiana di Scienze Forestali 34:211–242

Gabbrielli A (1985) Selvicoltura toscana nel '700. Parte seconda. Annali Dell'accademia Italiana di Scienze Forestali 34:179–226

Gabbrielli A (1997) Le trasformazioni del paesaggio forestale in Toscana: un tentativo di sintesi storica. Accademia italiana di scienze forestali, Firenze

Gabbrielli A (2006) Le vicende storiche e demografiche italiane come cause dei cambiamenti del paesaggio forestale. Annali Dell'accademia Italiana di Scienze Forestali 55:133–166

Galletti P (1988) Bosco e spazi incolti nel territorio piacentino durante l'alto Medioevo. In: Andreolli B, Montanari M (eds) Il bosco nel Medioevo. Clueb, Bologna

Gallo MA, Jovino F (2001) Vicende storiche del bosco in Sila Greca. In: Agnoletti M (ed) Storia e Risorse Forestali (History and forest resources). Accademia Italiana di Scienze Forestali, Firenze

Galloni P (2000) Storia e cultura della caccia. Greentime, Bologna

Giumelli G (1932) Nozioni di selvicoltura. Vallardi, Milano

Gori Montanelli L (1939) Selvicoltura e alpicoltura. Lega, Faenza

Gualdi V, Tartarino P (2006) Altre riflessioni sulla gestione, su basi assestamentali, della foresta mediterranea europea. L'italia Forestale e Montana 61(3):478–487

Imberciadori I (1953) Campagna toscana nel '700: dalla Reggenza alla Restaurazione, 1737–1815. Accademia dei Georgofili, Firenze

Jones P (1974) La storia economica, in Storia d'Italia, Annali 2. Einaudi, Torino

Le Roy LE (2009) Histoire du climat depuis l'an mil. Flammarion, Paris

Menicacci M (2002) Il ceduo a sterzo nell'Appennino Pistoiese. In: Ciancio O, Nocentini S (eds) Il bosco ceduo in Italia. Accademia italiana di scienze forestali, Firenze

Moreno D (1990) Dal documento al terreno. Storia e archeologia dei sistemi agro-silvo-pastorali. Il Mulino, Bologna

Moreno D, Poggi G (1996) Storia delle risorse boschive nelle montagne mediterranee: modelli di interpretazione per le produzioni foraggere in regime consuetudinario. In: Cavaciocchi S (ed) L'uomo e la foresta secc. XIII–XVIII: Atti della ventisettesima settimana di studi, 8–13 maggio 1995. Le Monnier, Firenze

Orlandini O (1833) Trattato di boschicoltura o elementi d'arte forestale. Telemaco Giani, Firenze

Pazzagli C (1989) Il paesaggio degli alberi in Toscana. La campagna tra pianura e collina. In: Bevilacqua P (ed) Storia dell'agricoltura italiana in età contemporanea. Spazi e paesaggi, vol I. Marsilio, Venezia, pp 549–542

Piccioli L (1923) Selvicoltura. Seconda edizione riveduta e ampliata. Utet, Torino

Piussi P (1980) Il trattamento a ceduo di alcuni boschi toscani dal XVI al XX secolo. Dendronatura 1:8–15

Piussi P (1982) Utilizzazione del bosco e trasformazione del paesaggio: il caso di Montefalcone (XVII–XIX secolo). Quad Stor 49:84–107

Renucci P (1974) La cultura, Storia d'Italia, Annali 2. Einaudi, Torino

Riccioni G (1994) Il mio bel giardino: storia del paesaggio del Mugello e della Romagna toscana. Cassa rurale e artigiana del Mugello, Firenzuola

Rombai L (2002) Storia del territorio e paesaggi storici: il caso della Toscana. Storia e futuro 1:1–20, p 15. Preprint at http://www.storiaefuturo.com

Rozier F (1832) Nuovo corso completo di agricoltura. XXIV. Tip. della Minerva, Napoli

Salbitano F (1988) Per uno studio sulle modificazioni del paesaggio forestale: il caso del monte Catria. In: Andreolli B, Montanari M (eds) Il bosco nel Medioevo. Clueb, Bologna

Salvi G (1982) La scalvatura della cerreta nell'Alta Valle del Trebbia. Note Dalle Fonti Orali, Quaderni Storici 49:148–155

Sereni E (1961) Storia del paesaggio agrario italiano. Laterza, Bari

Sereni, (1972) Agricoltura e mondo rurale, in Storia d'Italia, Annali 1. Einaudi, Torino

Sereni E (1981) Terra nuova e buoi rossi. E altri saggi per una storia dell'agricoltura europea. Einaudi, Torino

Toubert P (1983) Il sistema curtense: la produzione e lo scambio interno in Italia nei secoli VIII, IX e X. Storia d'Italia, Annali 6. Einaudi, Torino

Trifone R (1957) Storia del diritto forestale in Italia. Accademia italiana di scienze forestali, Tip. Coppini, Firenze

Zagnoni R (1997) La coltivazione del castagno nella montagna tra Bologna e Pistoia nei secoli XI–XIII. In: Foschi P, Penoncini E, Zagnoni R (eds) Villaggi, boschi e campi dell'Appennino dal Medioevo all'età contemporanea. Atti delle giornate di studio (21 luglio, 6 agosto, 14 settembre, 17 novembre 1996), Società pistoiese di storia patria, Pistoia

Zanzi Sulli A, Di Pasquale G (1993) Funzioni delle matricine nei cedui nella teoria selvicolturale del XVIII e XIX secolo. Rivista di Storia Dell'agricoltura 1:109–121

Chapter 5
The Woods for Shipbuilding

Abstract The needs of civil and military shipbuilding have profoundly influenced Italian forests over a long historical period. As with other cases, the management of forests for shipbuilding intertwined with other economic functions of the forest, in some cases complementary, such as grazing in the forest, in other cases competitive, such as the production of charcoal or poles. There is no doubt, however, that the demand for timber for naval fleets historically represents one of the most important activities that has influenced the forest landscape from north to south of Italy. As for military needs, a growing influence on forests is observed with the progressive assertion of Rome's dominance over the Mediterranean Sea. Around the second century, the fleet had steadily reached 280 ships, between triremes and quinqueremes. Venice represents the most important case for shipbuilding and the management of the forests in medieval times. In addition to the oak (50%) used for the keel and outside planking, 40% of the necessary timber was made up of conifers, especially Norway spruce (*Picea excelsa*) and larch (*Larix decidua Mill.*), while the remaining 10% was made up of beech, used for the oars. The conifers came from the Alpine areas, especially the Upper Belluno valley, Friuli and Trentino, while the oak forests were found in the plains. The need for the curved pieces necessary for the keel of the galley assigned to the grazed oak forests, characterized by a low density, a major role. Faced with the needs of its military fleets, Venice made a huge effort to census, preserve and manage the oak forests of its domains. Many forests were thus identified, confined and placed under the special protection of the Senate. Venice did also an estimate of all forest resources in its domains to establish their periodic control, combining it with a management aimed at increasing production. The first land register was made in 1489 and it was the first of about 40 land registers made by Venice until 1796 in which the boundaries, exposure, perimeter, number and health conditions of the oaks was reported. Also the Grand Duke of Tuscany made a forest inventory in 1750 for the needs of the military navy. The south of Italy was equally intensely exploited for shipbuilding. Interesting in this regard is the extensive use of oak that is made at the arsenal of Naples in the second half of the eighteenth century, in which oak constitutes 82% of the total. This signals an abundance of oaks in the south compared to the centre-north where oak forests suitable for the navy seem to be scarce and competing with the need to produce acorns for the grazing animals.

M. Agnoletti, *History of the Italian Forest Landscape*, Environmental History 19,
https://doi.org/10.1007/978-3-031-81631-4_5

5.1 Trees, Groves and Woodlands: The Centuries of Venice

The needs of civil and military shipbuilding have profoundly influenced Italian forests over a long historical period. As with other cases, the management of forests for shipbuilding intertwined with other economic functions of the forest, in some cases complementary, such as grazing in the forest, in other cases competitive, such as the production of charcoal or poles. There is no doubt, however, that the demand for timber for naval fleets historically represents one of the most important activities that has influenced the forest landscape from north to south of Italy.

As for military needs, a growing influence on forests is observed with the progressive assertion of Rome's dominance over the Mediterranean Sea. Around the third century BC, the Republic seems to have a first small military fleet composed of a few dozen vessels, mainly triremes, but already in 256 BC the Romans fought a naval battle against the Carthaginians with 160 ships, and in a second battle they used 330. Around the second century, the fleet had steadily reached 280 ships, between triremes and quinqueremes, a sign of the improvement of construction techniques that at that time already allowed the realization of imposing works. The trireme built by Hiero, king of Syracuse (270–215 BC), made with timber from Mount Etna—at that time rich in forests (Salvini 1754; Caro 1826; Munro 1867)—, had the hull (i.e. the part in contact with the water) covered with metal plates to protect the wood from the action of salt water and was launched with some machinery made by Archimedes (Tiraboschi 1822). It seems to have been even larger than another famous ship, that built by Philopator, Egyptian pharaoh of 204 BC, 300 ft long (about 100 m) and with a thousand rowers, on which a wooden palace of cedar and cypress was built.

The evolution of Roman construction techniques is evident from the descriptions of the famous Roman ships of Nemi, a lake in the Lazio region, that highlight sophisticated construction techniques, even in mechanical parts, such as bronze pumps and taps and revolving platforms on iron spheres. The two vessels, built in the period of Emperor Caligula between 37 and 41 AD, probably for ceremonial or leisure purposes, are of considerable size, with 75 m in length and 30 in width, and were made with the use of different species of timber, such as fir, Aleppo pine and oak (Quercus petraea and Quercus robur). It seems reasonable to think that the supply of the necessary timber for such demanding works could not have been carried out without forests destined for this purpose, or without people assigned to their management. The Roman armies were however able to set up fleets even at the time of need, as demonstrated by the 800 ships built by Julius Caesar on the northern French coast to invade Britain in 54 BC.

Military shipbuilding in Italy underwent an interruption with the fall of the Roman Empire and the barbarian invasions, almost all of which occurred by land, with some exceptions such as the invasion of the Saracens in the eighth century. However, they resumed decisively in the Middle Ages thanks to the maritime republics and ended in the nineteenth century, with the definitive assertion of iron ships. The possibility of better understanding the characteristics of the forests destined for shipbuilding and their management is offered to us in particular by the documentation available between the fourteenth and seventeenth centuries concerning Venice. The republic of Venice implemented a naval construction and forest management organisation that was unparalleled in the medieval period in Europe, as well as the supply of timber for civil construction and energy purposes.

The timber necessary for naval construction requires both coniferous and deciduous forests. In the naval field, this distinction had significant importance, especially in relation to their use for the construction of the hull (the submerged part of the hull), the superstructure (the part that remained dry) and the masts. Essentially, the main distinction was between timber suitable for resisting contact with water (thus the planks of the hull and the keel, almost always oak) and the rest (which could allow multiple solutions). In fact, all the main European navies, from the North Sea to the Mediterranean, assign a dominant role to oak species, but we should not think of too rigid distinctions. As we will see, the use of conifers for the hull is well documented given the good resistance of many species to prolonged immersion in sea water.

From a technological point of view, the main problem was resistance to severe deterioration due mainly to the attack of molluscs, crustaceans and fungi, among the most important of which are certainly the teredo and among the crustaceans those of the *Limnoria* genus. The teredo, known also to the Greeks and Romans, were considered a serious problem for their ability to build many tunnels that could completely compromise the wooden structure. Resistance to attacks was linked not only to the intrinsic durability of the adopted wood species, but also to salinity, water temperature and sunlight, so much so as to justify significant differences in the lifespan of vessels sailing in tropical areas, in the Mediterranean or in the northern seas.

An important element to evaluate both the complexity of naval construction and the problem of finding suitable timber is the fact that until the development of iron joints, around the eighteenth century, the framework of the vessels—essentially the frames and the stern and bow posts -, the bulkheads, plus a fair number of other assortments were made up of curved pieces, for which the natural curvatures of tree branches were sought to be exploited. The Romans already knew the superior quality, for mechanical resistance purposes, of timber already shaped in nature compared to those produced by axe masters (Ucelli 1940).[1] In Venice, they were generically defined with the term "stortami" (crooked wood), but had further denominations such as "volta da coscia" and "volta da gamba". In Liguria, the axe masters of the shipyard were called "maestro di garibo", a term of Arabic origin meaning arch or curve, just as with this name were called the wooden templates used to identify in the forest the trees to be felled, but a similar term was also used in France to describe the forests suited to produce curved pieces, called *bois de gabarit* (Boudriot 1999). These construction requirements greatly influenced the characteristics of the forests sought by the axe masters for naval construction.

In the Venice arsenal, the oak species used for the framework and for the external hull were mainly the pedunculate oak or famia oak *(Quercus peduncolata A.DC) and* the true oak *(Quercus sessi—liflora A.DC),* which constituted about half of the timber used in the Venetian galleys. The English oak, a lover of soil moisture and little light, was found mainly in the woods located in the plains and along the Veneto-Friulian estuary and, having the wood most resistant to water, was preferred in shipbuilding, but was also the most sought after for pastoralism because of its sweet acorns very much appreciated by animals raised outdoors. The sessile oak, less demanding of soil moisture, was more widespread in the hilly reliefs (Euganean Hills, Berici Hills, Asolano, Montello, Feltrino); however, the frequent substitution of this species with the downy oak and the difficulty of distinguishing hybrids between the two species must be considered. The presence of sessile oak and English oak is reported in various Italian areas, from the north to the south of the peninsula, but the need to find timber for the navy interested all of Italy, leading to a capillary exploitation of these woods, regardless of the distance from the shipyards. The good quality of the timber of these oaks for other types of civil constructions and carpentry activities, the usefulness of the oak groves for grazing and the tendency to reduce the plain oak groves for the cultivation of the land, as in the case of the reclamation of the Po Valley, or their transformation into coppice woods, generated strong pressure on these woods due to competition between civil and military uses.

In addition to the oak (50%), 40% of the necessary timber was instead made up of conifers, especially Norway spruce (*Picea excelsa*) and larch (*Larix decidua Mill.*), while the remaining 10% was made up of beech, used for the oars. The larch could reach 20–25% of the conifer quota if it was possible to obtain it or even more, given its robustness and better resistance to moisture. The conifers came from the Alpine areas, especially the Upper Piave river valley, Friuli and Trentino, but the trees also came from the Tuscan-Romagnolo Apennines, while the beech came from the Belluno Prealps. The commercial and especially the military navies of the sixteenth century saw the dominance of galleys as the main type, with more advanced construction techniques than those of the classical age, being longer than the Roman triremes and better built (Tursini 1940).

The large Venetian transport galley of the early sixteenth century was considered the most beautiful transport ship in the world (Naish 1993), it was 40–50 m long and could carry about 250 t of goods and about 200 crew members between rowers and gunners, effectively representing the product of a period considered as the peak of the Venetian arsenal (Concina 1987). For the Venetian military fleet, dominated in the middle of the sixteenth century by thin galleys and to a lesser extent by the galleazza and the bastard galley, it is possible to calculate the necessary timber thanks to the descriptions provided by the work of Frederic Lane (Lane 1934). Calculating the volume of the pieces and trying to estimate the assortments in which the measures are incomplete, adding the processing losses, we have that for the construction of a large galley about 490 m^3 of oak, 400 m^3 of spruce and larch (25%) and 100 m^3 of beech for the oars were needed, for a total of about 1000 m^3 related to the commercial trunks used. By this term we mean the assortment of raw timber sent to the arsenal from which the necessary pieces were then obtained, whether they were boards or curved pieces, with an estimated processing loss of around 20–25%, which would in turn be increased by another 25–30% when calculating the branches and tops of the original trees used, bringing the total of raw timber to about 1300 cubic metres. These are of course approximations that must take into account that the processing of a curved piece to make a keel or a rib is subject to very low yields, due to a waste higher than that to produce a board, also because the ribs were all handcrafted with the axe and hatchet, while the boards were produced with the frame saw (Agnoletti 1997). Similarly, the difference between the volume of the commercial trunk and that of the branches in conifers could be around 28–30%, while in a broadleaf it could even reach 50%, given the greater proportion of branches compared to the trunk (Stigliani 1952). In any case, the estimate seems to be confirmed by a comparison with that reported by Boudriot for the *ordinary galley* of 1736, for which estimated about 450 oaks were needed, and about 100–120 more for the *patron galley* (Boudriot 1999).

Faced with the needs of its military fleets, Venice made a huge effort to census, preserve and manage the oak forests of its domains. Many forests were thus identified, confined and placed under the special protection of the Senate or the Council of Ten, forbidding them to be damaged under the threat of extraordinary penalties. In this way, among many other forests, the fir forest of Somadida in Cadore in 1463, donated to the Republic by the Community of Cadore; in 1471 the oak forest of Montello along the Piave; in 1548 the oar forest of Cansiglio in the Belluno Prealps; in 1567 the fir trees of Caiada, always in Belluno; and in 1580 those of Visdende in Cadore. Within a century, the Republic bans the forests of the main species necessary for shipbuilding, alongside this initiative, aimed at ensuring control of forest resources, also the creation of land registers, i.e. forest inventories. This is a work that has probably no equal on the European continent in that period, never repeated by any country on this scale, which aimed to obtain an estimate of all forest resources in the Venetian domains and to establish their periodic control, combining it with a management aimed at increasing production. The first land register was made in 1489 and concerned the forests of Friuli. 32 forests were surveyed for a total area of 477 ha, and it was the first of about 40 land registers made by Venice until 1796 in which the boundaries, exposure, perimeter, the health conditions of the oaks. A second land register was made in 1536, and in this case each forest was not only confined, but also surrounded by a ditch and equipped with bridges and bars at the entrances, prohibiting any use that occurred without authorisation from the magistracy. After a few decades, in 1568, a master shipwright of the arsenal, Nicolò Surian, was commissioned to create a new land register in which all the oaks useful for the arsenal were counted, measured and marked, not only in public woods but also in private woods, and the country plants. In this land register are listed, individually, the oak trees suitable for planking and the curved pieces for the keel ribs. The land registers allow a fairly accurate estimate of the location and characteristics of the Venetian hinterland forests under the dominion of the Serenissima, which ranged from Vicenza to in Istria (Fig. 5.1).

Analysing the collected data, it emerges that towards the end of the XVI century the number of oak forests scattered in the Venetian plain was about 750–800, with an average extension of about 10 ha for each forest. The only exception to these averages was the Montello forest, which had an extension of about 6400 ha. If we add the forests of Montello and Montona in Istria, we reach a total of about 15,000 ha. In these forests, the Surian land registry had calculated the presence of at least 765,926 oaks, excluding the small ones, of which two-thirds were considered suitable for the

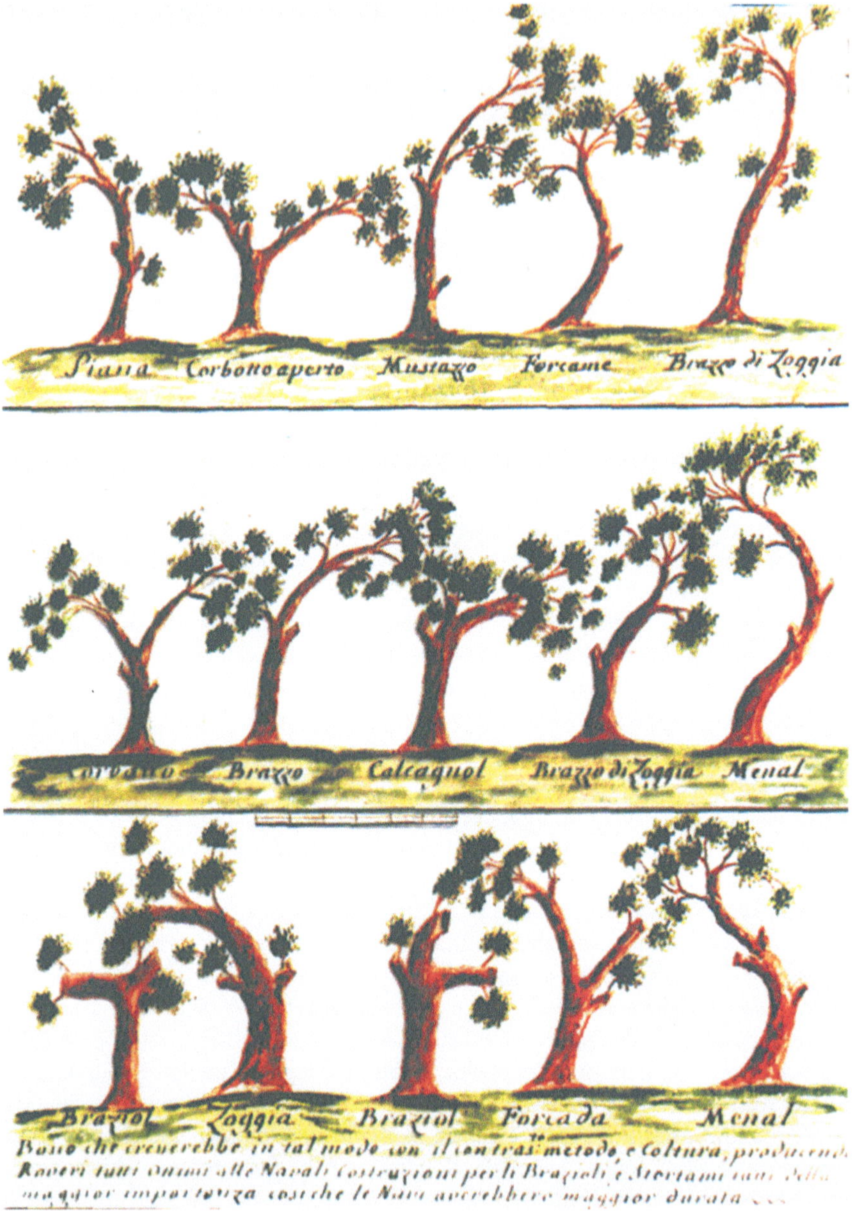

Fig. 5.1 1777 archival document illustrating the different shapes to give to the trees for the construction of the different parts of the galleys (State Archive of Venice)

arsenal's uses. As we can see, these are small size forests, often used for grazing, which testify to a landscape dominated by cultivated areas, in which the forests are a sporadic presence.

Considering the distribution and characteristics of the forests, and the needs for timber for shipbuilding, already calculated earlier, we can address the problem of the quantity of forests necessary for the construction of a galley. This requires an assessment of the density per hectare and the volume of timber of the plain oak forests and the fir trees of Cadore in the XVI–XVIII century, from which most of the assortments of conifers of the Venetian shipbuilding industry came. As for the former, evaluating the cubic capacity of commercial trunks to provide straight pieces for the hull, an average density of the plain oak forests of about 100 plants per hectare, and a volume of about 1 m^3, meaning by this the useful trunk obtainable from a whole oak (Susmel 1994), we find that to build a large galley about 490 oaks had to be felled. The extent of forest that would have been necessary to subject to cutting was therefore at least 4–5 ha, but since in the forests there were also other species besides oaks, unsuitable for shipbuilding, the total area affected was probably higher. Then there would be to calculate the differences between the forests suitable to produce curved pieces, therefore with a density of plants per hectare rather low, and those for the "straight" oaks, suitable to produce boards, which were tried to be managed in a different way from the silvicultural point of view, favouring for the latter a greater density to facilitate the vertical growth of the trunks, as illustrated in the 1777 termination. In reality, the densities of the individual forests were very variable, oscillating between 20 and 180 trees per hectare, therefore able to provide the variety of assortments required but in an uneven way, without the possibility of having forests specialised for one or the other assortment of timber required.

The estimate of the conifer forests necessary for a galley is in a certain sense more uncertain, since we do not have land registries available for the XVI century. We can therefore adopt as a reference point the estimate of 200–300 plants per hectare, with a unit cubic capacity of about 1 m^3 per tree, underlining that such values could vary a lot since such forests had different provisions, also depending on their greater or lesser purity with respect to the presence of other species such as beech and larch (Susmel 1951). To obtain the about 402 m^3 of fir and larch necessary for a galley it would therefore have been necessary to fell 1–2 ha of forest with clear cutting, eliminating all the plants present. In fact, the rules of selective cutting implemented in the Upper Piave Valley involved selecting plants by choosing those of suitable sizes, even though the history of those forests shows strong variability in the intensity of the cuts, which could also remove up to 50% of the provision. A wood harvest of 4–5 m^3 per hectare could express a realistic value of a balanced use in a low-productive forest, and in this case, the extent of forest to be subjected to cutting for the construction of a large galley could be around 80–100 ha (Patrone 1953). If these are the extents of forest necessary for a galley, it is more difficult to establish the quantities of timber necessary for the maintenance of the fleet. For Venice, it is possible to estimate the needs for the fleet of military galleys put to sea between 1504 and 1602, when 911 galleys were built, that is 9 galleys per year on average equal to about 3626 m^3 of raw oak timber. In some years, like 1602, the need reached about 56,840 m^3 of oak, even though we must consider that in the warehouses of the arsenal large quantities of already sorted timber were stored to build galleys quickly, according to the needs of the repeated wars against "the Turk". To understand the efficiency of the Venetian arsenal, consider that a fleet of 100 light galleys and 10 large ones could be set up in just two months in the middle of the sixteenth century, thanks to a production structure similar to an assembly line that required the storage of at least 90,000 m^3 of already sorted timber (Fig. 5.2).

From each warehouse located around the careening basin, the pieces were taken to assemble the galleys that passed from one warehouse to another, then they were put in the water once the keel was set up and continued until the exit of the arsenal where they arrived practically complete.

A fleet of 100 galleys seems to be the one considered useful not only by Venice, but also by the other military fleets that controlled the Mediterranean at that time. In 1572, about 100 galleys were counted in the service of the king of Spain, considered sufficient for defence and offence, including the control of maritime traffic, even though the Duke of Terranova suggested extending the fleet to at least 250, perhaps in memory of the similar amount of ships put to sea by the Holy League for the victorious battle of Lepanto in 1571.

At the time of the Contarmi census, in 1740, the number of oaks present in the Venetian domains had reached about 850,000, but this increase was not in any case sufficient for the needs of the navy. Making an estimate of the average usable

Fig. 5.2 Oak with a crown suite for the production of curved pieces used for the ribs of the keel of the galleys. Specimens with this very expanded crown shape could only develop in sparse forests, like those used for grazing

volumes and unit increases, we see that in the sixteenth century the annual production obtainable from the small forests mentioned earlier was around 10,000 m^3 of timber. If we add to this the mass presumably coming from the almost 8000 ha of the Montello and Montona forests, we see that the total production available for Venice was around 20,000 m^3 of timber per year. If we consider the needs of the Venetian navy, both military and civilian, Using Lane's numerical consistency as a basis, we see that in the fifteenth century, at least 50,000 m^3 of coniferous timber and a slightly higher quantity of broadleaf timber, mostly oak, were needed each year. If we were to consider only the military navy, we see that for the 120 ships built in 1504, at least 51,000 m^3 of timber, including conifers and broadleaves, were required. This is a value just below the average quantity of timber needed for the construction of galleys in the sixteenth century, estimated based on data available for seven years. The scarcity of oak groves led Venice to progressively import timber from abroad, while for civilian shipbuilding, they even came to purchase, in the seventeenth century, entire ships that ended up costing less than those produced at home (Agnoletti 1996).

The limited extent of the oak groves and the non-optimal tree structure for the production of necessary timber suggested to Venice to start managing the reserved woods with techniques aimed at improving their conditions for the needs of the arsenal. Initially, it was about converting the coppice woods, which were the majority, into high forests, then they moved to adapt the structure of the high forests to the needs of the navy. This trend is masterfully exemplified in a famous termination of 1777 in which the forms that the woods had to assume for the production of straight oaks, that is for making boards, and those for curved pieces, to be obtained through treatments intended to produce sparse woods and dense woods, with numbers of tree individuals even four times higher, but always to increase the number of oaks, is illustrated. Over the years, interventions were started that effectively obtained results, transforming the coppice woods into high forests and increasing the presence of the oaks that rose in about a century from 100 to 200 plants per hectare.

Another objective pursued by Venetian silviculture was to reduce the other tree species in the high forests and increase the overall presence of the oak. The oak and the hornbeam were indeed favoured by the management adopted in competition with other species present and cited in the land registers (ash, maple, elm, willow, alder, turkey oak, chestnut, downy oak); in the case of Montello, the chestnut and the beech were reduced leaving space for the oak alone. The techniques adopted over time modified the original specific composition, increasing the oaks at the expense of other species but causing problems to the natural regeneration of the oak, which does not regenerate itself in monospecific woods, and therefore causing a new progressive decline in the number of oaks. In general, the selective cutting represented the basis of the Venetians' silviculture; in this way, they tried to maintain a multi-aged structure of the wood that could periodically produce specimens of adequate size. Despite this, at the beginning of the 1600s, the conversion of the original coppice forests into high trunk forests was completed. However, by the 1700s, the forest was almost everywhere a high trunk, one-third of which had a normal uneven-aged structure and two-thirds were affected by even-aged structures, never reaching the final structure planned. On the other hand, the pressure of the populations on the forest resources

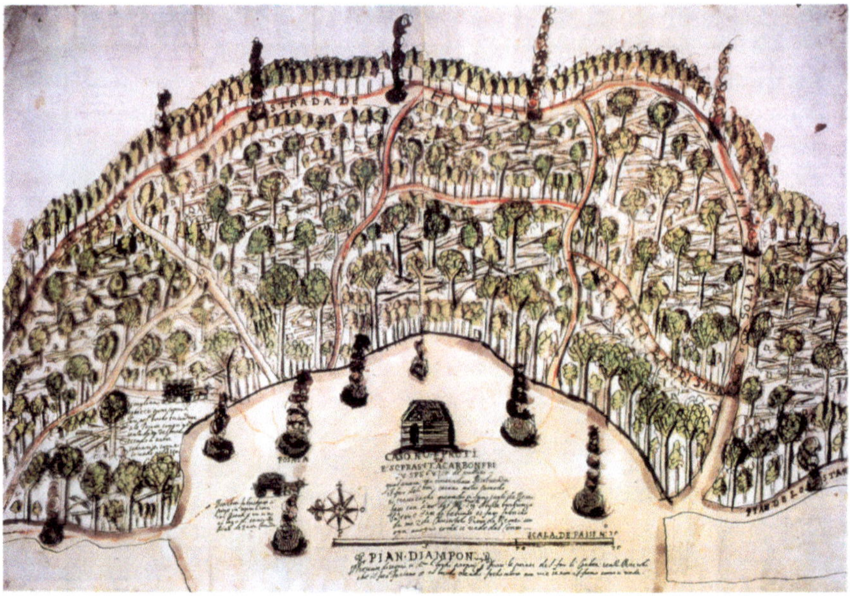

Fig. 5.3 The management plan of the Cansiglio forest for the production of oars whose surveys were started in 1638. In the drawing, which represents the Pian di Campon area, the forest is observed surrounded and crossed by forest roads to allow the transport of the felled trunks. The beeches are partly standing and partly felled, demonstrating the use of the selective cutting to manage the forests. Around the forest, the charcoal burners are observed with which the residues of the beeches not useful for other purposes were transformed into charcoal, at the bottom in the centre the house of the charcoal burners (State Archive of Venice)

of the plain was continually growing and the resentment towards the limitations and obligations imposed on the oak forests had to dramatically manifest itself with the fall of the Republic, with the arrival of the French, with cuts and deforestation that affected almost the entire territory (Fig. 5.3).

5.2 The Forest for the Oars: The Cansiglio

One of the most interesting examples to understand the management of the forests carried out by Venice for naval constructions is that of the beech forest of Cansiglio. As highlighted by Spada, with the acquisition by Venice of the Cansiglio in 1550, a first "general conterminazione" was carried out, that is, the definition of the forest boundaries, which identified an area not exceeding 12,000 ha (Spada 1995). In reality, a certain indeterminacy on the boundaries continued also with the boundaries of 1622, 1660 and the estimates of the following century. Estimating an average quantity of oars for a thin galley equal to 150 and 60 for the large ones, taking into account the proportion between large and thin galleys and also considering the different length of the trunks useful for making galley and galeazza oars, we can indicate about 2.5 m^3 the dendrometric volume of a beech of sufficient size (Giordano 1953). The data takes into account an average number of oars obtainable from a trunk equal to 4 and not 6, as was achieved from the best trunks, considering the considerable difficulties mentioned by almost all authors to obtain such production, almost always due to technological defects. Therefore, a need for about 800–900 m^3 of beech per year and a number of beeches of about 340–400 mature adult individuals per year for the arsenal can be admitted. Not included in the count are other pieces in beech that could enter the construction of the galley, but of which we have no news, even if compared to the Genoese—who used it also for the holds and the planking (Gatti 1999)—it does not seem that in Venice significant use of this species was made other than for the oars. Compared to the extension of the pure beech forest alone, equal to 3700 ha, the need for beech would have led to an average use of 0.25 m^3 per ha per year, about one beech every 12 ha. This is a very modest amount for the characteristics of a beech forest of that extension in good silvicultural conditions which, also considering the 1352 ha of mixed beech and fir forests present, could have easily supported much higher annual withdrawals. In light of this estimate, the technical, organisational and economic effort made by Venice in the Cansiglio is therefore of particular interest, which is not matched by what was done by other maritime republics as Genoa or Pisa (Puccinelli 1996), especially considering the much greater supply difficulties of the Genoese arsenal. Another element not considered in this estimate is the natural perishability of the oars, which lasted less than the average life of a galley, and possible breakages, very frequent especially in naval clashes.

The ban on the use of the Cansiglio forest exclusively for the arsenal comes almost a century after that of the oak forest of Montello occurred in 1471, in a period when the mechanism of reserved forests was already quite utilised in Europe, as evidenced by the edict of Charles IX in France in 1561 (Du Monceau 1800). Already in 1564, Gerolamo Foscarini, captain of Belluno, among various proposals for the management of the Cansiglio forest, suggested that "cutting in the woods should not be done where it was cut the year before, but rather rotating the forest cuts". A few years later, in 1592, Francesco Soranzo presented a first report with a proposal for a cutting plan, suggesting dividing the forest into 100 plots, as was done for the oak forest of Montona in Istria. From the analysis of the report, it is clear that the origin of management plans is located in a period prior to the initiatives for the Cansiglio and is initially linked to the problem of the plain oak groves, which are far more at risk, given their reduced extent in relation to the needs of Venice.

According to the writings of Theophrastus and Pliny, it seems that the first forms of management plans in the classical sense arose with the cultivation of the incense tree, whose forests were the first to be subjected to a regular management system and utilisation, as well as to pay taxes to the sovereigns and tithes to the temples. The rotation of cuts was also established by the Romans, as we have already seen. In other words, the so-called technical cut was identified, which will be the keystone for the management of forests for many more centuries. For high forests, such as acorn oaks, selective cutting was used instead, with a minimum care period of five years, in order to thin out the forest and promote fruit production.

In the case of the Cansiglio, the method of dividing the forest into annual cuts was adopted, which in some way reflected what in scientific terminology will later be defined as "planimetric-partitive", establishing a cutting shift that for the beech was set at 100 years. It is interesting to note that almost in the same period, in 1554, even in Saxony, the division of high forests into plots and the succession of similar cuts always proportionate to the productivity of individual species was prescribed, in homage to a general growing interest in the value of wood production that would lead Germany to become the homeland of economic silviculture. This is a process of increasing attention to the wood resource in a period when wood was not always the most important product of the forests as happened for acorns for pig feeding and for resin, which in the sixteenth century was often the main commodity traded on the Adige river from southern Tyrol to Verona (Fig. 5.4).

Soranzo's descriptions of the Cansiglio woods highlight the poor execution of previous cuts, the excessive presence of dead and shaded plants, the ineffectiveness of selective cutting for beech that impacts too much on the best subjects releasing many hopeless trees. It is therefore suggested to divide into 100 plots and to clear-cut one plot per year. The same proposal is repeated in the report of the rector of Belluno Giovanni Dolfin in 1613, while the poor conditions of the forest are confirmed by the final report presented by Rector Paolo Querini to the Venetian Senate in 1632. These continuous alarm signals are finally accepted by the Senate, which in 1638 sends Zorzi de Christofolo with the task of undertaking the implementation of a management plan based on the needs that emerged in the previous periods, summarised in the lack of real knowledge of the forest biomass, forest utilizations based on selective cutting according to the needs of the moment, the withdrawal of the best trees that had caused

Fig. 5.4 A large Venetian galley from the 1600s (Naish 1993)

an imbalance in the diameter classes, the lack of a rotation plan of cuts, the presence of only two-three trees per hectare with a diameter greater than 50 cm.

Zorzi was a master of the oarsmen, that is, a supervisor of the manufacture of oars for the Venetian military fleet on behalf of the arsenal. The assignment given to him by the patrons and providers of the arsenal shows how the figure of the forester, a specialised technician who would "respond" to the needs of a client and apply a model of forest management on a scientific basis, was still absent at that time. In fact, at least two centuries will have to pass before this happens, with the consequent rivalries between the navy and the forest administration for the exploitation of the forest: in a certain sense, the forestry administration was added to the various entities that had a say on the woods.

Zorzi began the surveys in Cansiglio on 31 August 1638, with the collaboration of the "Captain of the forest", four "oarsmen", that is, oar makers, and some "woodsmen", that is, loggers expert in felling and transporting timber. The "Cadastre" made consists of a register of 28 sheets for a total of 56 pages, with sheet sizes of 43 × 31 cm, and with 20 watercolour illustrated tables, in which all the 16 plots in which the forest is divided are represented. Each plot is therefore represented with a colour cartography and described with the boundaries and a survey scale indicated on the map. On the page next to the watercolour table, the number of trees "good at present for making oars" and the number of those of "undermeasure", usable after a variable period from 10 to 25 years, are indicated. Taking as good the measure of extension of the forest cited by Magoni for 1830, therefore of the only forested surface equal to about 5400 ha, compared to the total 7000 ha of the forest, it is seen that on average each plot identifies a surface of 337 ha. These are plots of enormous extension compared to modern management plans even for woods treated with selective cutting, perhaps justifiable with the scarcity of trees that are actually identified as suitable for making oars, but which certainly do not follow criteria linked to the need to have a size corresponding to the annual growth rate as in the modern management methods. During Zorzi's surveys, beeches with a circumference greater than the "brazadura", that is, greater than 55 cm in diameter, and those undermeasure, which would have become mature within at least a decade, were marked. This is a standard period for uneven-aged beech forests, a form of forest management that theoretically allows obtaining large-sized trunks and using at the same time the main yield and intermediate yield, also favouring the association with the fir, as can be seen from the conspicuous presence of this species associated with beech in Cansiglio. This is also the classic method used for the fir forests of the nearby Cadore valley, for which commercial customs set a price in relation to a technical age of cuts, independent of the maximum wood increase, so much so that the progression of the price paid did not increase proportionally to the increase in volume for trunks with diameter exceeding 12 ounces (34.8 cm), that is, the tree considered mature for commercial use. However, this is a form of management that requires the execution silvicultural treatments, as "thinning", to allow the plants to reach maturity, while maintaining an adequate distribution of age classes. Otherwise, over time, forests with uneven-aged structure by groups and not per tree foot are obtained, as often noted for the uneven-aged forests of Belluno not subject to correct silvicultural care. However, in Zorzi's surveys, not all the smaller diameter subjects forming a substantial intercalary mass often devoid of a real future were considered, so an inventory was not carried out as in the case of the oak groves of the plains. From the data indicated by Zorzi, the cutting of a total of 8259 beeches is observed, equivalent to an average of 516 plants for each intake, therefore an average of 1.5 beeches per ha, while the sub-measures add up to 7329, equivalent to 1.3 plants per ha. In total, for each hectare, we therefore have less than three beeches capable of being usable within 10 years, a very small quantity for a forest that at maturity should present at least 200–300 tree per hectare and annual increments of at least 4–5 m^3 per hectare.

The voluminous management plan of Cansiglio shows a well-articulated presence of service roads to facilitate logging, which further subdivide the plots into particles where the execution of a "thinning" testifies to the execution of a selective cut. At the same time, charcoal production is also carried out in charcoal pits located outside the plots, an activity of not negligible economic value for which a hut is also built clearly visible in the centre of the drawing. The possible explanation for the reduced quantity of usable beeches is linked not only to the poor conditions of the forest, but also to the peculiarity of the production of oars compared to the normal assortments of work and construction timber that can be drawn from the woods. In this regard, the cadastre presents a section with the description of three types of beech in relation to the production of oars and a radial section of a trunk where the mechanism of annual growth and the methods of measuring age are demonstrated. In the forest, there were plants over 200 years old that, however, present problems for the stuffing of oars; in fact, it is underlined in contemporary texts how woodcutters do not like to work these trunks. From a technological point of view, when a tree is too old, there can be the formation of a central area of the heartwood called "red heart" , an indication of a deterioration of the wood with defects that compromise its resistance and duration.

Regarding the processing, it is remembered that this was done with the squaring axe, that is, a tool to square the timber, that is, to create flat surfaces from round trunks. The squaring axes were divided into various models that can be divided into axes for forest work and axes for the carpenter's work. The main difference between these two types of tools lies in the length of the handle: in fact, the manufacture of beams, in the forest, provided for a positioning of the operator with the trunk close to the ground and therefore the use of a long handle. This type of tool could have symmetrical cheeks, sharp cutting edge on both sides and straight handle, and did not allow to achieve perfectly smooth surfaces. For the carpenter's work, however, the axes often had one flat side, an asymmetric edge and the handle inclined relative to the axis of the blade to allow the operator to stand away from the trunk, on one side of it, and achieve perfectly smooth surfaces. This second type of tool generally had a short handle that required positioning the trunk raised from the ground for work that required greater precision, and it is precisely the one used for the subsequent processing of the oars, which were then smoothed with the plane (Agnoletti 1997) (Fig. 5.5).

The presence of very large trunks that begin to deteriorate makes the work of the woodcutters more difficult both for the splitting into "quarters", and for the subsequent processing with the axe. In the beech forest, there could be beech trees

Fig. 5.5 Location of the oak forests for the venetian navy

with a twist in the grain, which does not develop straight as required for the oars. The twist of the trunk, therefore a helical course of its fibres, can be right-handed or left-handed, and in Cansiglio this defect represents a rather widespread characteristic of the beech. From these specimens, Zorzi explains that barely two galley oars are obtained, but even for these, the difficulty of the work is noted: it is clear in fact that the splitting with the axe and the division into quarters, from which the oars are then obtained, are made more difficult and less effective by the helical twist of the trunk. This defect was often compounded by a discordant inclination of the spiral of twist, with the result that any longitudinal section made produces timber with fibres never parallel to the section of cut and with the split with a wedge or the axe, wavy surfaces are determined, with consequent problems for subsequent processing to obtain the oar. The emphasis on the galley oars confirms, however, the diversification of the production of oars that were drawn from the forest, implying the possibility of producing different types for different boats, but also the low production. As we have pointed out it rarely reached six oars per trunk, on average it was around four for mature and in good condition trees, but often dropped to two.

Other beech specimens could be very branched, they are not calculated, but for the most part they are good for oars, according to Zorzi. The new reference to the possibility of making oars for the great galleys underlines again the importance of this boat for the military fleet, an impression confirmed by the marginal notes relating to the cutting of the catch n. 1 in 1638 from which it is underlined that 2010 oars are obtained, including 200 for the large galleys. This means a quantity sufficient to arm 12 thin galleys and 3 large galleys, enough for a production of a couple of years of the arsenal. The type of beech exemplified highlights one of the most common problems for construction timber, that is the presence of branches along the trunk that not only reduces the usable portion, but also makes the "stuffing" of the oars difficult due to the presence of knots normally considered weak points of the wood that compromise its solidity. As for the construction of galley oars, it required larger beeches and with technological characteristics that are difficult to be found in Cansiglio, and in fact the proportion of 10% of galley oars obtained from the cutting of the catch n. 1 in 1638 indicated in the register confirms that, although the overall extraction in terms of woody mass was very reduced, the qualitative requirements made the forest's production very scarce.

On the same page of the plan is drawn the radial section of a beech trunk, with the intention of showing how to calculate the age by counting the growth rings. According to Zorzi's surveys, the beech reached a technical maturity, in relation to the manufacture of the oars, around 100 years, starting a regressive phase from 200 onwards. Then. In fact, according to Spada, current surveys show that at 100 years old, the beech tree of Cansiglio reaches a diameter of 34–36 cm, therefore significantly less than the felling diameter set at 55 cm. The plan identifies a provision and an annual recovery, but in reality its application was only partial. In the years 1639, 1641, 1643 and 1644, multiple cutting areas were used simultaneously, then going on to cut also in those used previously, but releasing specimens for protective purposes, especially against the wind. From 1644, regular records of forest use cease, while towards the end of the next century the slow transformation of the woods from beech to fir will begin, following the cessation of oar production and the growing need for fir. A new type of forest, due to a different role of the wood resource, and also to a different fate of the Republic. The Cansiglio forest remains a significant example of one of the first management plans with which forest production was organised and the continuous shaping of the forest characteristics that are adapted to changing productive needs.

5.3 From Galleys to Galleons

From the sixteenth century, the galley gradually loses the central role that this boat played, from a commercial and military point of view, being supplanted by galleons and other types of ships. Compared to a galley, the "boats" and "ships" had a more complex structure, as well as the galleons and then the multi-deck military vessels that carried up to over 100 cannons and which will constitute the backbone of the military navies of the great European powers between the seventeenth and eighteenth centuries. The presence of a more composite typology compared to the galleys was reflected in a greater number of wood assortments required, but also in much higher volumes of timber. In fact, if about 1000 m^3 of timber were needed for a large galley, for a galleon it started from a minimum of about 5000 m^3. Also for this reason, Venice gradually loses its technological supremacy in shipbuilding in favour of nations like Holland, whose fleet appears to be composed of at least 10,000 vessels in 1697, compared to about 3600 in Venice. In a 1676 treatise by Claude Caron, carpenter of the French royal navy, 76 different types of wood assortments are listed but the masts of the three-deck vessels stand out, 78 m high and made in five different pieces of fir, in turn made by assembling four to six different trunks, having to reach a diameter of 1.2 m. Between 1718 and 1784 in Venice, when deep innovations in shipbuilding have already taken place but the city's role as a maritime power has also greatly diminished, we have a production of 56 line ships: 29 of the first rank, 12 and a half of the second, 4 of the third and about seventy smaller vessels, between xebecs, corvettes and galleys. For the calculation of the necessary timber, the data reported by Boudriot for France are helpful, especially for the fact that they present a rigorous classification of the military vessels of 1746, associating with each the quantities of necessary timber. This data also offers useful distributions between the curved and straight woods necessary for construction, showing a variable percentage between 25 and 35% of "bois de gabarit" necessary, even if in those years they had already begun to replace the "natural" curves with iron joints (Boudriot 1999).

Also interesting in Caron's treatise are the estimates related to the quantities of oaks, which for first-rate ships with 110 cannons and 3261 m^3 of necessary timber, indicate a need for 3500 oaks. According to these estimates for the 29 first-rate Venetian ships, taking a slightly lower quantity predicted for those of the third order, we would have a total of 89,900 oaks: 32,500 for those of the second rank, 9600 for those of the third, and 34,300 oaks for the rest of the fleet equivalent to a large galley. Calculating the usual cubic meter per oak we would therefore have a total of 166,300 m^3, which divided by 66 years gives 2519 m^3 per year: a demand compatible with the timber resources described by the Contarini census carried out in the eighteenth century, but which only concerns this type of timber, which as we have seen represents more or less half of the necessary timber. In reality, shipbuilding in the lagoon city is now in crisis and the arsenal no longer represents that model of efficiency of the past, having difficulty in launching large ships due to the too shallow seabed and its own organisational structure, denouncing construction costs double or triple compared to private shipyards (Lazzarini 2001). From 1785 to 1792 production more than doubled and the fleet's ships increased from 29 to 50, but the quantity of timber does not seem to be the main problem, but rather the quality, with the use of poorly seasoned material causing serious damage to the lifespan of the boats, and the organisation of transport. However, these are not new comments, as the same complaints were expressed at the end of the sixteenth century about the quality of the galleys, which were said to last no longer than 18–20 years as before, but only 3 or 4 (Lane 1934).

As for Genoa, quantitative data are only available for the eighteenth century. A contract cited by Luciana Gatti for the construction of an 18 m long ship allows us to roughly estimate the total amount of beech and oak necessary for the planks used for the construction of the hull, of which a series of measurements are provided by which it is possible to attempt an estimate (Gatti 1999). Considering the 300 oak planks and the 400 beech ones, we have a total of about 75 m^3, to which we must add a few tens of cubic meters for other indicated pieces, but it is obviously a fairly small boat (Giordano 1953). Considering that the merchant hulls of the late Middle Ages weighed little more than the load they had to carry, and that they required for construction a quantity of rough timber double such weight, we can however estimate the timber for the Genoese fleet according to some inventories cited by the same author. Around 1778 the Genoese fleet was equal to 181 units, for a total of about 27,611 t, with about 70–80,000 m^3 of timber necessary for its construction (Gatti 1999), while according to the French estimate of the 80s of the same century the total capacity of 42,130 t, distributed over 643 units, would have required 120–130,000 m^3 of timber. We have no news on the structure of the forests from where the Genoese drew their timber, although we know that in the fifteenth century the supplies came from Dauphiné, from the Maritime Alps behind Nice, from Corsica and from Tuscany (Puccinelli 1996). Finding the quantities and qualities of necessary timber was however a problem of considerable importance, for the solution of which several years of waiting might be necessary, given the scarcity of oak forests in the traditional supply areas.

To the scarcity of oak forests is therefore probably to be attributed the situation of the Genoese shipbuilding industry of the seventeenth century, where alongside oaks, significant quantities of beech are used, while the fir seems to play a very secondary role. In the supply of "square" necessary for an 18 m ship of 1699, it is observed that oak represents 43% compared to the remaining all beech, a significant difference given that almost all the lists of timbers used for the Genoese navy show considerable amounts of beech, in Venice used instead only for the oars. The use of beech as a species for the outer planking could certainly not be said to be optimal, as also confirmed by the French shipbuilding industry, which resorted to it only in periods of oak shortage or in special cases, in fact, beech boards cost half of those of oak in the seventeenth century. Beech was widely used in the construction of the Boulogne fleet with which Napoleon planned the invasion of England (Ballu 2000), but this species could certainly not stand out for great resistance to seawater. Its use nevertheless gave rise to numerous inconveniences, as described for a Genoese ship of 1670 in which the low percentage of oak used in construction is indicated as the main cause of the hull deterioration problems detected already on the first voyage (Gatti 1999). The scarce presence of oak is undoubtedly due to a lesser availability of woods of this species—a structural fact of Ligurian shipbuilding—which always had to be found in places very far from the shipyards; in this sense, Tuscany seems to be a privileged place both for firewood and for timber for shipbuilding. The timbers were in fact delivered to the mouth of the Serchio at Pietrasanta and Pisa. However, in the following centuries, Genoa maintained a shipbuilding industry, importing timber from other countries, even when iron became the main shipbuilding material, while in Venice shipbuilding was reduced to the construction of small boats for local transportation and fishing.

5.3.1 The Coastal Oak Forests of Tuscany

Some documents relating to Tuscany allow us to get an idea of the structure of the woods suited for shipbuilding, even if the Tuscan fleet was little compared to the

Genoese and Venetian ones, attesting to 5–6 galleys in continuous service. In the Grand Duchy, in the middle of the eighteenth century, the backbone of the galley was made of oak but, unlike Venice, the dead work could also be of domestic pine (Pinus pinea) or maritime pine (Pinus pinaster), the internal structures were instead of fir like all the masts. Elm was the protruding parts of the aposticcio, poplar the rowers' benches, the hatches and the bulkheads, walnut instead the stern castle. The granducal authorities also reserved the woods for the Pisa military arsenal; since 1559 the Cerbaie woods of Fucecchio were banned for oaks; to reserve the maritime pine woods of Monte Pisano a law was passed in 1569; instead for elm and ash timber a ban was issued in 1594. For beech and fir, large areas of the Apennines mountains were reserved from which such timbers were floated on the Arno and Serchio rivers; for example, the "Via dei Remi" (road of the oars) was called the route that brought the beech from the Garfagnana valley to the Pisa arsenal (Puccinelli 1996). The Pisa arsenal supplied itself with pine wood on the Monti Pisani, west of the city, still today characterized by the presence of extensive formations of maritime pine.

In the sixteenth century, problems of scarcity of suitable timbers began to be detected and relations were established with the papal domains, coming to import timber from the forests of Terracina and Cisterna, until the pope also vetoed it so as not to compromise the supplies of the Church State. Along the Tiber, an intense traffic of timbers was practiced, and at the two ports of the city, At Ripa Grande and Ripetta, boats docked that transported more than 50% of timber: according to Sansa, by the end of the sixteenth century, the eternal city consumed at least 110,000 m^3 of timber per year (Sansa 2003).

Interesting, for the necessary characteristics of the woods, are the measurements required for the masts of Tuscan galleys. For a flagship galley, the following were needed: a main mast of 24 m in height, with a diameter of 54 cm at the base and 34 cm at the top; a forecast of 16 m in height with a diameter of 39 cm 3 m from the base and 24 cm at the top; a main yard of 26 m in length with a diameter of 36 cm 7 m from the base and 16 cm at the top; a foremast yard of 24 m in length with a diameter of 31 cm. The ritual of choosing the trees in the forest took place in the presence of a large delegation of technicians, and the cutting workers had to climb the trees to verify the measurements and that the trees did not have visible defects (Gabbrielli 1990).

However, while firs and beeches were quite available, although the measurements required significant specimens, it was certainly the oak forests that were the weak point of the Tuscan navy. The opportunity to have a sort of "snapshot" of the characteristics of the oak forests available for the navy is given by a forest inventory carried out in the middle of the eighteenth century to assess the consistency of oaks suitable for naval constructions along the coast of the Grand Duchy. In the period considered, Tuscany is now firmly under the control of the House of Lorraine, but although the current legislation assigned to the grand duke jurisdiction over all forests deemed useful to the State, there was a strong shortage of oak timber for the construction of ships for the grand ducal fleet. To assess the presence of this type of timber and plan the use of this resource, the grand duke therefore ordered the carrying out of a survey on all the forests existing along the coast, which were visited by a special commission made up of axe masters, navy officers, arsenal managers and forestry officials. The results of the survey provide an original and detailed picture of the characteristics of the coastal forests that will not be analysed in such depth in the following centuries (Fig. 5.6).

The survey was completed in about three months, between mid-April and early July of the year 1750. The territory under examination went from the border of the Republic of Lucca, including the municipality of Pietrasanta, to the Papal State, practically today's Lazio region. The considered strip went from the beach inland for a depth of about 15 km, from which the first hilly reliefs were practically excluded, which made the transport costs unsustainable for a convenient use. It was indeed planned to transport the trunks to the sea by loading them onto ships, using draught animals and labour both provided by the local populations. The visited areas were divided into three departments referring to the southern, central and northern part of the Tuscan coast. The identified and visited oak forests were 27, but only in 16 of these were trees suitable for naval constructions found.

The purpose of the survey was to identify specimens of oak (Quercus petraea and Quercus peduncolata) suitable for naval constructions, in particular for the manufacture of the keel, the frames and the outer planking. Not indeed, pines were considered,

Fig. 5.6 Location of the oak forests for the navy in the Tuscan inventory of 1750

used for the structures of bridges, and holm oaks. However, the elm was considered, reduced to just 208 individuals throughout the territory examined. The inventory does not distinguish between the two species of oaks, but it is probably the English oak (Quercus peduncolata) that is most considered, bearing in mind the characteristics of the woods on which we have the most accurate data, namely the Migliarino and San Rossore area. In addition to the number of trees present, the diameter and height of at least 25% of the counted trees are estimated, but both parameters are evaluated with the technological perspective with which individual trees were viewed by the ship's masters. The minimum diameter considered was 28 cm, while for height, the total was not considered, but only that of the trunk deemed useful for obtaining planks. Therefore, the length of the stem was calculated up to the point where it forked or up to where it was judged suitable to produce such an assortment.

Contrary to Veneto, in Tuscany, no type of silviculture aimed at obtaining oak forests suitable for shipbuilding was practised, except for attempts made by the Grand Duke towards the second half of the eighteenth century. Instead, reliance was placed on what was found in "nature", so the situation described in the inventory is the result of the agro-silvo-pastoral practices and the forest utilizations carried out in the area. We must note that while the Venetian land registers do not take into account the number and quality of trees with an abundance of curved timbers, this inventory separately considers such individuals, signalling three categories of increasing branching in the list of trestle feet. This allows us to also have information on the density of the populations, albeit indirectly.

The woods of the first department were not all delimited, so we only have the extension of 13 woods out of the 16 considered. However, if we use the average data of trees/ha of the woods of which we have the surface (first department) we arrive at a total extension of about 44,000 ha. In the first department, we would thus have 38% of the total surface of the visited woods, in the second 10%, in the third 52%. The total number of inventoried trees is 64,802, but it was decided to immediately fell 19,321 individuals, equivalent to 29% of the counted trees. In the delimited woods (34,594 ha, equivalent to 78.7% of the total surface) we have one tree per hectare with suitable characteristics, with oscillations between 1 and 6, so these are mixed woods with the sporadic presence of maritime oaks. The average diameter is around 70 cm, with minimums of 28 cm and maximums of 185 cm. The average height is about 8 m, varying between 20.30 m and 1.2 m. The average volume of usable timber is about 2.56 m^3 per tree (Giordano 1953). Applying to the woods of the first department the average data resulting from the analysis of the oaks actually measured in the other departments, the total volume of the marked oaks is equal to 180,433 m^3. Calculating an average percentage increase of 1.2 m^3 per year, it follows that each year the available woody regrowth did not exceed 2165 m^3 in the entire area investigated (Susmel 1994), a very modest quantity even for just a few ships. If we consider what seems to be the best forest present on the Tuscan coast south of Pietrasanta, namely that of San Rossore extending for about 9000 ha, we see that within it there were 11,753 trees suitable for the navy, with a density of just 1.5 trees per ha, and that on the entire coast there were no more than 65,000 trees suitable for this purpose, much less than those present in the Venetian hinterland.

Considering the inventoried oaks, the distribution of diameters appears particularly deficient in the classes under 55 cm, compared to what we might expect in forests of this type, while it appears more similar to a normal trend the distribution of the diameter classes greater than 55 cm that progressively decrease in frequency. Even considering that the surveyors only marked trees with a diameter greater than 28 cm, it is evident that the forest population as a whole appears characterised by the lack of young trees, a phenomenon confirmed by the descriptions of the time. The news about San Rossore, the location to which 87% of the trees whose diameter and height have been measured refer, explains this phenomenon with the burden of grazing livestock and large wild game that prevented the regeneration of the forest, but also with the intense forest utilizations responsible for the reduced density.

If we observe the relationships between heights and diameters, the characteristics of the forest appear even more particular. In general, the heights vary between 1.2 and 20 m, while in various diameter classes, ranging between 35 and 95 cm, we find a large number of individuals with heights less than 5 m. Even considering the technological perspective with which the surveys were carried out, and any errors or approximations in measurements, it is evident that this situation is not explainable by thinking about problems related to the ecology of the species or to local environmental conditions, but rather to management practices. The technique of pollarding the trees, cutting them at varying heights from the ground, creating what could be defined as an "air coppice" is indeed evident. In this way, the production of forage branches was provided and it was prevented the bite of livestock from damaging the shoots of the stumps. This technique was also used by charcoal burners, who did not like too large trunks to make charcoal and cut the trees at half height using only the branches. This method was also used to try to recover the plants whose underground part of the crown appeared "dried up", a common phenomenon in San Rossore, where most of the inventoried plants did not have good technological characteristics being located in areas frequently or continuously swampy. As the diameter varies, the population does not show a proportional increase in heights. This phenomenon in turn is also reflected in the height-volume and diameter-volume ratios, there is in fact a much higher correlation between diameter and volume than between height and volume. This means that essentially it is the increase in diameter that determines the increase in volume, a phenomenon explainable both with the pollarding and with the presence of isolated individuals in sparse forests. From the inventory data it is evident that the Tuscan coastal forests are totally insufficient for the needs of the grand ducal marina, but it also confirms the presence of a structure substantially shaped by the needs of grazing. It is worth noting that the instructions given to the commission, included the recommendation to consider that not all the oaks suited for shipbuilding could be considered for cutting, an appropriate number oaks should have been left for the production of acorns needed to feed the pigs grazing.

The historical evolution of many coastal areas, not only Tuscan, populated by oak forests is quite interesting as many of them are replaced by reforestation of umbrella pine between the middle of the 1600s and the middle of the 1800s, following the rhythm of coastal reclamation. Once the drainage works were completed, the pine was sown along the coastal dunes so that the pine forest constituted a strip of protection

Fig. 5.7 The Pietrapennata forest, in Calabria, in a drawing by Edward Lear from 1847 (E. Lear, Diary of a walking trip. Reggio Calabria and its province)

of agricultural crops against the wind and against the movement of the sands. This is the origin of the approximately 6800 ha of Tuscan coastal pine forests, in which the production of pine nuts became particularly important towards the beginning of the 1900s, while today they mainly play a landscape and recreational role. The absence of appropriate management practices, that requires to remove the regeneration of different species growing under the shade of the pine and create open spaces among the pine trees, condemns these historical forests to be replaced by mixed oak forests, which spontaneously regenerate in place of the pine, as is happening in one of the most famous and ancient umbrella pine forests in Italy, that of Ravenna (Pasquali 1988). As we will see this is just one of the many examples of the conflict existing in this historical moment between the idea of renaturalization versus the conservation of the Italian cultural forest landscape (Fig. 5.7).

5.4 The Forests for the Arsenal of Naples

The south of Italy was equally intensely exploited for shipbuilding. Interesting in this regard is the extensive use of oak that is made at the arsenal of Naples in the second half of the eighteenth century, in which oak constitutes 82% of the total, while there is a total absence of beech and a minor role of fir which, even together with the larch and pine, brings the total quota of conifers used to just 12.5%. This signals an abundance of oaks in the south compared to the centre-north where oak forests suitable for the navy seem to be scarce. The areas where the Neapolitan arsenal was supplied—especially Abruzzo, Basilicata, Campania and Calabria, in particular Sila and Aspromonte—were rich in forests but usable with great difficulty (Gangemi 2001).[2] The list of qualities supplied by the various locations shows the absolute prevalence of oaks, which justifies the news that such forests were also highly sought after by France, Genoa, Tuscany and Malta. The total in cubic metres of timber brought to the arsenal in those years is 23,699, equivalent to about 11,849 m^3 put into operation, and therefore 8887 t of tonnage using the proportion of 1:2, which equated to about 3 first-rank ships. This is not a particularly aggressive navy, an impression confirmed by the fact that in the same years the arsenal only launches 7 galleys, 4 xebecs and a brigantine. A greater propensity for the construction and maintenance of the road network to facilitate the transport of timber becomes more noticeable in the last decades of the eighteenth century. In promoting the revival of such activities, it seems that the influence exerted on the monarchy by Giovanni Acton, who was given by Ferdinand IV in 1778 the task of reforming the navy of the Two Sicilies, immediately shows a keen concern about the location of forests supplied with trees useful for shipbuilding and therefore the conditions of the roads necessary to facilitate the transport of timber to the arsenal of Naples (Ostuni 1999). From the surviving documentation of an inquiry into the forests of the Kingdom promoted by Minister Acton in 1779, much information is gathered on the characteristics and consistency of the Kingdom's forest resources in relation to the provinces of Abruzzo Citeriore, Terra di Lavoro, Principato Ulteriore, Basilicata, Calabria Citra and Calabria Ultra (Gangemi 1996). The estimate certifies the presence of just over 2,106,000 trees, of which 1,760,000 in Calabria Citra alone, which included the Sila mountains. An extremely interesting element that can be detected from the inquiry is the presence of different forest species. The most cited species are certainly the oaks, with a presence that varies from 29% in Calabria to 56% in Terra di Lavoro, and which are described both in the form of forests and scattered trees in the fields. The situation appears very diverse in terms of species, with many chestnut trees in the Calabrian provinces, in Further Principality and in Terra di Lavoro, while hornbeams and maples seem to be more present in Abruzzo and Basilicata, where together with Calabria Citra there is the highest percentage of ash and hop-hornbeams, while beech trees are found everywhere. The presence of fir and pine trees, and of elms especially in Basilicata and Ulteriore Principality, is very significant.

The descriptions of the woods also report the concurrent uses to shipbuilding to which all the woods were nevertheless subjected. Therefore, the widespread presence of coppice woods for charcoal, heavy timber and poles, staves for barrels, hunting reserves, woods serving the Royal Foundry and the Powder Magazine are mentioned. The production of charcoal seems to be the main purpose of the beech forests, while for the chestnut groves their irreplaceable function for the nutrition of the population is mentioned, as well as the universally widespread grazing of pigs, the so-called "nerj". Additional uses are the production of manna from ash trees and especially pitch on the Sila, to which extensive pine forests are destined, as already observed for the Roman period.

For all the woods considered useful for the navy, the considerable difficulties of transport are reported: the best road network is found in the area around the capital where, in addition to the main routes, there was a fairly dense network of roads between the different municipalities and connections with the royal roads, which indeed, converged on Naples. More problematic, however, are the possibilities of transport relative to the more internal areas of the Kingdom, where the thinning of the main communication axes was accompanied—we think of Basilicata or Calabria—by a location of the woods in mountain areas often defined, by the contemporaries themselves, as "horrid" and "inaccessible". The sources do not fail to report the presence of "trattoj" (or "tratturi"), that is the ancient routes for centuries beaten by livestock transhumant and, in some cases, also passable by carts. In general, however, the routes that connected the woods with the most important roads were limited to difficult paths barely visible and in need of demanding restoration work, especially bridges to overcome ditches and valleys, with paths that did not always allow the transport of timber by "dragging" with oxen. Numerous woods were only reachable with pack animals, donkeys or mules, or even exploitable exclusively through the hard and limited work of men with shoulder transport. The image shows that Calabria is certainly to be counted among the provinces that, due to the consistency of forest cover and quality of timber, offered the greatest contributions to the construction of the war navy of the Kingdom of Naples. The region had one of the most varied and species-rich forest landscapes of the peninsula, and this impression is clearly perceived also from the writings of Edward Lear, traveller. An early nineteenth-century English traveller who extols the variety of the Calabrian forests, particularly the oak woods, often of gigantic dimensions (Lear 2003). Despite the fact that throughout the eighteenth century, but at a faster pace after the great famine of the 60s, the forest area had undergone a contraction in favour of the expansion of cereal crops, beeches, oaks (especially oak) and conifers (fir but, in particular, the larch, locally called "zappino") were abundant and transported overland to Naples or, more frequently, to the coast before being shipped to the arsenal of the same capital and, after 1783, to the new arsenal built in Castellammare (Gangemi 1999) (Fig. 5.8).

The Sila plateau had long been the most important forest estate of the Kingdom, providing both timber and resin: pines and firs were considered by some owners the only useful trees, while the rest of the woods were to be transformed into agricultural cultivation. In the two-year period 1788–1789 alone, hundreds of men and thousands of animals were engaged in multiple logging campaigns from Sila to Aspromonte, and in various places in Sila work was being done to fell and transport timber to the marinas of Rossano, Crocchia and Bruca. From the Aspromonte woods, oak timber was "lowered" to the marina of Gioia. The archival sources relating to these particular aspects of the supply of raw materials for military shipbuilding testify to the complexity of the operations of cutting and transport, defined as "Carrea de' legnami" and practiced in many woods of the Kingdom. When it was impossible or inconvenient to rely on land routes, from the islands of the Gulf of Naples or from the Amalfi and Sorrento coast, the "boat owners" and the crews of the boats were entrusted with the task of completing the transport by sea from the coasts where, in the almost total absence of ports, the timber flowed into simple collection points, up to the arsenal. From these brief hints it is well understood how, although in a

Fig. 5.8 Cutting areas and embarkation places, in Calabria Ultra and Citra, for timber useful to the war navy in the years 1788 and 1789 (Gangemi 2001)

technically less advanced way than in the northeast of Italy, the woods of southern Italy were nevertheless largely influenced by the needs of naval construction and that the oak woods suitable for naval construction were much more abundant.

From north to south of the peninsula, when oak could be counted on in abundance, this was the preferred species, while other types of timber were sought if such a resource was scarce; despite this, coniferous timber could still have good characteristics. In fact, conifers could be chosen from numerous species such as white fir, spruce, larch and various species of pine, given their function of support for the internal structures and the possibility of using them in contact with water, while oaks seem to be the only other ones used. There are important exceptions to these indications in the European field, as in the Portuguese case described in a sixteenth century treatise in which the cork oak assumes a main role for the frame, in replacement of which the holm oak and, it seems, even the olive tree could be used. The explanation for a choice somehow not in line with those of other European shipyards, as the Portuguese themselves observed, seems due to the different quality of the oaks present in Portugal, where the climate and the nature of the soil made these species unsuitable for naval construction, while the cork oak and holm oak woods are said to have provided abundant quantities of curved pieces. This is an observation of some interest, partly confirmed by the spread of cork oak and holm oak wooded pastures in the Iberian peninsula, with the development of trees with canopies suitable for shipbuilding, alongside another peculiarity concerning the use of chestnut for both the frame and the planking, albeit with some precautions related to the defects of this timber, which in Italy was always discarded for such use (Rieth 1999). However, it would be wrong to think that nations that did not possess large forests could not develop large fleets; in fact, for some countries, the opposite seems to be true. The cases of the Netherlands in the XVI–XVII century, which had no internal forest resources but possessed the largest navy in the world at that time (Silvy-Leligois 1962), and then of England, which certainly did not have adequate resources at home given the importance of its fleet, are emblematic of the ability to solve the supply problem, but also of the fortune of being able to count on the proximity of almost unlimited timber resources in the Scandinavian countries and in the Baltic republics, where extensive oak felling is reported in the territory of today's Poland.

Notes

1. The cutting of any timber with an axe or a saw produces a loss compared to the initial volume due to the waste (the sawdust) produced by the passage of the blade, or to the waste due to the wood removed to give the desired shape to a piece.
2. According to Maurizio Gangemi, Calabria was one of the most important regions for the supply of timber for the Neapolitan arsenal.

References

Agnoletti M (1996) Aspetti tecnici ed economici del commercio del legname in Cadore (XIV–XVI secolo). In: Cavaciocchi S (ed) L'uomo e la foresta secc. XIII-XVIII: Atti della ventisettesima settimana di studi, 8–13 maggio 1995. Le Monnier, Firenze, pp 1025–1040

Agnoletti M (1997) Ordinamento e catalogazione di alcuni attrezzi da taglio di uso boschivo. Annali di San Michele 9(10):62–74

Boudriot J (1999) Chènes et vaisseaux royaux. In: Corvol A (ed) Forèt et Marine. L'Harmattan, Paris

Caro A (1826) Eneide di Virgilio, vol I, Libro III. Società tipografica de' classici italiani, Milano

Concina E (1987) Venezia: arsenale, spazio urbano, spazio marittimo. L'età del primato e l'età del confronto. In Concina E (ed) Arsenali e città nell'occidente europeo. La Nuova Italia Scientifica, Roma

Du Monceau D (1800) Del governo dei boschi, Tip. Andreola, Venezia

Gabbrielli A (1990) Abeti, galee e galeazze nella marina toscana del XVI e XVII secolo. L'italia Forestale e Montana 45(4):302–312

Gangemi M (1996) Un'inchiesta sui boschi del Regno di Napoli nella seconda metà del Settecento. In: Cavaciocchi S (ed) L'uomo e la foresta secc. XIII-XVIII: Atti della ventisettesima settimana di studi, 8–13 maggio 1995. Le Monnier, Firenze

Gangemi (2001) Tra strade «impervie» e boschi «inaccessibili». Aspetti e problemi del trasporto del legname nel Mezzogiorno settecentesco. In: Agnoletti M [a cura di], Storia e risorse forestali, Accademia italiana di scienze forestali, Firenze

Gatti L (1999) Navi e cantieri della Repubblica di Genova (secoli XVI-XVIII), Brigati, Genova

Giordano G (1953) Manuale pratico di cubatura dei legnami e dei soprassuoli boschivi, Hoepli, Milano

Lane F (1934) Venetian ships and shipbuilders of the renaissance. The Johns Hopkins Press, Baltimore

Lazzarini A (2001) Domini, tecniche, organizzazione: il trasporto del legname dal bosco del Cansiglio a Venezia fra XVIII e XIX secolo.In: Agnoletti M (ed) Storia e Risorse Forestali (History and forest resources). Accademia Italiana di Scienze Forestali, Firenze

Lear E (2003) Diario di un viaggio a piedi. Reggio Calabria e la sua provincia. Laruffa, Reggio Calabria

Munro HAJ (1867) Aetna. Deighton Bell and Co., Cambridge

Naish GPB (1993) Navi e costruzioni navali. In: Singer C, Holmyard EJ, Hall AR, Williams TI (eds) Storia della tecnologia, vol III. Bollati Boringhieri, Torino

Pasquali G (1988) Il bosco litoraneo nel Medioevo da Rimini al Delta del Po. In: Andreolli B, Montanari M (eds) Il bosco nel Medioevo. Clueb, Bologna

Patrone G (1953) L'assestamento dei boschi disetanei del Cadore. L'italia Forestale e Montana 8(1):35–37

Puccinelli G (1996) Traffici di legname e vie dei remi nella montagna e nelle marine lucchesi, Istituto Storico Lucchese, Lucca

Ostuni N (1999) Le comunicazioni stradali nel Settecento meridionale. Edizioni Scientifiche Italiane, Napoli

Rieth E (1999) La sélection des bois selon le «Livro da fabrica das naos» (1570–1580) de Fernando Oliveira. In: Corvol A (ed) Forèt et Marine. L'Harmattan, Paris

Salvini AM (ed) (1754) Teocrito volgarizzato. Michele Bellotti, Arezzo

Sansa R (2003) L'oro verde. Clueb, Bologna

Silvy-Leligois P (1962) Origine et importance des bois acheminés par eau vers la Hollande aux XVIIe et XVIIIe siècles. Revue Forestière Franjaise 6:516

Spada G (1995) Il gran Bosco da remi del Cansiglio nei provvedimenti della Repubblica di Venezia. Ministero per le Risorse Agricole, Alimentari e Forestali, Roma

Stigliani R (1952) Lavorazioni boschive. Garino, Torino

Susmel L (1951) Struttura, rinnovazione e trattamento delle abetine del Comelico. L'italia Forestale e Montana 6:188
Susmel L (1994) I rovereti di pianura della Serenissima, Cleup, Padova
Tiraboschi G (1822) Storia della letteratura italiana, Tomo I. Della società tipografica de' classici italiani, Milano
Tursini L (1940) Note di architettura navale romana. In: Ucelli G (ed) Le navi di Nemi. Istituto Poligrafico dello Stato, Roma
Ucelli G (1940) Le navi di Nemi. Istituto Poligrafico dello Stato, Roma

Chapter 6
The Timber Forests in the Eastern Alps

Abstract The alpine forests located in the north-east of the country have been an inexhaustible source of construction timber since the Roman period; the valleys of the Piave and Adige were the backbone of trade with the plain from the thirteenth to the nineteenth century and still today provide the best construction timber. It is in the 14th–16th that the systematic exploitation of the forest begins also from the point of view of the creation of the technical and commercial structures necessary for forest utilizations. The production of construction timber was part of a complex of functions performed by the forests and regulated by the alpine communities. These are institutions that in the Upper Piave Valley organised themselves into the Magnificent Community of Cadore, in 1338. The most intense exploitation of these forests starts after the incorporation of Cadore in the Venetian domains in 1420. Studies of pollen diagrams have shown a significant human influence in the distribution of vegetation on the Italian Alps due to timber trade with the steady growth of conifers until the twentieth century. The climatic conditions, the orography, the hydrography, the lack of the roads in the alpine environment that characterizes the Piave Valley necessitated the development of a wide range of techniques and structures specifically designed to overcome these difficulties and bring the timber down to the plains by river driving. The increase in volumes of timber traded along the Piave river led to the gradual establishment of customs in the management of forests that took into account not only economic needs, but also the conservation of the forest, and which were applied and respected for several centuries. The management system applied in is selective cut, which originates as a system removing individual trees from a forest, in opposition to the cutting of all the trees typical of clear-cutting as occurring in the nearby province of Trentino. This system established that the tree was felled when it had reached maturity, understood as commercial "maturity", with the aim of providing certain assortments required by the market. The growth of timber trade in this area continued until the first world war bringing the Italian timber dealers to dominate the market, cutting forests also in the countries of the Austrian empire and importing to Italy. After the war Austria took the leadership in timber production directly exporting timber to Italy, this marked end of timber industry in the northeast of Italy and the birth of the Italian furniture industry as a result of the changes in the market conditions.

M. Agnoletti, *History of the Italian Forest Landscape*, Environmental History 19,
https://doi.org/10.1007/978-3-031-81631-4_6

6.1 Forests and Communities in the Upper Piave Valley

The alpine forests located in the north-east of the country have been an inexhaustible source of construction timber since the Roman period; the valleys of the Piave and Adige were the backbone of trade with the plain from the thirteenth to the nineteenth century and still today provide the best construction timber. The alpine communities with their rules ensured an exploitation that held together local needs and commercial needs, managing to maintain better than in the plain the consistency of forest resources. The most important forests in this area were the beech forests and coniferous forests, especially fir and spruce. The proportions between these species have been profoundly modified by human activity over time, and the current landscape is the result of a long historical process that has influenced the characteristics of the forests especially since the medieval period.

The production of construction timber for a long time was a marginal aspect compared to a complex of functions performed by the forests and regulated by the alpine communities. These are institutions that in the Upper Piave Valley organised themselves into the Magnificent Community of Cadore, in 1338, dividing the territory into "Rules", an institution of probable Celtic origin that consolidated following the Lombard invasions. These were composed of associations of families: the Community was governed by a Council composed of elected members, with further subdivisions at the territorial level such as the "Centenari", very similar to modern municipalities, that grouped more Rules. The most important area for analysing the evolution of forest resource management in this part of the Alps is certainly the Upper Piave Valley, an area subject to the control of the Magnificent Community of Cadore. From the fourteenth century, the trade in Cadore timber began to take on considerable dimensions and the passage of the Community of Cadore under Venetian rule in 1420 did nothing but sanction an economic and commercial union that had existed for some time. This was a choice also dictated by the need to defend themselves from the ambitions of the feudal nobility of the princes of Tyrol, and in fact the Ampezzo territory was annexed in 1518 in the Archduchy of Tyrol where it was destined to remain until the end of the First World War when the whole area became part of the Italian state (Fig. 6.1).

It is in this period that the systematic exploitation of the forest begins also from the point of view of the creation of the technical and commercial structures necessary for forest utilizations (Zangrando 1951). The first area that begins to be affected by the exploitation of the forest seems to be that of central Cadore, where concerns for the availability of timber are already felt in 1300 when the two centenari of Pieve and Valle decide to exclude some woods from community use ("vizzare" in the local dialect) to ensure the necessary timber for the construction or repair of houses, bridges or roads (Venturoli 1930). This type of interventions soon becomes a common practice and spreads throughout Cadore. They not only respond to the need to control the cutting of the forest, but are also a symptom of the growing importance of the timber trade and the increasingly vast economic interests connected to it that require greater public control over the forests. The main tool to achieve this control

Fig. 6.1 The various phases of work in the forest (du Monceau 1764)

is the practice of "vizzare", a term that corresponds to banning a forest from public use, reserving it for the needs of the community, a measure that begins to be used first by individual Rules and then by the Community of Cadore itself.

Until the mid-sixteenth century, bans are mainly operated by Rules and Centenarians, but new regulations inserted in the statutes from 1527 expressly prohibit individual Rules from reserving the woods without prior authorisation from the Council and impose a ban on cutting timber for commercial purposes in reserved woods (Zangrando 1961). These new regulations essentially sanction a general trend of this period which sees a progressive disempowerment of the Rules, therefore of the Community, as the main economic and political subject and an equally progressive concentration of power in the hands of the Council of the community. If on the one hand the Council's claim to the right to reserve the woods asserts the community's right of ownership, granting the Rules only the use of timber for social works, on the other hand it concentrates even more power in the hands of the Council itself. The Council, present in every phase of social life, was controlled by the oligarchy of notables who often used it to their advantage, taking advantage of their role. And it is precisely from the intertwining of interests between the notables of the Cadore Community and the timber dealers, often the same individuals, that the fate of the wooded areas is determined. A first element in which these connivances materialise is the leasing of community-owned woods, a practice to which the community is forced to face public expenses. To avoid excessive exploitation and allow the merchant to regulate the cuts over time, the tenant is granted the right to extend or renew the contract provided that it can be proven that the conditions of the forest have improved. In this way, leases are renewed for many years, but end up creating a sort of merchant's right over the woods in question (Venturoli 1930). These forms of tenancy involve merchants and public representatives in commitments that are stipulated in the form of private writings, but which in reality sanction agreements with clear commercial intentions (Fig. 6.2).

Fig. 6.2 The design of Iseppo Paulini from 1601 appears in a study on the Alto Bellunese presented to the Senate of Venice, which links the mountain to the problems of the progressive silting up of the lagoon. On the left, a forest before cutting and on the right, the same forest after cutting. On the right, the forest is released on the top of the mountain to protect the soil from erosion, below the forest with some trees left standing and others felled, as in the selective cutting's felling, in the valley bottom the cultivated fields (State Archive of Venice)

The most glaring manifestation of the intertwining of interests to which the woods are subject concerns the commercial rental of reserved woods. This custom, now consolidated in the sixteenth century, is used by Rules and Centenari which, indebted for grain, for salt, for public works, borrow the necessary money from the official or the notables of the centenary, Renting the marked forests for their own use, thus opening the forest to speculation. According to the statutes of the Cadore Community, the reserved forests cannot be rented, nor can commercial logging be carried out there, as they are intended only for the use of the regular community. Violating this rule not only contravenes a statute, but goes against repeated provisions of the community council. The custom established in the second half of the sixteenth century, however, sees the councillors themselves, who within the council deliberate against the cutting of the marked areas, privately appropriating the banned forests. An emblematic figure of this state of affairs is that of Tiziano Vecellio, notary and councillor, but also a timber merchant with his own sawmills in Ansogne and a famous Italian painter of this period (Zangrando 1958). Some documents from the period show, for example, how Vecellio is able to manage on a personal basis the reserved forest of Ombrio belonging to the Rule of Candide by drawing up private contracts with the various members of the Rule. In doing so, these latter deprive themselves, for very modest sums, of the timber intended for domestic use, giving free rein to the commercial exploitation of the forest. To carry out these operations, Vecellio generally goes into partnership with other notables and it is not surprising, being in Comelico, to see him alongside Andrea Gera, an official of Comelico Superiore, a member of one of the most powerful families in the area (Eicher Clere 1988). The massive use of the reserved forests for commercial purposes undoubtedly constitutes the economic basis on which the political power of the notables rests and one of the reasons for the intense exploitation of the Cadore forest resources in this period.

6.2 From Beech to Fir and Vice Versa

After an initial greater attention to the beech, used mainly as firewood, the cuts increasingly focus on the fir. However, by affecting the density of the fir trees, the natural regeneration of the beech is favoured, which, having come to clutter the clearings created by the cutting of the fir, greatly exceeded in number the natural regeneration of the fir. This is a normal ecological process linked to the different needs of the two species. For this reason, at the beginning of the sixteenth century, the fir market had greatly reduced and a new trade in beech was beginning to flourish, which became the object of new cuts, so that in less than two centuries the fir trees grew again and the trade in construction timber resumed, letting the beech trade decline. The transformation of the landscape described here appears as the consequence of intense cuts that do not destroy the forest but modify its composition. In fact, studies of pollen diagrams in the post-glacial period have shown how, after the immigration of the white fir and beech into the Eastern Alps during the Boreal period for natural reasons, they then spread simultaneously during the Atlantic period. Subsequently, until the beginning of the era of the most intense forest uses, spruce, white fir and beech reach approximately identical distribution, although in some places phases of beech prevalence are found in the Sub-Atlantic period (500 BC–1500 AD). In practice, these studies confirm a significant human influence in the distribution of vegetation on the Italian Alps after the year 1000 (Kral 1989).

Regarding the sixteenth century, it is worth noting that the problem of timber supply for mining and metallurgical activities in Cadore, Agordino and Zoldano, which were economically more significant than construction timber at that time, had already become significant (Cucagna 1961). The forests that bear the brunt of this situation are undoubtedly those of beech, which provide firewood for the workshops. Despite attempts to curb the problem by reserving beech forests, divided into forests for domestic use and supply forests for the workshops, the forests of this species tend to decrease (Venturoli 1930; Sala 1937; Schupfer 1912). If in these centuries beech areas are reduced to procure fuel, in the following centuries the situation changes for opposite reasons. In the nineteenth century, there is a tendency to strongly prefer spruce over beech due to the latter's lower commercial value. There are recommendations and official provisions from the beginning of the century in which it is expressly ordered to "ring" the beech or even to "uproot" it in the forests of Somadida and Caiada (Coletti Di Lucia 1988). Ringing is one of the oldest practices for felling trees, especially large ones, for which the use of axes was considered too demanding. The bark was indeed cut all around the perimeter of a tree, interrupting the transport vessels of the nutritive flows, xylem and phloem, from the roots to the crown, thus leading the tree to desiccation and then death. This allowed for easier felling or burning of the tree. In the second half of the nineteenth century, favouring conifers over broadleaves even became a specific direction of the forest administration of the Kingdom of Italy (Piccioli 1885).

The phenomenon of alternation in the species that make up a forest as a result of commercial logging is not an isolated case. As we have seen, a case similar to Cadore concerns the forest of Montello, initially composed of chestnuts and beeches, which became an oak forest after being banned for the construction of the galleys of the arsenal of Venice (Di Bérenger 1965). Particularly interesting, both for the different source and for the many points in common with Cadore, is what is described for the Vosges, a major forest region located in the southeast of France where the forest was treated with occasional cutting, commercial exploitation was intense and sawmills were equally numerous. In this area, in the sixteenth century, the prevalent species were oak and beech, while in the following century the situation seems to change in favour of spruce. As hypothesised for Cadore, here too it seems to be the need for fuel for the metallurgical sector that determines the change (Boithias and Brignon 1985) (Fig. 6.3).

Fig. 6.3 An early seventeenth-century Venetian sawmill along the Piave, drawn by Iseppo Paulini (State Archive of Venice; from Agnoletti 1998)

6.3 Chutes, Dikes and Waterways

The climatic conditions, the orography, the hydrography, the lack of the roads in the alpine environment that characterises the Piave Valley necessitated the development of a wide range of techniques and structures specifically designed to overcome these difficulties and bring the timber down to the plains. Techniques to which it is worth paying some attention, given their fundamental role not only in making resources available, but also in modifying their qualitative and quantitative characteristics, influencing their conservation and use. One of the main characteristics of the technical organisation of this sector was the large amount of time between the request and the actual supply of the material. The purchase of cutting rights, the planning and implementation of forest utilizations, the transport of goods to their destination, was a process that could extend for a period even longer than 12 months, due to the relative rigidity in the sequence of operations. As with the agricultural sector, where the various phases of work marked the life of the mountain populations, forest work had its seasons. Although they had greater flexibility in choosing the moments of their execution, operations could only be carried out under certain conditions. Felling had to be carried out by the end of summer, before the winter snowfalls, which were essential for transport in the forest. During the winter, in the presence of snow and especially ice, logging had to be carried out, as it would have been much more laborious or impossible to drag the timber from the felling site out of the forest on bare ground. This operation was facilitated by the construction of wooden chutes, a structure made with the same felled trunks that descended along the sides of the mountains and into which the timber was started helped by the surface made slippery by frost and snow (Agnoletti et al 1986). These were costly structures made by expert personnel that were part of a complex system of internal forest transport, structured to facilitate reaching a watercourse or the rare roads (Occhi 2000). Similarly, when the thaw and rains increased the flow of the streams, it was necessary to carry out river driving, i.e. the transport of timber along the rivers, before the periods of low water when even the hydraulic sawmills placed along the rivers could not work to saw the raw timber for lack of water.

Until almost the entire nineteenth century, the possibility of carrying out the timber trade in a profitable way was linked to the existence of a watercourse towards which to channel the trunks felled in the forest, so the rivers constituted the main axis for the development of transport. We know that the Romans organised flotations along the main Italian rivers and therefore probably also on the Piave. The "dendrofori", technical corporations of "timber makers" who were particularly concerned with the river and land conduct of timber, already existed around 716 BC, and traces of their activity have been found in Belluno, Feltre and various other areas of the Alpine arc and central Italy. The secondary hydrographic network of the Alpine arc presented abundant, fast and almost everywhere embedded waters, with strong variations between maximum and minimum throughout the year, allowing only loose river driving. On the contrary, the hydrographic characteristics of the main rivers, especially the Adige and Piave in terms of traffic volumes, allowed their use as stable commercial routes, used for rafts and boats used for the transport of timber and other goods.

As a result of this hydrographic situation, the organisation of work generally envisaged in the upper part of the valleys the loose driving of trunks, with the realisation of various types of accessory works. In the lower part, at the confluence with rivers or lakes, other works were instead created to block incoming logs and divert them along the banks or into the ditches feeding the water powered sawmills for processing, after which the sawn assortments continued their journey tied in rafts. There were therefore two distinct phases: loose river driving and driving in rafts, which, since the sixteenth century in Trentino, were regulated respectively by the Trifft Ordnung and the Flòss Ordnung, regulations that organised such activities. In this area, the course of the rivers naturally identified two different markets for the produced timber: the Venetian and the Lombard, fed respectively by the eastern and western valleys of Trentino, while the Piave Valley and all the rivers of Friuli mainly fed the Venetian market (Agnoletti 1998) (Fig. 6.4).

In the tenth century, Emperor Berengar made agreements with both Arnulf of Germany and the bishop of Belluno regarding the rights and duties for the transit

Fig. 6.4 Representation of the landscape of the Alto Bellunese made by Iseppo Paulini at the beginning of the 1600s. The importance of pasture compared to the presence of the forest is observed (State Archive of Venice)

of timber in Trentino and Friuli. On the Adige in 1181, there is talk of a "field" in Egna for tying the timber from the Val di Fiemme along the Avisio into rafts, confirming the early assertion of this production area in the Trentino area (Dal Ri 1886). The agreements of Bishop Eberardo of 1155, on the other hand, provided for the construction of a commercial port in Riva del Garda (on the Garda lake) to collect timber coming from the Sarca river. The free flotation progressively extended, especially from the fifteenth century onwards, causing a growth of disputes between merchants and local populations for the damage caused by the disorderly race of the logs downstream, which is accurately reflected in the protests of the various communities, with calls to court and sentences to settle the disputes. As for the Piave, the act that clarifies the modalities of flotation is from 1308, but we know that this trade was common practice also in the previous centuries. With the passage of Cadore under Venetian rule in 1420, and the consequent intensification of river traffic, the need to regulate such an important sector as that of trade on the river became increasingly pressing on the part of the Serenissima.

The flotation was carried out by specialised personnel and presented many risks, often having to operate in the water to free the masses of logs that formed in the riverbeds. Once the final destination was reached, some workers also had to retrace the stretch of the river backwards to recover what was left behind, so both the employment of labour and the costs were significantly high (Occhi 2000). The operations were coordinated by a manager who, in addition to organising the work, was responsible for it towards the merchant or the communities that organised the transport, such manager was present in all the main torrents where this activity took place. Among his competencies were also recognitions along the banks of the watercourses, to study the problems of the route and the protections to limit the damages. Other important figures in this activity were the "counters", personnel assigned to count the logs and separate them by property according to the marks imprinted on the trunk with specific tools, an operation that was carried out on the banks of the river where the timber was gathered; in Tyrol it was forbidden to "water" the timber not marked. The Trifft Ordnung of 1646 ordered not to insult the counters, which speaks about the delicacy of the operation and the social climate in which it took place, given that the operation often also served for the imposition of duties.

If the force of the waters of the torrents was not enough to drag the trunks, dikes were erected to create artificial floods capable of transporting the timber downstream with their water force. These dikes, could be permanent or temporary and were present in many locations in the Alps, constituting a structure widely used and described in documents since the XV and XVI century (Agnoletti 1996). The permanent dikes, rarer than the temporary ones, acted by opening a door in the middle of it that caused the sudden release of a mass of water sufficient to transport the timber accumulated under the dike, they were operated by specialised personnel and had a considerable cost. They were complex structures, with internal rooms for the operation of the opening mechanisms, very different from the temporary dikes that were more commonly used to float the timber along the minor torrents. If they were built by merchants, the costs fell on the budget of the commercial activity, but if they were made by the holders of the cutting rights, the construction costs were compensated by the collection of duties (Occhi 2000).

Important structures were also the "cìdoli", consisting of barriers placed across the watercourses, intended to stop the floated trunks. Below the upper part they had a rack, or comb, composed of vertical pylons so close together that they did not allow the passage of the trunks, which was however possible by lifting the rack, or the individual pylons, by means of a winch. In the XV century there was a rack to stop the timber of the Avisio river at Lavis, while another was present at Fonzaso around the XVI century to collect the timber coming from the Cismon river, but the most important of those known in Italy was the cìdolo of Perarolo di Cadore, reported since the XIII century and which remained active until the XIX century.[1] These structures were always combined with the "roste", wooden channels which had the function of directing the trunks towards the feeding channels of the water sawmills. They underwent periodic rebuilding and their construction represented one of the most frequent items of expenditure in water transport, which was provided by merchants in partnership or individually, if these served to feed their own sawmills.

The river driving in rafts instead involves the transport of timber in the form of rafts, but not all types of timber could be used because of their specific weight. If necessary, rafts were made with conifer trunks on which some hardwood trunks that could not float individually because of the different specific weight were loaded, and this always happened for the heavy oak trunks used for shipbuilding.[2] In the case of the Adige river, we have evidence of its use as a river route since Roman times for transport on rafts; the "ratiarii", those in charge of conducting the rafts on the rivers, were organised into corporations, as were the "dendròfori" who dealt with timber transport by land. During the early medieval period, the Adige already played a strategic role as a waterway, as numerous agreements regarding transit and duties on goods demonstrate. In 1182 the prince Bishop of Trento was granted by the emperor the collection of navigation duties (Andreazza 1989), while the statutes of the "Art of the rafters" of Verona from 1260 list in detail all the most important stops along the river. In this century, the river is a rapid artery of traffic leading to Venice, connecting Venice with Verona, Tyrol and present-day Germany. In the second half of the sixteenth century, the importance of the Adige as a river route for European trade appears to be well established. A document from 1596 describes the itinerary of goods coming from Belgium to Italy, indicating Antwerp, Brussels, Cologne and then Tyrol via the Adige (Canali 1939) (Fig. 6.5).

To the east of the Adige, all the river network instead directly led to Venice. The Brenta was one of the waterways of strategic importance, in fact Bassano del Grappa had a port for rafts and several canals connected the river with Padua, Venice, Chioggia and the Adige. However, the Piave was the most important river for the transport of timber to Venice, which even the "German" merchants used to transfer timber from the valleys north of Cadore, dragging them with oxen and winches over

Fig. 6.5 The "Cidolo" of Perarolo in a photo of the late nineteenth century. It is a barrier built across the Piave watercourse in 1668, intended to stop the floated trunks

the Alpine passes and making agreements to have ports on the river. On the Piave, the raftsmen joined in a corporation only in 1462, about forty years after the union of Cadore with Venice, but their activity was already established in the thirteenth century, when we have documents about the disputes between the Bishop of Belluno and Venice for the duties on raft goods.

The interest of the merchants of the plain in the Alpine timber resources is constantly growing from the thirteenth century, so much so that in the sixteenth–seventeenth centuries the mountain environment is at the centre of multiple commercial interests and considerable economic fervour, linked to the exploitation of forest and mining resources, but facilitated by the abundant availability of hydraulic power. The merchants, distributed at the nerve centres of the watercourses in conjunction with the timber transit areas, played the role of intermediaries with the commercial circuits of the plain, maintaining contacts with the holders of exploitation rights, represented by the local communities and the German-speaking territorial princes. In this context, the economic world, first and foremost the Venetian one, with its nobles and financiers, found in the timber resource an important opportunity to invest their fortunes, an activity facilitated by the progressive expansion of mainland domains. The Venetian policy of the fifteenth century therefore tends to ensure greater control also of those areas that had significant timber resources. As already mentioned, Cadore was annexed to the Republic in 1420, but also in southern Tyrol the penetration took place in the same period; after the Vallagarina the Venetians pushed as far as Folgaria, to use the timber of those forests and establishing close commercial relations with the port of Calliano, which began to serve the arsenal directly. The Venetian control also extended westward, reaching not only the Giudicane Valleys, but also Brescia in Lombardy, especially Val Trompia and Val Camonica, establishing relationships that lasted even after the Republic's withdrawal from these territories following the wars with Austria (Valenti 1907).

The choice of Venice as a reference point, even for those traders coming from other areas who opened their "shops" in the city, was justified not only by the growing size of the Venetian market, but also for financial reasons. The immobilisation of capital required by the forest-wood supply chain was considerable, and the long time between the anticipation of capital and its fruition forced merchants to seek the substantial capital needed where it was available. From this point of view, there is no doubt that Venice was a privileged place to find money, not only in the Italian panorama, and it is equally true that the economic rise of many families of timber merchants from the 1500s onwards shows the effectiveness of this strategy. These reasons, in addition to the inherent slowness of the reproducibility of forest resources, forced operators to build stable relationships over time, through methods that could vary in the case of local communities or territorial princes.

In the Tyrolean area, relations with the German-speaking nobility and the archducal institutions were linked to the loan mechanism, with payments at the various Tyrolean forest and customs offices where they secured logging licenses, a custom that ended up building important ties with the Tyrolean state, for which timber was a significant source of income, continuous and constant. In the sixteenth century, 74% of the revenue from Tyrolean customs came from timber duties, an important source of income, therefore, but also a double-edged sword. With the growing importance of trade and the role of loans for the archducal coffers, the role and political importance of merchants increased in step, so much so as to obtain investitures and noble titles that even allowed them to reach positions such as that of archducal adviser held by Giovanni Battista Someda at the beginning of the 1600s Occhi (2000). This situation allowed merchants to assume prominent roles in the alpine communities, often superior to those of the local feudal lords who sometimes found themselves losers in legal disputes (Fig. 6.6).

Although among the changing fortunes of the individual families that have succeeded each other in the alpine area, the merchant class colonised the entire alpine arc. east, gradually taking into its hands all the timber trade, a process that will proceed in crescendo until the First World War (Agnoletti 1998). In many cases, these are monopoly positions held by a few families, such as the Angeli, along with the Petricelli and Maccarini, who in Fonzaso in the first half of the 1600s control the timber industry and trade, with the sole worry of circumventing the bishop's customs control. If, all in all, contact with the Austrian authorities was quite simplified, due to the full sharing of the "money-logging licenses" exchange, relations with local communities were more complicated, especially in cases where these had autonomous governing bodies and established regulations for the use of timber resources.

6.4 The Selective Cut

The increase in volumes of timber traded along the Piave river led to the gradual establishment of customs in the management of forests that took into account not only economic needs, but also the conservation of the forest, and which were applied and respected for several centuries. Although non-Cadore traders were considered foreigners by the communities—so much so that it was expressly forbidden for them to bring labour from other areas to work within the community's borders—, the rules they established influenced the entire forestry sector, representing not only one of the original features of the Cadore technical system, but also one of the most important models for Italian silviculture.

Cadore has always been considered the homeland of the unevenaged forest, that is, a forest characterised by the simultaneous presence of plants of very different sizes and ages. The management system that applies to this type of forest is selective cut, which originates as a system removing individual trees from a forest, in opposition to the cutting of all the trees typical of clear-cutting. This system established that the tree was felled when it had reached maturity, understood as commercial "maturity", with the aim of providing certain assortments required by the market. When cutting, therefore, plants equal to or greater than a certain size were selected, established based on commercial criteria, but interventions should also have been carried out to improve the quality of the forest. This method was opposed to clear-cutting used in some areas of southern Tyrol and in Austria, where entire portions of forest were

Fig. 6.6 In this drawing by Roswitha Asche, the phases of timber transport on the Piave to Venice from the Cadore mountains are schematised. To the letter “a” a wooden chutes that carries the trunks to the river for free flotation, to “b” a wooden dike, to “c” some *menadas* guiding the trunks on the river, to “d” the Perarolo cìdolo, to “e” a Venetian water sawmill, to “f” and “g” the transit of some rafts along the Piave, to “h” a boat for transport, to “i” the river ascent with animal towing, to “1” the arrival of the rafts in the Venice lagoon (Agnoletti, *Foreste e segherie nel Trentino,* cit.)

felled eliminating all the trees. With this system, the creation of even-aged forests was facilitated since the regeneration occurred in the same period and the trees reached maturity simultaneously, giving trunks of similar sizes. In truth, the clear-cutting system, towards the 1600s, also appears to be employed in Tuscany by the benedictine monks of the Abbey of Vallombrosa, who just then begin to draw decent profits from the commercial use of the fir forests owned by the Abbey. However, selective cutting has been the most widespread silvicultural method in Italy for high forests, especially in Venetian domains.

With the Cadore customs, the mature tree, the one ready to be cut, was the one having 12 oz in diameter, which was equivalent to a Venetian foot (34.8 cm) measured at the height of the chest of a man (1.3 m); to this class (XII) a fixed economic value was attributed from which the values established for the other classes depended, as shown in Table 6.1 (Giordano 1953). What is important to observe is that the price paid was not proportional to the actual volume of the trunks, but depended from the value of the boards obtainable from the respective diameter classes,[3] devaluing those of small dimensions and overestimating those obtained from larger dimensions. Commercially, this system gave a great advantage to merchants who were buying: in fact, it was enough for a trunk not to reach, even by a few millimetres, the measurement of the higher class to be paid as the lower class. The actual volume was thus ignored and the size underwent a significant depreciation.

The undisputed merit of this method was to set simultaneously rules for the management of forests and a quick system for assessing the commercial value of the boards obtainable from the trunks, in fact, the diversification into diameter classes implied the presence of uneven-aged forests, composed of trees of different ages and sizes (Da Rios 1922). The 12-oz trunk was practically the one corresponding to the tree considered "mature", having a diameter of about 34.8 cm above the bark at 1.30 m from the ground. In principle there was no convenience in using a forest too intensively by reducing excessively the trunks of large dimensions and leaving subjects of too small measures, since the waiting time to have mature trunks would have been very long and small trunks were paid little. For the coniferous forests of Cadore, a curing period was foreseen, therefore a waiting period for the next cut, of about 10–12 years, which was the average time required for the trees belonging to

Table 6.1 Merchant customs for the evaluation of timber in Cadore

Commercial classes	Actual volume in cubic metres	Variation % of the price paid
VIII	0,245	0.25
X	0.375	0.50
XII	0.550	1
XV	0.8	1.5
XVIII	1.1	2
XXI	1.44	2.5

The commercial classes in the first column are expressed in Venetian ounces. The percentages in columns 2 and 3 refer to the 12-oz trunk, a trunk 34.8 cm in diameter measured at chest height

the diameter class immediately lower than the mature one (10 oz) to reach maturity, that is 12 oz. The commercial system described for Cadore went on until the First World War when the Italian army, having to buy timber for war operations, imposed the switch to the decimal metric system, a transition already sanctioned by a law of 1890 but never applied. The selective cutting system is practically the most suitable method for maintaining forest characteristics closer to the natural structure, resulting in a forest landscape characterised by trees of different ages, diameters and sizes.

6.5 Forests and Timber Trade Between the Eighteenth and Nineteenth Centuries

In the eighteenth century, Cadore, Camia and the Asiago plateau are the areas of the Venetian domains where the forests regularly provide large quantities of timber. With the difficulty of finding materials, in particular oak timber, the production costs of ships have quadrupled and other nations, particularly the Netherlands, take the lead. The Dutch have revolutionised this industry by organising more efficiently both shipbuilding and the procurement of raw materials from the Baltic countries, managing to produce ships at lower prices (Lane 1978). For its part, the arsenal increasingly uses material from Cadore. In the eighteenth century, the trunks traded annually along the Piave reach the figure of 200,000–300,000, while every year about 1400 rafts sail the Piave towards Venice (Agnoletti 1996). In addition to the increase in trade with abroad, we must not forget also the growth of internal needs. Around the fifteenth century, the population of Cadore is between 6000 and 10,000 units (Zanderigo Rosolo 1982), in 1557 it reaches 13,217, while around 1790 it becomes 22,814 (Fontana 1972). This increase cannot but also reflect on the wood uses for internal consumption, both in timber and firewood. It should also be remembered that all of Belluno, in this period, seems to suffer from a great scarcity of timber and Cadore is the area that provides almost all of the firewood (Vecchio 1974).

The end of the century certainly represents a critical period for the woods, especially those in the Venetian hinterland. With the arrival of Napoleon in 1797, there is a frantic rush by private individuals and municipalities to cut (the latter even sell the woods by issuing permits to uproot the trees). The boundaries of the woods reserved for the arsenal are removed and in general there seems to be a period of total freedom, or rather anarchy, at least until the promulgation of the forest law of 1811 (Di Bérenger 1965). With the fall of Napoleon and the return of the Austrians, there is a general increase in the timber trade and a further intensification of commercial uses. The progressive expansion of the sector is confirmed by a series of observations related to the birth of new commercial firms, the increase in the transit of timber on the Piave and the construction of new sawmills. The beginning of the 1800s sees the formation of an important alliance between the Lazzaris family and the Wiel family: the progressive expansion of the family's activities leads the Lazzaris firm to assume a prominent role in the timber trade, and in 1860 it will represent the most important industry in the Cadore sector.

The events of this company are very interesting to understand the evolution of the timber trade, not only in Cadore (Pavan 1988). In 1821, the good progress of business convinces the Lazzaris to deal, in addition to trade, also with the processing of timber. For this, they build a factory in Perarolo with sawmills and carpentries. A few years later, large timber depots are built in Covolo, Spresiano and Treviso. Next to these depots, shops are also opened for sale in order to meet the needs of construction timber and firewood of the plain. The general management of the company is instead located in Venice. These developments outline the business strategy of the company. The operations of cutting, processing and transporting timber are concentrated in Perarolo. The sale in the plain is carried out in the branches of Covolo, Spresiano and Treviso, all located along the Piave. The timber shipped to Venice is instead introduced into the main Italian and foreign commercial squares by the local agency.

The market growth recorded in this period also favours the construction of new sawmills, which are traditionally almost all located in the stretch between Perarolo and Longarone, the most important area for this activity. In Ospitale di Cadore in 1825, as many as twelve are built by Candido Coletti who, being an expert technician, with some adjustments manages to also improve the quality of the sawn timber. From the second half of the century, the data relating to timber production are continuously growing. In 1841, about 2000 rafts of planks, 800 boats made with beams, 500 "fanghere" composed of beech arrive in Venice from Cadore. In 1871, the rafts will be 3000. Other data provided by Wessely (Wessely 1860) indicate that the floated timber is around 300,000–400,000 trunks in 1859, but as we will see these figures are destined to increase.

It is difficult to make precise calculations on the exploitation of forests in Cadore, but if we accept an approximate estimate we can attempt an evaluation using Wessely's data related to surfaces. Calculating the timber coming from Alto Cadore, Auronzo district, taking as an average the cut from 10 oz and increasing this quota with illegal cuts and internal consumption, we arrive at a value fluctuating between 7.3 and 8.3 m^3 per ha of annual production. This value is almost the same indicated by Susmel (Susmel 1952a, b) as production for the Comelico fir trees belonging to the most productive woods, while the average productivity was around 4–5 m^3 (Patrone 1953).[4] It should be noted, however, that Susmel's data refers to the period after the second world war, in which usage has greatly decreased and therefore the conditions of the woods are certainly better than they were in the previous century. In addition to this, we must think that in the mid-1800s the characteristics of the felling, logging and transportation of timber certainly do not allow for the full exploitation of the woods, so the cuts are concentrated in the most easily accessible areas.

A reflection of the progressive intensification of forest utilizations seems to be observable also in some changes in the technical characteristics of water powered sawmills. In general, we can say that the set of technical changes that occurred in that period, which also included the reduction in the thickness of the blade and the spread of the teeth, seem to reflect the concern to save wood, unlike what happens in Austria, for example, where this concern is much less. Around 1860 the trade of Cadore timber is still in a rather flourishing phase. Guernieri calculates that in this period the people employed in cutting, transporting and processing timber in the province of Belluno are more than 10,000, about a seventeenth of the entire population (Guernieri 1862). Despite this, both in Venice and throughout the Mediterranean—the traditional outlet for timber traded by the Venetians—the competition of Austrian firms was increasingly felt. Cadore timber seems to possess better technological characteristics (Costantini 1864) but also higher prices. In this regard, it should be noted that in Venice Cadore timber pays duty while this is not the case for that coming from other parts of the Austro-Hungarian empire.

From a report of a business trip in 1858 on behalf of the Lazzaris company, we learn that the market situation for Cadore timber is still good. In the Bologna and Ferrara province, a lot of Cadore timber is used, its merits and qualities are recognised, but in Tuscany the forests of Camaldoli and Casentino provide all the assortments required at lower prices. At the port of Livorno, large quantities of timber arrive from Trieste and from Fiume, also these at lower prices than those of Cadore. A certain stagnation in the timber trade is observed in Rome, while a strong consumption of Cadore planks is noted in Palermo, although some loads coming from Fiume cost less. Cadore timber is also widely used in Catania. The greatest demand, however, is noted in Malta, which is the hub of the African market. Larch and fir are much sought after and the Arabs want only Cadore fir boards. It seems that in Malta attempts to introduce timber from North America and Austria have had no success. Great use of Venetian timber is also made in Corfu, especially in construction, confirming the tradition linked to the timber trade from Venice to the east.

In the second half of the 1800s, however, difficulties begin to increase. In November 1871 the new Fortezza-Villach (Austria) railway is inaugurated. This line and the new connections to Trieste allow Austrian timber to reach Venice quickly and at competitive prices. In addition to the price, the products coming from Austria have two advantages: the greater speed with which they reach the market and the absence of the dark colour that floated timber takes on after being immersed in water for a long period. We must not forget that from the moment of felling to when the Cadore timber arrives in Venice can take more than a year, and moreover, one cannot be certain of the delivery times. This makes it necessary to have large warehouses whose costs are obviously reflected on the Prices. The negative repercussions of this situation are not long in coming. In the following years, the prices of Cadore timber suffer a sharp drop: from 7 to 9% from 1875 to 1878, from 13 to 17% in 1879, from 25 to 30% in 1880–1881 (Fabbiani 1973).

The growing difficulties of the Cadore economic model can also be inferred from the crisis of the hydraulic sawmills. For these, in addition to the unfavourable economic situation, natural events also play a negative role: in fact, in the disastrous flood of the Piave in 1882, at least forty are destroyed, and it is symptomatic that many of them are not rebuilt. It is interesting to note the behaviour of the most important company on the Piave, the Lazzaris company, during this period. After a careful analysis of the situation, the company's management decided to move the bulk of production to Spresiano (TV) where a very modern steam sawing plant was built in 1883. The choice of Spresiano is dictated by various reasons. First of all, the company's warehouses are located here, then the proximity of the river allows to continue receiving raw materials from Cadore, but above all this location is well served by roads and railways to market the products.

It is during this period that the economic model linked to the exploitation of the forest in the Upper Piave Valley begins to enter into crisis. Without a doubt, it is also a difficult time for the entire Venetian mountain, characterised by strong emigration that involves various sectors in addition to that of wood. Until then, the substantial integration and organisation of forest cuts, river transports, sawmills, had allowed the system to withstand competition, even if some elements such as sawmills and the river were beginning to no longer be up to the times and the forest was heavily exploited. For many centuries the relationships between these factors had remained substantially unchanged, but now the system shows the inability to change to adapt to new needs. Thus, with the development of roads and railways, the river that represented the unifying pole of this economic model is no longer up to the situation and tends to put the other sectors in crisis as well.

6.6 The Major Changes of the Twentieth Century

At the turn of the century, the timber market is in further expansion thanks to the growth of construction, railway constructions and shipbuilding. From 1900 to 1910, it seems that between 500,000 and 600,000 logs per year have left Cadore, almost double what Wessely indicated in 1858 (Fabbiani 1973). As observed for the Somadida forest regarding forest utilizations, these are often more or less overt forms of clear-cutting. In some areas, however, there is a real change in the form of government: for example in the Val Fiorentina, part of the Cordevole river basin, clear-cutting had been introduced since the previous century, and not on small surfaces but on cuts extended even 30 ha (Sief 1960). The major changes underway in the organisation of the timber trade are also evident from some data relating to flotation along the Piave. From 1905 to 1914 A total of 1,476,332 logs are floated. On average, therefore, only 164,000 pieces are floated per year out of the approximately 500,000 per year previously mentioned. The arrival of the train in Perarolo on 5 June 1913 definitively marks the abandonment of floating, which then completely ceases with the construction of dams for hydroelectric purposes. This period represents a new moment of difficulty for the timber trade, and in particular for that of Cadore. In the 1911 financial report of the Lazzaris company, one of the reasons for the crisis that the company was going through is indicated in the inactivity of the sawmills in Perarolo. The same company now mainly procures timber in Austria, but also the Rettenegg plant in Styria (Austria), where the company in 1908 had purchased a plot of forest and built a sawmill, is in difficulty.

Only with the outbreak of the First World War is there a certain recovery of the Cadore timber trade. However, in 1930, timber exports appear to have been reduced to about 90,000 m^3 per year (more or less 280,000 trees of 10 oz) (Pais Tarsilia 1930). The sawmills do not exceed fifty units, and many of them are electric. In these years, the first scientific studies aimed at ascertaining the real characteristics of the Cadore forests are also carried out. The most important of these works is undoubtedly that of Da Rios, which appeared on the occasion of the first Italian forestry conference, held in Florence in 1921 (Da Rios 1922). Only in the early years of the post-war period are increasingly detailed studies carried out on the forests of Cadore. In general, it is admitted that the composition of the forests has been strongly influenced by the commercial preference given to red fir (Susmel 1952a, b). This trend has resulted in the reduction of beech and a strong reduction in the percentage of white fir, both species that, from an ecological point of view could occupy larger areas. The high yield of some areas compared to others with lower yield, even in the same forest, are explained by the fact that certain areas are more easily accessible because they are better served by roads or logging tracks compared to others that are more difficult to reach. This has led in the past to more intense uses. An interesting aspects that studies on the forests of Cadore highlight is the presence of small evenaged forests, of different ages, alternating with large evenaged forests. These two types of forest landscape, hardly well definable in their boundaries, occupy almost all of the Cadore area and it can be said that they are equivalent in terms of extension, while unevenaged forests does not exceed 1% of the total forest surface (Patrone 1975). The history of the forests of Cadore shows that, although subjected to intense cuts, in reality the forests have never been destroyed and above all that these forests do not currently have a structure that can be defined as natural, but are the result of many centuries of forest utilizations operated to support local populations and the commercial activities of Venice.

6.7 The Transformations of the Forest Landscape in Trentino

Trentino, the southern province of Trentino-Alto Adige region, allows us to study in more detail the changes in the Alpine forests for a period subsequent to what was analysed for Cadore, also considering the events of the timber industry that played a decisive role for the characteristics of the forest landscape. In terms of forest management, it is observed that in the provincial territory historically the presence of the coppice forest, in addition to the high forest, is well documented and played an important economic role. For the Garda area, the coppice wood is mentioned since the eighth century BC, as a complement to agricultural activities, this form of forest management also persisted in the Roman period and in subsequent centuries.

By the end of the 1800s, the growth of the timber industry led to a gradual decrease in coppicing in favour of high forest, especially conifers. Coppicing represented a low economic value production linked to agricultural activity, even though its main use was obviously to produce firewood and charcoal, which was preferred for its qualities over conifer wood. This form of forest management was characterised by great versatility, being able to adapt to the most diverse needs, producing both poles for vines and fruit orchards, and construction material, as well as fodder, in addition to fuel. With the reduction of coppice areas, which accelerated in the first decades of the 1900s, the need for firewood of the populations shifted towards the waste products resulting from the use of coniferous forests. The decline in the need for firewood and charcoal was also contributed to by the Trentino industry. In fact, with the introduction of electricity at the end of the 1800s, the energy problem for industrial activities was definitively solved. Industries no longer had to rely on steam, which required huge amounts of firewood and charcoal, but could be powered by electricity. Despite this, coppicing still held a position of some importance in the Trentino forests. The situation remained this way until the Second World War, during which large quantities of firewood were used.

In the following years, the production of firewood suffered a decline throughout the national territory due to the progressive spread of compressed gas and other petroleum products. In Trentino, the presence of cheap firewood, also resulting from the use of high coniferous forest, somewhat limited this phenomenon. It is clear, however, that the replacement of fuelwood created further unfavourable conditions for the cultivation of coppice, if only for the export that was made towards the Po Valley. In the 1950s, coppicing is now considered a secondary importance forest due to its low economic value, and if simple coppices decrease, coppice with standards disappear. For these forests, an improvement is hypothesised in those years mainly through coniferisation, therefore a further increase in conifers at the expense of broadleaves, with a view to increasing the economic value of forest production. In the 1970s, with the creation of the first Trentino Forest Map, it is observed that coppicing now only constitutes 25% of the forests, while coppice with standards is not even mentioned.

The reduction of the coppice forest went entirely to the advantage of the high forest, which saw its importance grow thanks to the increase in the economic value of construction timber and the fall in the value of firewood. Beyond strictly economic aspects, the decrease in coppicing marks a changed relationship between man and territory. Coppicing, as a form of forest management, implies a greater presence of man, also for its role within agriculture. The decrease in this formation involves the progressive decline in the importance of agricultural activities and the reduction of human presence, the shift of the economic centre of gravity towards other productive sectors, and the social changes of which the abandonment of coppicing is an indicator.

6.8 From Broadleaves to Conifers

The increase in the extent of high forests was therefore accompanied by the progressive increase of conifers at the expense of broadleaves. The preference for resinous trees is already evident in the sixteenth century. The Ferdinandine ordinances regarding the woods in Primiero and Tesino in 1558 express concern to safeguard above all the Schwarzwald, that is, the coniferous forest reserved for trade. When we talk about commercial timber in Trentino, we can only refer to the two most important species from this point of view, spruce and larch, since the white fir was almost ignored. At this time they are already appreciated and sought after, as also appears from the commercial documents of the beginning of the fourteenth century, relating to the purchase of batches of timber in Val di Fiemme, by Veronese merchants. The preference expressed for the red fir is a normal fact, which is observed by also analysing the transport on rafts that was done along the Adige (Fig. 6.7).

The larch - another very sought-after species but also much scarcer—was known and appreciated already by the Romans, as we have seen; its spread, often in the phase of repopulation of deforested lands following fire or grazing, goes through alternating events. The extension of pastures in high-altitude territories, combined with the desirability of its wood, constituted a threat to its spread, but its ability to recolonise the areas degraded and abandoned by grazing has ended up constituting an advantage. In the fifteenth century and in the following centuries, there seems to be a certain tendency to defend this species from excessive exploitation. The rule cards of Deggiano of 1494 expressly prohibit cutting larch. The bishop of Trento in 1551 advises not to use fir and larch for fencing works, but rather the avezzo (white fir). In the ordinances of Charles VI of 1735, the monetary penalties for those who cut more than allowed are one florin for each larch, while for the red fir 30 carantani are paid. In addition to being used as construction timber, this species was also used for the collection of resin, an important economic activity in the forests: for each larch carved to make turpentine, 6 carantani were paid. Another very common use of larch was for the support poles for vines, given its great resistance to rot. To obtain this type of assortment, many young plants were cut, often with damage to the density of the forests.

In general, the forests in these centuries showed a greater presence of species compared to the current situation. In 1586, in Val di Sole, in the territories of the rule of Dimaro, the woods are described as consisting of larch, spruce, scots pine, white fir with the presence of alder and birch. Woods very different from the current ones where the larch forests have more space, especially in the band between the valley floor and 1300 m of altitude. Val di Sole is still a special case, given that the larch is in its *optimum* constituting 41–43% of the valley's forests. Its exceptional spread compared to the rest of Trentino is probably of anthropogenic origin, and due to grazing. In the middle and lower Val di Sole, however, the Swiss pine is missing, also disappeared due to anthropogenic causes. The greater presence of broadleaves is sometimes confirmed by the possibility of making comparisons between the current situation and documents from the past. Some orders from 1537 and 1693 prohibit the cutting and export of chestnut timber from Meano and Sardinia. In both these areas, the Trentino Forestry Map, in the 1970s, does not mention chestnut groves even as mixed formations. Until towards the end of the nineteenth century, there are no data for a temporal comparison regarding the presence of conifers and broadleaves (Fig. 6.8).

From 1892 onwards, however, it is possible to trace, at least in broad terms, the trend of this ratio which appears very significant. In fact, a strong decrease in broadleaves in favour of conifers is evident, which reaches its peak around the 1970s.

Fig. 6.7 Remains of a stone canal, for the logging of timber in Val di Fiemme. These artefacts still punctuate the forest landscape of the north side of the valley (photo Mauro Agnoletti)

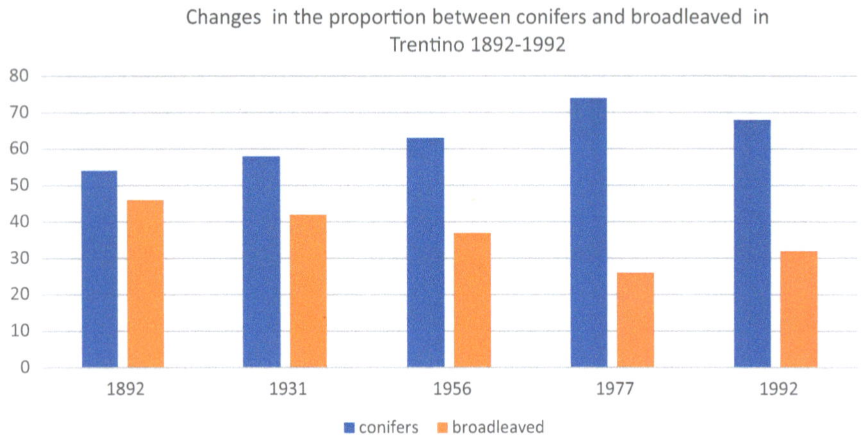

Fig. 6.8 The changes in the proportion between conifers and broadleaves from 1892 to 1992 in Trentino illustrates the great changes undergone by the forest landscape in terms of species to meet the greater demand for construction timber coming from conifers

This fact leads us to make some considerations. We know that the spruce was the main species of the stands and that its economic maturity was generally reached around 70–90 years, although the cutting cycle could vary, due to the different growth rate, and even exceed 120 years. We must therefore think that the origin of the high percentage of conifers recorded today must be placed in a period that goes from the end of the nineteenth to the beginning of the twentieth century, when the preference for conifers is also facilitated by favourable market conditions. This, however, represented a need that had already matured from the nineteenth century and was pursued with determination also by the Austrian forestry administration. In the second half of the century, promoting resinous trees becomes a precise direction also of the forestry administration of the Kingdom of Italy. These trends represented the influence of agronomic type forestry of German origin, which preferred monospecific plantations of conifers. These trends were in turn spread by forestry technicians who dealt with the management of the forests of Trentino since the first half of the nineteenth century.

In any case, more than technical teachings, which had a limited practical impact, economic convenience considerations were valid. Following the increase in the road network, it became convenient to exploit forests that were previously too far from the markets. From the moment it is no longer necessary to reduce all types of forest indiscriminately to firewood, but it begins to be technically possible and economically convenient to trade construction timber, things start to change. The possibility of exporting sawn timber favours the spread of water sawmills and, since the greatest demand was for red fir planks, a greater attention for this species is obtained as a consequence. This also to the detriment of the white fir, which is considered inferior from a technological point of view. In this process of reduction of some species in favour of others, we must also consider the action carried out by man but not directly related to trade. The disappearance of the beech, for example, or its current very limited extent, is not only due to the timber industry. This species has instead been subjected to intense use by local populations who have always preferred it as firewood and for the manufacture of household items, without simultaneously thinking about promoting its renewal.

Another important factor that has influenced the composition of the Trentino forests was undoubtedly the reforestation work, which we will discuss more broadly for the post-unification period. Reforestation is already mentioned in the eighteenth century, in fact in 1735 the Ferdinandean ordinances allow work in the plots cut only for reforestation, while those who let their land become forested could freely dispose of the timber that would have been produced. The rest of the ordinances leave no doubt about the intention to promote not only the increase of the forest, but especially the growth of species suitable for trade. From the beginning of the 1800s, in addition to promoting forest regeneration, there is also a need to intervene to defend the territory with hydraulic and forest works, the execution of which is also codified through technical regulations. For the Val di Flemme, Francesco Meguscher, chief inspector of the Tyrolean and Vorarlberg forests, compiled a reforestation project as early as 1831, some sowing was done in 1832 and some plantations in 1836. From that period onwards, there is also the spread of forest nurseries that rapidly increase in number, producing mainly spruce and larch. The reforestation work also had legislative support, in fact the law of 1852 obliged to reforest the sections of forest cut in the imperial and communal forests. Resinous species were generally preferred, which were also increased in districts where the presence of conifers was already considerable. In Val di Fiemme, for example, from 1878 to 1898, at least 150 ha were artificially reforested, with 971,845 seedlings and 76 kg of seeds. More precisely, 776,745 3-year-old red fir trees, 188,880 2 and 3-year-old larch trees, 3300 4-year-old Swiss pine trees, 2000 4-year-old dwarf pine trees and 1000 beech trees were used, with a total expenditure of 6843.79 florins. The account of the sale of the plants from the Povo forest nursery in 1885 gives us a very significant report on the species produced. We have larch at the top with 108,200 plants, then black pine 90,500, red fir 62,150, acacia 61,100, ash or hornbeam 53,200, hawthorn 28,300, Scots pine 17,150, chestnut 7000, oak 6000, sweet chestnut 5650, Christ's thorn 5350, hornbeam 3500.

Another important moment regarding reforestation was after the First World War. As early as 1923, the War Damage Section of the former Trento Forestry Inspectorate was in charge of the reforestation works. To repair war damage, work was carried out in 97 municipalities, from 1923 to 1928, and in another 37 in 1929. 1,102,000 seedlings of red fir, larch, black pine were planted, but there were still works to be completed in another 50 municipalities. Other municipal-owned areas were also reforested, to compensate for the uses due to extraordinary cuts carried out in the years immediately after the war. In addition to this public action, private individuals also carried out reforestation, on a smaller scale, but always favouring conifers. The reforestation was therefore having a significant impact on changing the proportion between deciduous and coniferous trees, so much so that among the foresters themselves there were voices of perplexity, due to the rapid change that this type of choice brought to the woods of the province.

An analytical comparison between the percentage of deciduous and coniferous trees in the various forest districts is only possible with the data from 1892, which we have compared with those of the Trentino Forest Map. The comparison highlights a general increase in conifers at the expense of deciduous trees, an increase extended to all districts of the province, albeit with different percentages. In areas like the Val di Sole, the advent of roads and the Trento-Malè railway made it very advantageous to export timber and therefore increase production, and the Valsugana with its railway could do the same. The most significant increases concern the Val di Sole, with a + 30% of conifers, the Giudicane (+ 32%) and the Alta Valsugana (+ 40%), although due to changes in the district boundaries the data may be distorted. For the Val di Sole, the increase is also reflected in the number of electric sawmills from 1929 to 1955, which grew ninefold. The Giudicane Alps also saw a parallel increase in sawmills from 4 to 27.

The trend towards increasing the production of conifers, particularly red fir, was further increased by unification with Italy. From this point of view, the Trentino forests played a decidedly strategic role since they contained not only a large part of the Italian woods of this species, but also some of the best ones. The growth of conifers was noted with some concern in the 1930s, when it was observed that several coniferous forests located at their lower altitudinal limit had developed thanks to the gradual elimination of deciduous trees. In the years 1950–1953, the production of working timber was now totally composed (99.5%) of coniferous timber. Within this, red fir was the most used species. The remaining quota of deciduous forests was composed of 70% beech and almost all the rest chestnut. This situation also corresponded to the majority presence of the spruce, compared to other conifers, as a component of the woods. Guglielmo Giordano assigns this species 78% of the surface of all the high forests, while Sembianti assigns red fir 42%, larch 15%, Scots pine 10%, white fir 6%, other pines 5%. The two authors, however, share the value assigned to the total of coniferous forests (Fig. 6.9).

The detail of the production of timber in the various districts in 1952 also indirectly gives us an idea of the presence of the various species. Detailed surveys carried out by Cristofolini in 1963, based mainly on management plans, would give the following percentages for the Trentino forests: red fir 65.1%, larch 17%, white fir 12.2%, Scots pine 3.1%, Swiss pine 1.8%, beech 0.5%, black pine 0.3%. Even after the introduction of forestry on naturalistic bases, occurred a few years after the second world war, the timber industry has continued to influence the evolution of the Trentino forests, in fact the uses of spruce have even increased. The current situation sees a limited presence of broadleaves that represent a minimal percentage of the Trentino forests. Therefore, in terms of specific composition, it can be concluded that the landscape of Trentino has been largely modified by human influence.

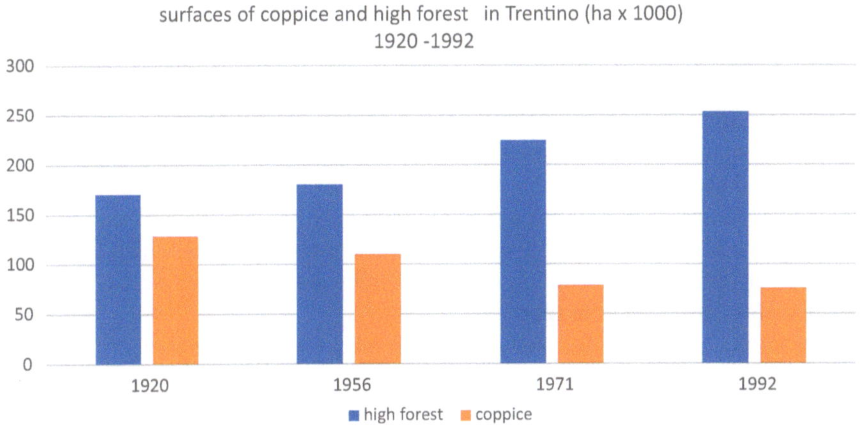

Fig. 6.9 The evolution of forest management in Trentino sees the decrease of coppice forest in favor of high forest. This transformation of the landscape occurred for reasons related to the greater economic importance of timber from high coniferous forests compared to firewood from coppice forests

6.9 Firewood and Timber

As far as the use of the forest is concerned, an interesting aspect concerns the relationship between the consumption of the population and the timber destined for trade. There were essentially two types of use that were opposed in Trentino: that for firewood and that for construction timber; and over time their relationship tends to change in relation to economic and social evolution. The decrease in the importance of firewood and the growth of that for work finds its point of trend reversal in the 1930s. From this moment the gap between the two values increases more and more, with a logical interval in the period of the second world war in which the needs of the population took over. At this point we might ask ourselves what was generally the destination of the timber that absorbed the largest share of uses. An answer can be found by summarising some elements already discussed, such as the prevalence of internal consumption, especially wood, compared to export; the important presence of coppice forest and the use of its production almost entirely for firewood; the prevalence of firewood production, compared to construction material, until the 1930s; the reduced percentage of construction material used for internal consumption, against the almost total use of firewood by the local populations. Having said this, we can conclude that the intense exploitation of the Trentino forests from the 1800s until the second world war is to be attributed to internal consumption and in particular to the consumption of firewood, more than to the production of construction timber destined for export.

Regarding the methods of managing the forests, the presence of "fratte" and therefore the clear-cutting of all the trees in a given plot, is documented in the medieval period, but we cannot say that this form of treatment was the most widespread:

selective cutting prevailed, as we have seen for Cadore. In the nineteenth century, despite the influences of German economic forestry, clear-cutting seems to have been practiced regularly only on 15% of the provincial surface, especially in Val di Fiemme. In the following years this trend increases: in 1955 30% (according to some 50%) of the forest surface presented an evenaged structure, mostly derived from clear-cutting and partly from other silvicultural methods. The rest of the forests had the typical characteristics of uneven-aged forest only on 15% of its extent. In their majority, the forests presented a very heterogeneous structure in which the traces of irregular felling, carried out on the bigger trees, located in the most convenient positions for cutting and transport.

The felling operations, focused especially on big trees, without bothering to also eliminate those decaying and to distribute the cuts to maintain the uneven-aged forest, had the general result of reducing the average diameters of the trees present in the forest. The larger trees had always been much sought after, and the timber orders for the Sacco raftsmen in the 1700s explicitly report the indication “in the largest possible sizes” for both squared and round beams. At the beginning of the 1800s, however, there are already the first observations on the fact that the Trentino forests appear degraded and there are complaints, above all, about the disappearance of ancient forests, meaning forests that presented trees with significant diameters, about one meter.

The spread of the road network had allowed to reach even areas where cuts had never been made, or anyway where occasional felling had been carried out and not too intense, and that could now be exploited more regularly. This phenomenon and its effects find some correspondence in certain technological aspects of the sawmill industry. As already observed for the Cadore, in the early 1800s the intensification of uses had led to the reduction of the average diameters of the logs to be sawn. The reduction in diameters, at the sawmill level, resulted in an adaptation of the machines to saw smaller logs, going from 70 to 80 cm in diameter to 40–50. This observation seems to go hand in hand with the intensification of uses that precisely in the early years of the century reached the highest levels (Fig. 6.10).

6.10 The First World War and the End of an Era

The effects of the First World War on the timber market allow us to understand some fundamental aspects concerning not only the decline of production and trade of timber in Italy, but also much of the dynamics that affected the Italian forest landscape, well beyond the narrow territorial area of the province of Trento.

The outcome of the first world war had dramatic consequences on all aspects of the social and economic life of the north-eastern regions. The collapse of the Austro-Hungarian empire and the birth of new national states completely changed the political framework of reference and the economic effects of these changes were not long in coming. On the other hand, the Italian state, having taken possession of the so-called New provinces, soon made its influence felt. Trentino suddenly found itself

Fig. 6.10 1829 drawing. Lumbermen working in the bed of the Cordevole river in order to facilitate loose driving (Trentino)

in directly involved in the Italian political and economic situation, with consequences also on the timber industry. In those years, Trentino had already achieved an important role as a supplier of timber to Italy. In 1915, its exports constituted about 1/8 of all the timber imported from Austria, which in turn more than doubled its export to Italy between 1900 and 1913, but the events following the world war played a distinctly unfavourable role in the Trentino timber industry with a real upheaval of the entire timber market, which had a significant effect both on the forests and on the industry.

At the production level, there was a decrease of about 36%; but in the face of a reduction in exports of about 22%, the real collapse affected internal uses which suffered a drop of 62%. The reasons for the crisis are various and must be linked to the Trentino situation, to the Italian one, and to the foreign market. The war years had greatly reduced manufacturing activity with a contraction of production in the entire agricultural sector. The devastated area covered a third of the province and the cultivated agricultural area had been greatly reduced. During the war—especially for military needs—and in the immediate post-war period there was certainly an excessive exploitation of the forest, which had been used above its own productive capacity. Many forests had also been damaged by war operations, trenches, bombardments, fires. Wood production in 1919 reached 715,000 m^3 and the following year was even higher, exceeding the forest's growth rate. The surge in production was motivated mainly by internal needs, which had pushed beyond any previous limit due to the war and post-war situation.

The new fact of this period is that Trentino became part of the Italian market, with all the consequences that this entailed. Since the second half of the nineteenth century, the timber problem had taken on a fairly defined appearance for our country. Italy was a heavy consumer, but could not produce all the wood it needed. Before the war, against a consumption of round timber of about 3,200,000 m^3, only 500,000 m^3 were produced in Italy, while 80% of the import came from the Austro-Hungarian empire. From 1895 to 1913, the empire had significantly increased its export to our country, going from about 330,000 to 1,050,000 t, three quarters of which arrived by rail and the remainder by sea. In total, this was about 2,863,000 m^3 of timber, of which the vast majority (more than 80%) was sawn timber (Serpieri and Segala 1915). The main areas of origin were those closest to our borders: Carinthia, Tyral, Vorarlberg, Carniola, Croatia, Slavonia, Styria, Bosnia. Among these, all of Carniola's export was directed to Italy, while Carinthia and Styria exported only a third to our country (Segala 1915). The incorporation of Trentino and the new provinces within the national borders did not change the situation much since these regions could only contribute 500,000 m^3. Of course, this increase meant doubling national production, and bringing it to about 1,000,000 m^3, yet despite this, only a third of internal needs could be met.

The war and its consequences had, however, changed another important aspect of the timber trade, which is necessary to understand the events following 1918. Given the scarcity of domestic production, even before the war many Italian companies worked abroad, buying timber or even working it on site, to then import it into Italy. Almost all the import of timber from Austro-Hungarian empire was in the hands of Italians, and the value of our industrial heritage in those areas was estimated at over 200 million pre-war lire (Serpieri 1919). Before the war, many modern industrial plants had been installed in Rovereto, Mezzolombardo, Bolzano, Bressanone, Brunico, Vipiteno, in Val Pusteria and in other locations along the Brenner border and also in Austria. Among all, undoubtedly the Feltrinelli company, destined to become one of the first Italian multinationals, was the one that had created the largest commercial organisation and carried out the most important business. Just think of the purchase for 18 million crowns of large forests in Hungary and Romania, to be exploited in 25 years (Segala 1915) (Fig. 6.11).

Italian traders bought the production in bulk directly from Austrian forest owners, and this caused resentment among local timber industrialists. In fact, almost all operations related to cutting, transport, sawing, seasoning, assortment of timber were subordinate to capital, to direction and largely to Italian labour. For this reason, as early as 1902, pressure had been exerted by Austrian industrialists, at the Ministry of Agriculture in Vienna, to dislodge Italian dominance. However, the forest owners had always refused to accept these proposals, fearing to lose their best customers (Micoli 1923). The only thing they had been able to obtain was the increase, in 1912, of the transport tariffs of round logs by railway, to limit the growth of the sawing industry in Italy and Germany, to the detriment of the Austrian one. With the end of the war, however, things changed; for some time now, the local press had been advocating the need to keep the production and export of timber completely in Austrian hands, warning that Italy would have to adapt to the prices imposed by local producers (L'alpe 1918). Italian companies therefore found increasing difficulties in operating in Austria, with a consequent double serious damage to our economy. Not only did trade gradually pass into the hands of Austrian companies, but since the Italian ones worked with Italian labour, jobs were also lost. The events of these years will practically mark the fate of the Trentino and Italian timber industry.

The economic difficulties in which Austria was struggling in the post-war period had a strong impact on its timber industry. The strong devaluation of the crown

Fig. 6.11 The forest of Paneveggio in Trentino is quite famous for the quality of the tone wood used for musical instruments. One of the most famous violin maker of the world Antonio Stradivari, who lived between the sixteenth and the seventeenth century, used to collect spruce wood from this forest

and the general problems led the Austrian government to adopt important measures, which consisted in monopolising the export of timber, which was one of the main resources of the nation. In Vienna, the Holzausfuhrstelle was created with the task of establishing prices in relation to the exchange rate of the crown and the foreign market. All producers had to adapt to these prices and were paid directly by the government, which, however, retained a part of the proceeds. At the end of hostilities, in 1918, after the peak reached in the same year, prices had a sharp reduction and, in 1919, Austrian exports of sawn and squared timber decreased a lot compared to 1913. (77.2001) (Vitale 1921). For this reason, and due to the need for timber for reconstruction, intensive logging was carried out in Italy to also take advantage of the inevitable rise in prices that was expected to follow this scarcity.

With the ongoing devaluation, the price of Austrian timber had become very competitive, not only with Italian timber, but also with that of the countries of the former empire that had become dangerous competitors, such as Czechoslovakia and Yugoslavia. Austrian exports to Italy, taking advantage of this favourable situation and the increasing demands of our market, therefore regained ground, reaching 247,500 tonnes in 1920, and at the same time Austria gradually raised the prices of sawn timber from 160 to 320 lire/m^3. In 1921, perhaps also due to a stagnation in demand due to the reduced needs of reconstruction, but also in the context of a drop in the price of wood at an international level, the Austrian government suddenly decided to reduce the benefit for the treasury, reducing the price of timber to 200 lire and at the same time bringing the exchange rate from 12 to 18 crowns (L'Alpe 1921). This decision had immediate serious consequences on Italian production. By being able to sell timber at the border at a price of 200 lire/m^3, the Austrians effectively nullified the economic value of many of our forests and extended the areas with a negative balance, where the costs of use exceeded the value of the timber produced.

As we have said, as a result of the war and the situation created in the timber market, intensive logging was carried out. The mirage of easy profits due to post-war reconstruction had led to logging being undertaken by makeshift companies, which were then absorbed by larger ones. The sector had attracted workers from other industries, who became woodcutters, carters, sawyers, fuelling an excessive growth in the number of employees. The collapse of prices as a first consequence meant that large quantities of accumulated timber remained unsold. The situation was serious everywhere: in Trentino, as in other areas, municipalities had put timber up for sale that found no buyers, and therefore the administrations found themselves without income to meet budget needs. The municipalities themselves had anticipated the costs of use, but were unable to sell the timber and recover these costs.

Italian traders naturally did not just watch what was happening across the border, in fact, they had begun to react as soon as the Holzausfurhstelle in Vienna had decided to lower the export price of sawn timber from 270 to 200 lire. The chambers of commerce, municipalities and associations of industrialists and traders in north-eastern Italy circulated various memorials to denounce the situation, asking the government to impose new customs duties (L'Alpe 1921). The Rovereto Chamber of Commerce did the same, asking in a document of 15/6/1922 for interventions on the level of railway tariffs, as we have already seen, and on the customs level. Interesting is the clarification (at point 2b) where it is asked to make possible the importation of round timber which was hindered by "prohibitive" railway freights on foreign routes. This factor deprived the Trentino sawing industry of raw material, while the market was flooded with imported sawn timber, to the full advantage of the Austrian sawing industry. In practice, Austria, through a system of export taxes and railway tariffs, made the export of round timber to Italy economically disadvantageous. This triggered further resentment from Italian industrialists, who openly asked Italy to assert its rights of the victorious country, returning the market conditions to the pre-war situation (Micoli 1923).

On the Austrian side, a policy was clearly implemented aimed at the Italian market, which on the other hand was subject to strong imbalances. It was easy to conclude that the ideal solution for our sawmill industry and our forest production would have been the increase in import duties as had already happened in 1887, but in reality the government's policy on this front was rather cautious. Traders had their faults, since in better times they had speculated both on imports from abroad which were in their hands, and towards the owners of Italian forests, and now they were asking for high duties on sawn timber and the exemption of round timber to saw it in their sawmills, now present in too high a number. In addition to this, if the duty protected domestic production, it would have disadvantaged our exports to Austria. Anyway, although not everyone agreed, the pressures from the industry were partially accepted, and in 1921 new customs tariffs were set, which exempted round timber and set a new duty for sawn timber (Serpieri 1921). In reality, the duty was very moderate and therefore insufficient for the purpose, moreover in this way those who imported round to work it in their own sawmills were favoured and the owners of Italian forests were disadvantaged. As was therefore predictable, despite this protection in favour of the sawmill industry, timber from abroad remained more convenient, especially the Austrian one.

In the following years, the problem of relations with Austria was the subject of various government decisions, which also concerned timber, with the application of the most favoured nation clauses, mutually adopted (Carloni 1923). Much of the battle was played on railway freights, which our railways did not intend to lower for internal transport. The prices of foreign timber and therefore its excessive competitiveness with our products were apparently considered the main problem, with accusations of unfairness to the policy of the countries exporting to Italy. Magazines such as "Il Legno" and "l'Alpe" clearly reported the ongoing debate and the issues at stake. The controversies were often direct and Austrian personalities also took part in person to counter the theses of Italian industrialists. On the part of the latter, it was claimed that the fluctuations in the price of Austrian timber were the work of a calculated speculative trend implemented by the government in agreement with the large companies in whose hands the timber trade had passed. In this sense, the industrialists expressly asked the government to implement a series of measures that would limit the damage of such a policy, and restore the national sawmill industry to competitive conditions.

In reality, there were structural problems, such as high production costs and the scarcity of raw materials, which did not favour a positive evolution of our situation.

Moreover, the government seemed unwilling to intervene decisively, perhaps not taking into account that Austria also desperately needed to sell us its timber. On the part of traders and industrialists, there was also a lack of the ability to implement and agree on choices that would favour the production of national forests. In fact, technological investment in the sector could have been improved to reduce high utilisation costs, reduce the number of sawmills and increase their production capacity, by setting up modern sawmills. During the war, considerable profits had been made, exploiting the dizzying increase in timber prices due to the military operations, but in the long run, speculative operations could not prove successful.

As can be inferred, the events described had a significant influence on the sawmill industry. During the period in which traders had gained control of all the importation of Austrian timber, the number of sawmills had greatly increased, both those in Italy and those built abroad. Immediately after the war, as the Austrian importation decreased, it was necessary to cut and saw the necessary timber in Italy, and so many new small plants were built, especially in Trentino. The growth of the sawmills was stimulated not only by the needs of reconstruction, but also by speculative intentions linked to the trend of prices. A decent amount of capital was invested in the establishment of sawmills and the construction of new means of communication and transport, causing an excessive growth of the sawmill sector (Institute of Agricultural Economics and Statistics, Ministry of National Economy 1927). According to some estimates, in 1921, in Trentino Venezia alone there were 1265 sawmills, of which 161 were multi-blade: a disproportionate growth, about four times higher than the productive possibilities of the region. This growth was also contributed to by the prices charged for sawing, which had grown in step with the prices of timber. In fact, in 1918 private sawmills were asking for a price five times higher than that charged before the war for each cubic metre of sawn timber. This increase could be justified in part by the increase in costs and depreciation, but in part it seems to be the product of a speculative attitude. The difficulties created by Austria in the export of logs, to remedy the disastrous state of its sawmill industry, directly penalised our sawmills. They saw the costs of raw materials rise, while the imported sawn goods arrived at proportionally lower prices. In fact, the transport of boards by railway was much more convenient than that of logs, due to the loss of processing to reduce the logs into boards in Italy, and the better use of space on the wagon. Despite this, the number of sawmills continued to grow throughout the decade following the war.

Since Trentino was now in Italy, it is logical that it too would feel the imbalances of its new market position, imbalances that became increasingly evident over time. After the price crisis of 1921 and the problems of excessive previous utilisation, the production of Trentino forests slowly rose, while the prices at the felling post were continuously increasing, even though the forests with negative cutting costs were reduced. The second half of the 1920s did not see an improvement in the general market situation, indeed further damage was done with the entry into force, on 1st October 1926, of the direct Austria-Italy tariff, under which imports from Austria enjoyed a reduction in railway freight. There were immediate protests from Trentino traders because, according to them, the new tariff caused at least a loss of 8 lire/m^3 to the timber produced in the province. The concerns were well-founded, so much so that the following year, probably due to the stagnation of building activities and competition, prices fell and the crisis lasted until the end of 1929.

The events of the Trentino industry in this period are now comparable to those of Italy. We have already mentioned the severe shortage of raw materials that weighed on this industrial sector, with a consequent increasing dependence on imports from abroad. As for imports, from 1922 to 1935, they constantly increased from 1,100,000 to 2,200,000 tonnes, only to decrease significantly in the following years. 84% of this import was squared, rough, sawn timber, all material therefore processed by sawmills. The problem of customs tariffs and those applied on various railway routes remained the same, so that those applied always ended up favouring the imported timber. The customs tariff of 1887 established import duties on timber quite high, but that of 1921, as has been said, and the subsequent one of 1934, did not improve things. Regarding railway transport, however, it should be remembered that in addition to the north, the tariffs also disadvantaged the south of Italy. To give an example, timber coming from Cosenza (in Calabria region) to reach Milan paid a tariff of 166.21 lire per tonne, while that coming from Postumia, and heading to Milan, paid 146 lire per tonne benefiting from the direct tariff Italy-Yugoslavia.

In 1932 the import of timber mainly came from European countries (83%), but among these Austria played a decisive role, especially for the industries of the north-east. Keeping in mind the old Austro-Hungarian Empire, we see that imports from Austria, Yugoslavia and Hungary represented, in value, 63% of the total of timber of European origin. Although it was no longer an empire, Austria still played an important role, with 25% of the total value of imports, while Yugoslavia covered 35%. It was mostly resinous timber, which met the competition from Yugoslavia, but which was of better quality and above all enjoyed special commercial concessions. Out of a total of 351,558 t of timber imported from Austria, 89% was made up of sawn and squared timber. The Austrian competition therefore continued to hit precisely that Italian and Trentino industry that had received so much impetus in the first decades of the century and which was now beginning to feel the crisis.

Between the beginning of the century and the First World War, the industries of the sector had experienced exceptional development. The sawmills and the large carpentry workshops doubled and the employees grew by 136%, but the greatest increase was had by the furniture industries which grew almost tenfold, with an increase of 335% in employees. The growth recorded in the post-war period continued, as by the 1930s the companies appear almost doubled compared to 1911. Dividing this industrial sector into timber production, primary processing (sawmilling) and in processing industry, we see that the first sector, in 1935, counted 6679 firms with a number of employees equal to 100,000 units. 55% of these companies were located in the north, 23% in the centre and 21% in southern Italy. Among the various regions, Lombardy stood out, followed by the three Venices and Tuscany; in the north east the most important province was Trento, followed by Bolzano and Vicenza. The processing of wood was instead carried out by about 89,600 firms, of which 6600 were industrial and 83,000 were artisanal. The industrial companies, i.e. with more than three employees, employed about 80,000 workers, while the artisanal ones were estimated to employ at least 200,000 employees. For some time, at least until the crisis of 1929, the two types of companies, industrial and artisanal, had integrated quite well. The industry was reserved for large supplies and the manufacture of products that required a lot of labour and mechanised processes, instead the artisanal companies dedicated themselves to small supplies that required careful work, generally on direct order of the client. A trend not very dissimilar from today's, where craft businesses tend to produce quality items rather than work in large quantities. The technological development achieved in the sector had led to the production of quality products, especially in the furniture sector, which could finally compete with foreign production not only in the domestic market, but also abroad, reducing imports and significantly increasing exports. Thus was born the Italian furniture industry destined to become a world leader at least until the arrival of Ikea, the Swedish furniture company in 1989.

The contractions in consumption that were recorded at the end of the 1920s led to a decrease in the activity of industrial companies, which reacted by laying off workers. This workforce was however absorbed by small businesses, which could also accept orders from large industry. This allowed craft businesses to resist better and to increase in number at the expense of medium and large industries. According to Battaglia, large companies, having greater constraints, especially in terms of working hours and wages, had higher production costs, even by 400%. This phenomenon also gave rise to a transformation of the industrial class into an entrepreneurial class more inclined to take on large orders and distribute them to small businesses, rather than working in their own facilities.

As for raw and semi-finished timber, however, the Italian market continued to be directly influenced by the foreign market, which satisfied a large part of the national requirement, automatically determining the price level to which our production had to conform. Sometimes the prices of foreign timber, despite transport, were so low as to be almost equal to the cost of Italian production. However, there was a substantial difference between the different types of industries: those producing raw wood and boards suffered directly from foreign competition, while those in the transformation sector, such as furniture factories, were in better conditions. This is why in this sector the national industry was able to develop more, while that of primary processing could not withstand the competition. This also explains the delicate situation of customs policy that could not increase duties too much on imported sawn timber, at the risk of seeing retaliatory measures taken on other products exported from Italy. This also signals an important reversal of trend that will lead to an increasingly reduced economic use of Italian forests and a progressive abandonment of their active management, in favour of increasingly massive imports.

6.11 Local Traditions and New Forest Policies

To the previous aspect, purely economic, another factor is added to explain the change in the forest landscape, that is very representative of the changes in the economic and cultural situation of the country. In the second post-war period, considering the state of the Trentino forests, it is observed that they suffer the effects of an exploitation model that had intensely utilized the forest. In 1953 the growth yield, that is the annual formation of woody mass given by the physiological growth of the trees, is estimated at around 546,000 m^3, while the uses were pushed around 571,850 m^3 as an average of the three-year period 1950–1953. Clear cutting was practiced on about 50% of the territory, while for the rest it was a selective cut that, as we have already seen, usually took the best plants. The regeneration of the forest was rather poor, the diameters of the trees rarely exceeded 50 cm and the woody mass was about 162 m^3 per hectare, far from the 250–350 considered normal.

As a result of this situation, at the beginning of the 1950s, there was a kind of rethink that involved some of the forestry service managers of the time. With the intention of modifying the silvicultural management practised up to that point, a new management strategy was developed that intended to enhance ecosystem values. New criteria were adopted that in some way wanted to refer to the principles of the so-called "naturalistic silviculture", a method used by part of the Swiss and French forestry world (Susmel 1973). Without a doubt, this was a choice that made a decisive break with the past, and therefore with traditional productive criteria: this vision tended to favour not only a protective function of the forest, but also a tourist-recreational and environmental one, thus meeting the new cultural trends that aimed to replace economic values with environmental ones (Casagrande 1990). In essence, clear- and artificial regeneration should have been abandoned, even though in some areas of the province this cut had been practised for centuries without particular problems.

The intention of the foresters was to rely on management methods that approached that of control, developed by Adolphe Gurnaud in Switzerland at the end of the nineteenth century. Not all the foresters of the time agreed with this approach, however. The period was characterised by intense debate, which involved all of Italian silviculture, and was marked by the important forestry congress of Florence in 1954. The debate continued throughout the 1960s, during which the foresters of Trentino set out lines of action that aimed to achieve mixed, uneven-aged forests by groups, to be implemented gradually. The main tool for achieving these objectives were the management plans to which all the forests of the province were subjected. The new silvicultural methods involved, in principle, the abandonment of clear-cutting and the widespread adoption of cuts intended to eliminate trees in bad conditions, both in even-aged and uneven-aged forests. The cuts were supposed to favour species such as beech and white fir, as well as other broadleaves, to mitigate the predominance of spruce.

In practice, clear-cutting was not definitively abandoned. There were indeed exceptions constituted both by geographical areas with particular characteristics and by forests with structures that were poorly suited to different techniques. In Val di Fiemme, for example, clear-cutting was maintained, also because the Community of had its own autonomy in terms of forest management, which had to be taken into account. In general, in even-aged forests, the surfaces of the clear cuts were greatly reduced, which were thus to become the embryos of future group-based uneven-aged structures. In uneven-aged forests, on the other hand, thinning was carried out to select the overstory and ensure regeneration, affecting the structure of the forest.

Fig. 6.12 Three rafters of Codissago, a village along the Piave river, are building a raft to transport timber using traditional techniques. Periodically the rafters descend the Piave up to the old port of the rafts in Venice to keep this historic tradition alive

The actual action of the foresters, in the years straddling the mid-1950s and ‘60s, was mainly to reduce the intensity of the cuts, to allow the reconstitution of the woody mass (Fig. 6.12).

In about a decade the forest utilizations went from 363,380 m^3 in 1953 to just 239,000 m^3 in 1964 (Pozzati 1990). The utilization was done following the rules drafted by the plans and under the control of the provincial forestry inspectors. This policy was obviously against many interests. In addition to those of the industry, which saw the available timber shrinking, there were indeed the immediate interests of the municipalities and local populations. To understand the effects and the application difficulties of this new direction, it is necessary to dwell on the mechanism by which the use of the forest had so far taken place.

Until then, the procedure followed for ordinary cuts provided that the representatives of the municipalities and the forestry authorities would meet annually in the so called “forest sessions”. On this occasion, the former indicated the woody mass they wanted to cut during the year, for the needs of firewood and work of the inhabitants. If the requests seemed excessive, local inspections were made, to evaluate the provision of the forest and the admissibility of the request. Extraordinary cuts could instead be requested by the municipalities, or by private individuals, for commercial purposes, to meet the needs of the budgets. In these cases, forestry rules were not much followed, therefore large quantities of timber were removed, through which the municipalities made income to cover the needs of the budgets. The new forestry policy intervened on both types of utilisation, with a very tight control of the quantities assigned to the cut and the execution of the same, in accordance with the management plan.

In the early years of application of this forestry policy, a large part of the utilisation were in fact only thinnings, which could not obviously interest the best trees, but mainly concerned predominantly lower quality trees. To requests of the local population to cut more, it was answered that there were no trees to fell and it was not uncommon for forestry inspectors, who went into the forest to mark the trees to be felled of a municipality, to find themselves facing heated resistance and disputes from those, from the mayor to the population, who did not want to settle. In some cases, there were episodes in which the inhabitants of a village positioned themselves in the forest next to the trees that each wanted to cut, at all costs, with understandable situations of embarrassment. It is therefore imaginable the resistance that this new policy encountered at all levels. On the other hand, it was happening not only a change in forestry management, but above all a cultural change that was modifying uses and very rooted customs (Fig. 6.13).

The possibility of implementing this policy was certainly a demonstration of strength of the provincial forestry administration, but undoubtedly it was facilitated by some more general factors. Firstly, the forest as a whole had quickly lost the characteristics of an important source of sustenance and work for the mountain populations, being replaced by other activities such as tourism and its induced activities. This resulted in less pressure from the populations, even for those secondary forest uses that were now limited mainly to the collection of firewood, which not even everyone did anymore as before, given the introduction of alternative fuels. The new management tended to adapt to a change in progress in society, less dependent on the forest and wood, and to the fact that the timber industry had now lost its battle against the foreign competition, both in the production phase and in the processing phase. The shift of the industrial axis towards the plain, close to the most important communication routes, meant that supplies of timber from abroad were increasingly easy and convenient. Local woods were only used in cases where they were convenient both for geographical location, and therefore the convenience of transport, and for the quality of the timber. An aspect to which the industry had become much more attentive, as from abroad it was possible to obtain valuable timber at more advantageous prices. After a few decades of the new forest management strategy, the provision of Trentino woods now touches peaks of even more than 800 m^3 per hectare, but without this having generated a greater importance of timber in the economy. Even Val di Fiemme, historically the most important area of Italy in terms of timber production, is no longer an important area from the point of view of the timber economy. The woods of Trentino are presented as a fundamental part of the naturalness of the environment to promote tourism, an idea supported also by the recent reintroduction of bears, but in reality, the composition of the forest species and the structure of the forest have been significantly influenced by man over the centuries.

Fig. 6.13 The Stua of Padola in the upper Piave valley. This dike was built to facilitate river driving of timber. It is still existing today

Notes

1. The birth of Perarolo, the most important commercial centre of Cadore, home to the main commercial sawmills and where the cìdolo was located, is set between the fourteenth and fifteenth centuries. The cìdolo was a wooden barrier placed across the river, created to allow the arrest of floated logs.
2. Fresh timber with a specific weight greater than 1 cannot float; this is the case for many species of broadleaf trees such as oaks, which were indeed loaded onto conifer rafts to reach Venice.
3. This class included all saw logs having, at the lower end, a diameter under bark of at least 12 oz (an ounce is equal to 2.9 cm), i.e. 34.8 cm. Logs with a diameter (at the lower end) less than 15 oz, equal to 43.5 cm, which was the next class (XV), were also considered to belong to

this class. Regardless of the actual volume, for the other diameter classes, values deduced from experience were adopted. Once the price for the standard 12-oz measure was established, the prices for each class were automatically determined, so if the class XII log cost for example 10, that of XV was worth 15, while that of X was worth 5.

4. On this point, some of the most important representatives of forestry at the time, such as Susmel and Patrone, were perfectly in agreement.

References

Agnoletti M (1996) Aspetti tecnici ed economici del commercio del legname in Cadore (XIV–XVI secolo). In: Cavaciocchi S (ed) L'uomo e la foresta secc. XIII–XVIII: Atti della ventisettesima settimana di studi, 8–13 maggio 1995. Le Monnier, Firenze

Agnoletti M (1998) Segherie e foreste nel Trentino dal medioevo ai nostri giorni. Museo degli usi e costumi della gente trentina, San Michele all'Adige

Agnoletti M, Tognotti E, Zanzi Sulli A (1986) Appunti per una storia del trasporto di legname in Val di Fiemme. Quad Stor 62:491–504

Andreazza A (1989) Sacco e gli zattieri. Tonelli, Riva del Garda

Boithias J-L, Brignon M (1985) Le scieries et les anciens sagards des Vosges. CRÉER, Saint-Dié

Canali G (1939) I trasporti sull'Adige da Fronzolo a Verona e gli spedizionieri di Sacco. Archivio per l'Alto Adige

Carloni P (1923) Recenti modifiche al regime doganale dei principali prodotti boschivi. L'alpe 7:142

Casagrande G (1990) Caratteristiche della selvicoltura naturalistica trentina. In: Provincia Autonoma di Trento, Teleferiche e selvicoltura, convivenza possibile? Atti del convegno, Moena, Trento, 24–25 settembre 1990

Coletti Di Lucia N (1988) I boschi riservati della marina da guerra austriaca. In: Caniato G, Dal Borgo M (eds) Dai monti alla laguna. Produzione artigianale e artistica del Bellunese per la cantieristica veneziana. La stamperia di Venezia, Venezia

Costantini G (1864) Il Cadore e i suoi boschi. Tip. Antonelli, Venezia

Cucagna A (1961) Le industrie minerarie, metallurgiche e meccaniche del Cadore, Zoldano e Agordino, durante i secoli passati. Università degli Studi di Trieste Facoltà di Economia e Commercio—Istituto di Geografia, Trieste

Da Rios GB (1922) Sulla composizione normale delle fustaie di resinose da taglio saltuario. Atti del I convegno tecnico-forestale italiano: promosso dall'associazione dei funzionari tecnici forestali e dall'istituto superiore forestale nazionale—Firenze 1921, Vallecchi, Firenze

Dal Ri G (1886) Notizie intorno all'industria e il commercio del Principato Vescovile di Trento nei quattro secoli precedenti il Concilio (1545) fino alla secolarizzazione (1803). Programma della I.R. Scuola Commerciale di Trento, A.S. 1886–1887, Trento

Di Bérenger A (1965) Studii di archeologia forestale. Tip. Coppini, Firenze

du Monceau D (1764) De l'exploitation des bois, Chez H.L. Guerin et L.F. Delatour, Paris

Eicher Clere P (1988) La comunità sregolata. Notai-notabili e potere locale nel Cadore del secondo '500. Tesi di laurea, Università degli Studi dì Venezia, Facoltà di Lettere e Filosofia, a.a. 1987–1988

Fabbiani G (1973) Auronzo di Cadore. Pagine di storia. Tip. Piave, Belluno

Fontana G (1972) Notizie storiche del Comelico (Cadore) dalle origini al 1866. Castaldi, Feltre

Giordano G (1953) Manuale pratico di cubatura dei legnami e dei soprassuoli boschivi, Hoepli, Milano

Guernieri A (1862) Cubazione dei legnami squadrati e rotondi, Colombo Coen, Trieste

Il mercato del legname e L'importazione austriaca (1921) L'Alpe 7–8:140

Istituto di Economia e Statistica Agraria, Ministero dell'Economia nazionale (1927) Prezzi e costi di lavorazione delle piante nei boschi italiani. Tip. di Mariano Ricci, Firenze

Kral F (1989) Le vicende del popolamento forestale sulle Alpi italiane. L'italia Forestale e Montana 44(2):107–131
Lane FC (1978) Storia di Venezia. Einaudi, Torino
L'importazione del legname in Italia e la Jugoslavia (1918) L'Alpe 7–8:179
Micoli G (1923) L'importazione del legname austriaco. L'Alpe 3:33
Occhi K (2000) Mercanti di legnami fra la contea del Tirolo e la repubblica veneta. Tesi di dottorato, Università degli Studi di Milano, Dip. di Scienze della storia e della documentazione storica, a.a. 1999–2000
Pais Tarsilia A (1930) Il Cadore e la sua ricchezza forestale. Castaldi, Feltre
Pavan D (1988) L'Industria del legno Rottolo Lazzaris diSpresiano (TV), dalle origini alla prima guerra mondiale. Tesi di laurea, Università degli Studi di Padova, Facoltà di Lettere e Filosofia, a.a. 1987–1988
Patrone G (1953) L'assestamento dei boschi disetanei del Cadore. L'italia Forestale e Montana 8(1):34
Patrone G (1975) Ricerche sulla fustaia disetanea del Cadore. Annali Dell'accademia Italiana di Scienze Forestali 24:29
Piccioli F (1885) Le principali foreste d'Italia. Nuova Rivista Forestale
Pozzati A (1990) Situazione delle utilizzazioni boschive in Trentino e tendenze. In: Provincia Autonoma di Trento, Teleferiche e selvicoltura, convivenza possibile? Atti del convegno, Moena, Trento, 24–25 settembre 1990
Sala G (1937) Il Cadore e i suoi boschi. L'alpe 2:240
Schupfer F (1912) Il Cadore, i suoi monti e i suoi boschi. Tip. del Senato, Roma
Segala G (1915) Le provenienze del legname importato dall'Austria in Italia. L'alpe 3:115–119
Serpieri A (1919) Per l'approvvigionamento del legname nel dopoguerra. Regio Istituto Superiore Forestale Nazionale, Firenze
Serpieri A (1921) Il nuovo regime doganale del legno. L'alpe 10–11:169
Serpieri A, Segala G (1915) La guerra e la crisi del legname in Italia. L'Alpe 10–11–12:334
Sief L (1960) Aspetti pedologico-forestali della Valle Fiorentina. L'italia Forestale e Montana 15(4):146
Susmel L (1952) Stato normale delle abetine del Comelico. Monti e Boschi
Susmel L (1952b) Trattamento delle abetine disetanee. Monti e Boschi 4:148
Susmel L (1973) L'opera dei forestali nella tutela dell'ambiente in Trentino Alto-Adige. Monti e Boschi
Valenti S (1907) La repubblica di Venezia alle Giudicane interiori. Archivio Trentino
Vecchio B (1974) Il bosco negli scrittori italiani del Settecento e dell'età napoleonica. Einaudi, Torino
Venturoli G (1930) Vizze e boschi del Cadore. L'Alpe 17(2):80
Vitale A (1921) La crisi di ribasso dei prezzi del legname resinoso. L'alpe 9:146
Wessely J (1860) Die Venetianischen Brettmühlen der Piavethäler. Vierteljahrschrift für Forstwesen, X Band, Heft 1, Wilhelm Braumüller, Wien
Zanderigo Rosolo G (1982) Appunti per la storia delle Regole del Cadore nei secoli XIII–XIV. Istituto bellunese di ricerche sociali e culturali, Belluno
Zangrando F (1951) Il porto del Piave. Tip. vescovile, Belluno
Zangrando F (1958) Le seghe dei Vecellio ad Ansogne presso Perarolo. Tiziano negoziante di legnami. Archivio storico di Belluno Feltre e Cadore
Zangrando F (1961) Note sulla storia giuridica ed economica dei boschi cadorini. Monti e Boschi 7:309

Chapter 7
Monastic Silviculture: Vallombrosa

Abstract The importance that some monasteries had in the management of forests over the centuries has led some scholars to define the concept of monastic silviculture, meaning long-term management activities, carried out with technical characteristics linked to the vision of the forest typical of the monastic orders. In their approach to the forest, the various orders showed substantial differences, which determined the history of some forests, especially in the Apennines mountains. The Italian debate has mainly concerned two of them: the Benedictine and the Franciscan. For the Franciscans, who belong to the mendicant orders, economic activities were not as important as contact with people, preaching and almsgiving. Consequently, they cultivated and managed the forest in relation to self-consumption needs and not so much to trade. The Benedictines, in fact, developed an approach to the forest very different from the Franciscans, They cultivated the forest keeping in mind the needs of their abbeys, without disdaining the timber trade. A real organisation of forestry activities in a productive function should be considered a prerogative of the Benedictines, who were able to derive significant profits from the sale of timber. This aspect seems to reinforce the role of the order as a symbol of a culture and silvicultural knowledge developed between the late Middle Ages and the modern period, which is alongside that of major political entities, such as the Republic of Venice, aimed at managing the forest with criteria very close to modern economic silviculture. In this perspective, the monks of the Abbey of Vallombrosa seem to assert themselves as precursors of a type of silviculture that only during the nineteenth century, particularly with the birth of the Italian forestry school founded in Vallombrosa in 1869, was attempted to develop and proposed also at a national level. Vallombrosa is located on the Pratomagno mountain (1000 m AMSL), 40 km from the city of Florence. The foundation of the monastery of Vallombrosa occurred in 1037 by San Giovanni Gualberto, which gave rise to the minor benedictine order of the Vallombrosans. Some events occurred after the foundation of the forest school in 1869 offers an interesting case, considering the modern feelings on forests and forestry among urban population and the early developments of landscape conservation in Italy based on aesthetical concerns.

M. Agnoletti, *History of the Italian Forest Landscape*, Environmental History 19,
https://doi.org/10.1007/978-3-031-81631-4_7

7.1 Monastic Orders and Forest Management

The role of monasteries has been widely documented by many authors regarding the promotion of agriculture, particularly in the early medieval period (Fumagalli 2003). Less known, however, is the role of monasteries in forestry activities. The importance that some monasteries had in the management of forests over the centuries has led some scholars to define the concept of monastic silviculture, meaning long-term management activities, carried out with technical characteristics linked to the vision of the forest typical of the monastic orders. In their approach to the forest, the various orders showed substantial differences, which determined the history of some forests, especially in the Apennines mountains. The Italian debate has mainly concerned two of them: the Benedictine and the Franciscan.

For the Franciscans, who belong to the mendicant orders, economic activities were not as important as contact with people, preaching and almsgiving. Consequently, they cultivated and managed the forest in relation to self-consumption needs and not so much to trade. The forest cuts carried out by the order were very limited, often restricted to decaying plants or carried out with the aim of preserving the forests, so as not to exploit them excessively, with management aimed at moderate use of forest resources. Francis, with his evangelical poverty, represented the "shadow of the church" and, like other outcasts such as outlaws, hermits, plague victims and madmen, found in the forest his refuge (*lucus asylii).* As reported by Thomas of Celano in the *Life of St. Francis,* when the brothers cut down the trees he forbade them to cut down the whole tree and ordered the gardener to leave the borders around the garden uncultivated, so that in time the green of the grass and the splendour of the flowers sing how beautiful the Father of all creation is (Paci 2002). The characteristics of the religious order then prevented the consolidation of significant agricultural and forest possessions and each convent only had land for self-consumption gardens, flower beds for the church and a small forest to produce firewood, some beams and planks, without transfers to third parties except for small acts of charity. There were no cultivation rules, except customary ones, nor needs to regulate commercial relationships, non-existent or entirely occasional.

The history of the Verna forest, around the famous Franciscan sanctuary of La Verna in Tuscany,—on the border between Casentino valley, the Upper Arno Valley, and Valtiberina, and for almost eight centuries the property of the Franciscan friars—, has been reconstructed mainly through the documents kept in the archive located in the Clausura of the convent: there are indeed scarce news in regulations, constitutions, provisions, to be related, when they exist, more to contingent facts than to an orderly and disciplined forest management, recognisable through the same rules. We have to wait until the end of the sixteenth century to find the first rule, a provision of the general minister of the order Bonaventura Secusio from Caltagirone, that imposes on the father guardian of the Verna to plant at least two hundred firs a year, before Pentecost. The provision, among other things, does not have much silvicultural importance, both because the fir forest was regenerating without problems, and because probably in this way it was intended to counter a criticism about the alleged bad state of the forest, complained about by some and due perhaps due to abuses by neighbours rather than excessive cuts. It should also be remembered that the practice of creating small plantations by uprooting and transplanting spontaneously grown seedlings, was certainly widespread throughout the Casentino area at least since the end of the Middle Ages (Borchi 2001) (Fig. 7.1).

In the last few decades, concepts such as the conservation of nature and the return to nature, ideas that can be linked to much of the environmentalist thought developed since the end of the nineteenth century, have been associated with Franciscan thought. These ideas perhaps found their highest expression in the encyclical of Pope Francis "Laudato Si" of 2015, which specifically mentions the thought of Francis of Assisi about nature conservation. Certainly, interpretations aimed at presenting Franciscanism as a philosophy of "non-intervention" in the forest are to be considered incorrect, as the forests owned by the monasteries were used, albeit mainly for self-consumption and for the local populations. Nonetheless, the Franciscans also resorted to the sale of timber when they needed it.

If one wants to study the impact of the monasteries' activity on forest management and the actual influences of monastic culture on the forest landscape, other orders, particularly the Benedictine one, should be taken into consideration. The Benedictines, in fact, developed an approach to the forest very different from the Franciscans: as the historian Franco Cardini explains, the Camaldolese and the Cistercians

Fig. 7.1 The original structure of the Vallombrosa monastery at the time of its foundation (State Archive of Florence). The landscape is characterised by the fir forest on the left of the enclosure and the beech forest

sought the solitude of the forests mainly "to domesticate it" (Cardini 1992). Respect for nature did not mean observing it in a meditative detachment, but guiding it for the benefit of the community and the order. They cultivated the forest keeping in mind the needs of their abbeys, without disdaining the timber trade. It seems, however, established that a real organisation of forestry activities in a productive function should be considered a prerogative of the Benedictines, who were able to derive significant profits from the sale of timber. This aspect seems to reinforce the role of the order as a symbol of a culture and silvicultural knowledge developed between the late Middle Ages and the modern period, which is alongside that of major political entities, such as the Republic of Venice, aimed at managing the forest with criteria very close to modern economic silviculture. In this perspective, the monks of the Abbey of Vallombrosa seem to assert themselves as precursors of a type of silviculture that only during the nineteenth century, particularly with the birth of the Italian forestry school founded in Vallombrosa in 1869, was attempted to develop and proposed also at a national level (Agnoletti 2000).

7.2 The Abbey of Vallombrosa

Vallombrosa is located on the Pratomagno mountain (1000 m AMSL), 40 km from the city of Florence. The foundation of the monastery of Vallombrosa occurred in 1037 by San Giovanni Gualberto, which gave rise to the minor benedictine order of the Vallombrosans, the forest is often mentioned among the landed properties attached to the hermitage which at that time was made up of a circular wooden enclosure, with a simple wooden church in the centre, and around it twelve larger buildings and twenty smaller buildings, also likely made of wood. Wood was therefore the main material used for the construction of the monastery: outside the enclosure, in fact, two clear patches of fir and beech trees can be seen surrounding it with almost equal extension. The donations of land rights made by some followers of the saint and by Itta, abbess of Sant'Ilario, belonging to the faction of the Conti Guidi, included various types of woodlands and other uses of the soil. If the prevalence of these two species therefore seems established near the hermitage, the presence and composition of the forest in the land possessions that in the act of donation include woods of turkey oak, chestnut and beech, as well as fields, pastures and vineyards (Gabbrielli and Settesoldi 1985).

The increase in possessions in the two centuries following the foundation, first through donations and then through direct acquisitions made on a large scale by the abbey itself, led to the constitution of a vast heritage that extended from the top of Pratomagno mountain to the Arno, within the current territory of the municipalities of Pelago and Reggello. Such acquisitions were made with great foresight, trying to consolidate isolated plots and increasing the value of the possessions with low-altitude agricultural lands, more profitable, but not neglecting the purchase of fir and beech woods.

Towards the 1300s, the traditional Benedictine organization which envisaged a vast dominical property cared for by the monastic community to which the leased lands were added (Salvestrini 1996)—was replaced by the sharecropping system. This reorganisation reduced the "dominico", the part of the owner, to pastures and the forest, while the more productive lands were subject to farming agreements, through which the sharecroppers directly managed the lands; but it also resulted in a different administrative structure. Around 1400, the Vallombrosa estate therefore appears divided into four administrative units, which grouped the farms located in the properties of Pitiana, Paterno, Ristonchio and Sant'Ellero, while a fifth administrative unit, that of Guarlone near Florence, was established towards the end of the fifteenth century. To the five properties were added the ancient plots conducted by the monks, located between the Secchieta and the hills of Valdarno, characterised by steep slopes and therefore less interesting for agricultural management, but more useful for grazing and forest crops. Most of the land left to the direct management of the Abbey, in addition to some farms located at the highest altitudes, was therefore characterised by a large spread of beech and fir forest, while other species composed the woods at lower altitudes.

The establishment of the monastery took place in a historical period in which the forest, from a large open territory to which everyone drew according to their needs, gradually became the object of increasing attention and control by the authorities, also due to the significant demographic increase that was recorded in that period (Duby 1970). This was especially true for the more easily cultivable lands of the plains and hills; but also the mountain territories, until then characterised by the free grazing of many animal species, were made the object of greater attention, especially for their forest resources. In the context of medieval agriculture the role of the forest and especially that of forest plants was but very different from the current one, as well as their distribution and extent. Indeed, despite current forest typologies operating classifications that link the species distribution to the nature of the soil, altitude and species characteristics, each historical period has seen the prevalence of woods characterised by specific structures, densities and compositions, determined by silvicultural practices linked to the social and economic characteristics of the time. Vallombrosa was certainly no exception to this model and the development of farms, with the consequent increase in demographic pressure on natural resources, also conditioned the characteristics of the woods. As for the extent, both the analysis of written sources and cartography suggest a reduced extent of the forest compared to the current one, with isolated patches in the midst of large extensions of pasture, often wooded pastures with the presence of isolated trees or small groups. Such trees, as well as the trees on the edge of cultivated fields, were often used to produce forage leaf, and for this purpose pollarded at a certain height from the ground to protect the shoots from the bite of livestock.

The formation that absolutely had the greatest extent was probably the oak grove (downy oak and turkey oak), followed by the chestnut grove. The chestnut tree appears mentioned in purchase contracts both in the form of single trees and as a wood, with a presence certainly higher than the current one. This species was held in high regard for its extreme versatility, producing construction timber (poles for vineyards, beams, boards, staves), food for humans and also tannin for leather tanning. For this reason, it was expressly protected within agricultural contracts in which it was imposed to improve and in any case not to cut the chestnut trees, also because part of the rental fees of the abbey were paid in chestnuts. Other woods of great importance for all of medieval Tuscany were the oak groves, fundamental for the production of acorns, so much so that in years of low production the pig farming suffered negative repercussions. In the land assets, both downy oak and turkey oak woods are described several times, to which a large number of place names correspond: in these woods, pigs were taken to pasture both by the tenants, and by the monks, who were in turn directly engaged in this activity. Another type of woods that had a considerable extent were the so-called "stipa" woods, which are a low scrub, often heather or other shrub species.

In light of this situation, it is understandable how in the first centuries of the monastery's life the beech and the fir had a rather limited role, both as a surface and as an economic value, and among the two it seems rather the beech the one subject to the greatest attention, also from a religious point of view. It is well known that the sacred plant to the monks was actually the beech, a choice linked to a miracle performed by the saint right in the hermitage: it is said that the tree lowered its branches to cover Saint John Gualbert who had no shelter (Gabbrielli and Settesoldi 1977). The holy beech, dedicated to him, is represented in a land register of uncertain age, probably eighteenth century, in which the oratory, the source and the beech are shown surrounded by a wall to protect it and propose it for the adoration of the faithful.

The beech forests were therefore protected by the Vallombrosans, so their presence was always indicated in the contracts and the trees constituted an exclusive good of the monastery, also forbidding the tenants to cut them. The beech was especially appreciated for making charcoal, but the practice of making cultivated plots and charcoal in the beech forests was expressly prohibited, as is the sale. The species that in this early period of the monastery's life seems to be in decline, and which in any case occupied a limited area, seems to be the fir, sporadically present even at lower altitudes than the current ones, around 500–600 m, but gradually pushed back to the higher altitudes along with the beech by the expansion of agricultural crops and chestnut trees (Solla 1893). It is therefore understandable how the management of beech and fir forests was carried out directly by the monks, since for the colonists they had limited utility, and it was certainly the beech that was held in higher regard for the production of charcoal and the manufacture of common use objects. The most widespread species according to the sources seems to be the chestnut, which was often mentioned in lease contracts. The monastery continued to purchase between the thirteenth and fifteenth centuries many chestnut groves on the mountain of Vallombrosa and in the Val di Sieve.

7.2.1 From Self-Consumption to Timber Trade

From what we have said, it does not seem that in the first centuries of life of the abbey the production of wood had a significant role, but over time both this and the role of the forest seem to become more important. The forest historian Antonio Gabbrielli reports a first contract of 1369, in which assortments of fir and beech are sold, and this is followed by that of 1389 in which along with the fir a slightly greater quantity of chestnut is sold (but the cutting would have had to occur over two years). However, these seem to be quite isolated cases, in fact until 1500 a fair part of the timber trade was conducted with the tenants and local woodsmen, within the framework of a typical agro-forestry-pastoral economy in which the sale of timber had a limited role. In these first sales the percentage of fir was always quite contained, we do not know if due to a still reduced interest in this species or due to the difficulties related to the costs of its transport. The acts relating to 1389 and 1574 concerning cuts of firs provided that it was the Florentine buyers to saw the trees and clear them within six months. In truth, during this period the Opera del Duomo—the institution created for the construction of the Florence Cathedral that owned extensive forests in the Tuscan Apennines—was already drawing significant quantities of fir from the forests of Romagna Toscana, and had also begun from the previous century a progressive action of reducing the beech to favour the more valuable fir. Probably the extent of the fir groves of Vallombrosa was not yet such as to allow regular trade. What strikes for their size are the thousand oaks belonging to the oak grove of Vallombrosa sold together with some beeches in 1519, a quantity not far from the eight hundred trees felled sixty years later. In this contract it is specified that the various types of timber (oaks, brooms, ashes, hornbeams) present in the surroundings of Tosi, where the cutting took place, were to be cut and not uprooted, with a clear reference to the need to maintain the forest in the face of advancing agricultural crops.

A sign of greater attention to timber production in this century, compared to firewood and faggots, is given by the presence of a water sawmill near the monastery, as can be observed from the Vallombrosa map of 1584. Usually, the cutting of the trees was assigned to master woodcutters, but the presence of a saw, in addition to the usual attention for the technological innovations by the monks—who also in other European monasteries made use of these machines—, shows the need to work fairly high quantities of wood, no longer linked to an occasional use of the forest resource (Reynolds 1984). Certainly, like the other hydraulic sawmills built in the woods of the Opera del Duomo, these were not the fast Venetian alpine sawmills Agnoletti (1998), but machines with large hydraulic wheels and a multi-stage advancement system that did not saw at high speed, but reduced the use of labour.

If the sixteenth century does not yet seem characterised by a significant role of the forest in the monastery's economy, with the following century things begin to change. The 1627 budget assigns to the sale of woods and firs in all territories belonging to the Vallombrosa congregation the value of 8861 scudi, bringing the amount of this item to 10% of all incomes. This figure placed the forest fifth after wine, grain, livestock and pastures, and oil as a source of income for the monks. Certainly something had changed in the importance that fir timber had assumed in the Tuscan economic context, and therefore also for the monastery which now regularly sent timber to Livorno, the new port city built by the Medici family, via floating down the Arno river, in addition to trading it locally.

A new interest in this resource by the Benedictine order also seems confirmed by the activity of the Camaldolese brethren, who from the seventeenth century also began a profitable commercial activity opening deposits together with the Vallombrosans in Arezzo, Florence and Livorno, while until then they had limited themselves to self-consumption and a few occasional sales. Specifically for Livorno, the timber registers kept by the Tuscan grand duke's administration report the shipment of at least 25,355 pieces, sent from Vallombrosa at the end of the seventeenth century, mostly made up of "firs", and destined for naval constructions, underlining the important role assumed by the monastery as a supplier of timber for the navy (Salvestrini 1999), for whose needs forests were cut in various parts of Tuscany as we already discussed.

The growing economic importance of the monks' timber trade, however, required greater organisation not only of forest utilisation, but also of silvicultural activities. Already in 1421, the monks had begun to purchase seedlings, that is young fir plants, to extend the cultivation of conifers, but it is in the 1600s that the surfaces are significantly expanded through frequent plantations, as happened in 1645 with the planting of at least 29,470 seedlings, while 16,000 were purchased on the Falterona in 1697. From this moment on, the management of the fir forests is configured as an important economic activity for which silvicultural techniques based on artificial regeneration are implemented, using seedlings of the same age, with regular planting in rows, with the aim of obtaining an even aged forests of a monospecific type. This activity was, on the other hand, justified by the evident objective of increasing the profits derived from the sale of timber, which happened towards the end of the eighteenth century, with a revenue due to the cutting of the fir forest and other woods equal to 35% of the total income. This is a value more than tripled compared to the previous data and linked to the silvicultural activities carried out a century earlier, since the cutting cycle ranged from 70 to 100 years.

It can therefore be stated that Vallombrosan silviculture assumed a respectable technical dimension already in the seventeenth century, there is no news about other communities in Tuscany that carried on this type of activity at the same level. Oslo the Opera del Duomo in Florence preferred to gradually eliminate the beech, "ringing" it, to make more room for the fir, rather than resorting to artificial regeneration, aligning with what was done by the Republic of Venice in its domains. It is perhaps with the latter that we could draw an interesting parallel, with the forestry of the monks representing a technical model opposite to that implemented by Venice, although motivated by the need to extend the reduced surface of its fir forests and referring to a few hectares. The transformations and interventions that gave Vallombrosa the appearance that the local landscape still has today, in relation to the dominance of the white fir, are however due to Abbot Luigi Fornaini (1756–1838) (Salvestrini 1996). This is, on the other hand, a historical period that sees a constant growth in the importance of coniferous timber, and even Venice now favoured the expansion of the fir, after a period in which the beech had received more attention. Thus, a gradual replacement of deciduous forests with coniferous ones is accelerated, slowly affecting both Italy and other European regions, and continues with increasing intensity until our century.

The action of the monks is therefore representative not only of a local phenomenon, but of a historical period in which the extension of fir forestry with the technique of clear cutting and artificial regeneration will gradually assume the dimension of a European phenomenon. This was also linked to the strategic importance of the wood resource in Europe and its scarcity, which now made it convenient to resort to imports even from the "new world" (Hatzenberger 2000). The importance assumed by the monastery of Vallombrosa in the forestry field, thanks to the cultivation of the fir carried out by the monks, will be at the base of the motivations for the foundation of the first Italian Forestry School, established in this location in 1869. After the establishment of the Royal Forest Institute of Florence in 1914, the Royal Higher Agricultural and Forest Institute in 1924 and finally the Faculty of Agriculture in Florence in 1936, Vallombrosa will play the role of the location of the field courses for the forest technicians of the Kingdom, after the second world war Vallombrosa will be continue to be the location of the field courses for the forestry students of the School of Agriculture of the University of Florence. The cultivation of the fir started by the monks was continued by the forestry school and the extension of the fir forests was considerably increased.

Some developments of the forest school offers an interesting case, considering the modern feeling on forests and forestry among urban population, as also for the early development of landscape conservation in Italy. Between the nineteenth and twentieth century Vallombrosa was a well-known summer destination for the holydays of the rich Florentine families who spent some weeks in the nearby location of Saltino, just a couple of km from the Abbey, where several hotels and restaurants were built. They were looking for some relief from the heath of the city in summertime, a tradition continuing today, although the number of hotels has greatly reduced, and people get there mostly for daily trips. The foresters of the school, during the first years of activity, adopted the same methods developed by the monks and later scientifically systematized by the German forest school, the leading one in Europe in those times, applying clear cuts to the fir forest. At same point the rich Florentine bourgeoisie started to complain about the poor aesthetic value resulting from this management method, also suggesting that a degradation of the forest was underway. They first addressed their complaints to the school and later, since nothing happened to the ministry in charge of forests. The ministry, in turn, reported these complaints to the board of the school, whose members where all foresters.

The board rejected these complaints stating that they were working not for pleasing tourists, but for building the Italian forest science and respond to the need of the country. In that period timber was the third imported good, after coal and wheat, and the government wanted to increase timber production and take care of the bad conditions of the mountain areas, developing afforestation plans. Nevertheless, noting that the forest school did not pay much attention to the complaints, at the beginning of the twentieth century Vallombrosa was included by law among the "climate stations" and, according to the regulation applied to these sites, the cutting of the forest was forbidden. Beside this episode, it is worth nothing that the reports of the meeting of the board of the school in those years, clearly shows that an important debate on clear cut versus selective cut was underway and the latter became the most widespread silvicultural method adopted in Italy. There was an ongoing debate in Europe in those times, criticizing the German economic approach to forestry and suggesting "close to nature forestry" of which the Italian foresters are well aware. However, it does not seem to emerge any reference to the philosophical discussion referring to the writings of Henry David Thoreau, John Miuir, in the US as well as to the thinking of John Perkins Marsh, who visited Vallombrosa, that we discussed in the introduction of this book.

Nevertheless, we can look at these events also as the prodromes of the development of landscape conservation in Italy, officially starting with the first law written by Benedetto Croce, the most important Italian philosopher, in 2022 during the fascism. The law, besides monuments and settlements, speaks of the protection of the so called "natural sceneries" including in this definition forests and agricultural areas. In the following decades, the absence of any consideration on the cultural origin of the Italian rural landscape will combine with nature conservation, developed in the 80s of the last century in Italy, leading to the definition of an "aesthetic-environmental" approach in landscape conservation laws. The strong limitations of traditional forest management forms, or any other form of management negatively affecting what is considered the ideal natural state of the forest, was the result of these development. Recently, in order to settle a debate lasted for more than 150 years, the central part of the forest has been turned into a forest museum, where clear cuts are allowed, while the surrounding area, mostly a mixed forests, has been declared "Natural Reserve".

References

Agnoletti M (1998) Segherie e foreste nel Trentino dal medioevo ai nostri giorni. Museo degli usi e costumi della gente trentina, San Michele all'Adige

Agnoletti M (2000) Fra agricoltura e selvicoltura: il ruolo del bosco e la coltivazione dell'abete a Vallombrosa fra XII e XVII secolo. L'italia Forestale e Montana 55(6):386–397

Borchi S (2001) Cenni storici sulla foresta della Verna. In: Agnoletti M (ed) Storia e Risorse Forestali. Accademia Italiana di Scienze Forestali, Firenze

Cardini F (1992) Guerre di primavera. Studi sulla cavalleria e la tradizione cavalleresca. Le Lettere, Firenze

Duby G (1970) L'economia rurale nell'Europa medievale. Francia, Inghilterra, Impero (secoli IX–XV). Laterza, Bari

Fumagalli V (2003) L'uomo e l'ambiente nel Medioevo. Laterza, Roma-Bari

Gabbrielli A, Settesoldi E (1977) La storia della foresta casentinese nelle carte dell'archivio dell'opera del Duomo di Firenze dal secolo XIV al XIX. Ministero dell'Agricoltura e delle Foreste, Roma

Gabbrielli A, Settesoldi E (1985) Vallombrosa e le sue selve. Nove secoli di storia. Ministero dell'Agricoltura e delle Foreste, Roma

Hatzenberger F (2000) The historic evolution of Haitian forest. In: Agnoletti M, Anderson S (eds) Forest history: international studies on socio-economic and forest ecosystem change. CABI International, Wallingford-New York

Paci M (2002) L'uomo e la foresta. Meltemi, Roma

Reynolds TS (1984) Medieval roots of the industrial revolution. Sci Am 251:122–130

Salvestrini F (1996) Il patrimonio fondiario del monastero di Vallombrosa fra XIII e XVI secolo: presenza ed utilizzazione del bosco. In: Cavaciocchi S (ed) L'uomo e la foresta secc. XIII–XVIII: Atti della ventisettesima settimana di studi, 8–13 maggio 1995. Le Monnier, Firenze
Salvestrini F (1999) L'Apport des Vallombrosains et des Camaldules à la marine toscane (1650–1720). In: Corvol A (ed) Forèt et Marine. L'Harmattan, Paris
Solla RF (1893) Caratteri propri della flora di Vallombrosa. Bollettino della Società Botanica Italiana

Chapter 8
Lowland Forests and Mountain Reforestations

Abstract The presence of trees in the countryside is a process that begins in Etruscan times and increases in the late Middle Ages. It accelerates with the great changes of the XVI–XVII century, which see the decline of Italy as an important commercial power due to the changed conditions of competitiveness of the economic system and its progressive transformation into an agricultural country, with a land investment in progressive growth. The processes of expansion of crops in the plain and in the hills, although crossed by various phases of crisis, would tempt to describe these landscapes as without forests, but it is certainly not so. The density of trees in some areas reaches values that do not suggest tree plantations, but rather real forests. In some areas there can be 50–70 plants per hectare, in others the double rows made the number rise even to 200 plants per hectare. From the 1920s onwards there is a progressive decline which goes hand in hand with the mechanisation and specialisation of agricultural crops that evolve towards monoculture for industrial purposes. Until about 1980, more than the half of the fuel wood in Italy was not produced from forests, but from trees in the field. This suggests that a different kind of forest, not the dense, homogeneous, forest cover we usually think of, but rather an agroforestry system, can play both an economic and environmental role. The work of reforestation and hydraulic arrangement of mountain basins represents the most important and problematic commitment undertaken by the Italian administration from the Unification of Italy to the period between the two wars, with the aim of increasing the country's timber resources, but also of reducing the hydrogeological instability of the mountains. Attention to this problem continues today. It is also interesting to reflect on this history considering the current scientific suggestions and political initiatives promoting extended afforestation for climate purposes. Sometime these plans lack of a careful considerations of the lessons learnt from the past, that, if not taken into careful consideration, may lead to the failure of these projects. The arrangement works of the Italian mountain river basins goes through different phases. In the first period the hydraulic arrangement of the riverbeds, basically the building of dikes, compared to afforestation was preferred. Later, we notice a more balanced approach assigning importance to both methods. In any case, most of the afforestation carried out by the state until the years after second world war was unsuccessful, with a large preference assigned to conifers creating low quality landscapes, almost no

M. Agnoletti, *History of the Italian Forest Landscape*, Environmental History 19,
https://doi.org/10.1007/978-3-031-81631-4_8

management occurred after planting and a very small number of new forests created. The doubling of the forest area due to agricultural abandonment does not seem to have solved the problems connected to reducing the hydrogeological instability of the mountains.

8.1 The "Ordered" Reforestation Between the XVI and XVIII Century

Deforestation for the cultivation of new lands, excessive exploitation due to grazing activities, the intensity of cuts for charcoal production and the effects of deforestation are a constant in narratives about forests, from the north to the south of Italy. The reduction of forests also constitutes the largely prevalent theme of much of the forestry and environmental historiography: some authors have indeed given a rather detailed picture for the 18th and early nineteenth centuries (Vecchio 1974). There are also processes that oppose these trends, such as reforestation, which can be studied from different but complementary points of view, such as the growth of the forest in abandoned agricultural areas, planned reforestation and the use of tree for agricultural purposes.

The presence of trees in the countryside, especially in sharecropping areas, is a process already described by Sereni in the *History of the Italian agricultural landscape*, mixed cultivations with vines and trees are already documented for the Etruscan times in central, south and northern Italy. With the fall of the Roman Empire there is a strong widespread decrease of agricultural activities, but during medieval times, with the spread of sharecropping and in the late Middle Ages the presence of trees and hedges in cultivated areas increases. The trend towards an increase in trees outside the forests, however, accelerates with the great changes of the XVI–XVII century, which see the decline of Italy as an important commercial power due to the changed conditions of competitiveness of the economic system and its progressive transformation into an agricultural country, with a land investment in progressive growth. The processes of expansion of crops in the plain and in the hills, although crossed by various phases of crisis during the seventeenth and eighteenth centuries, would tempt to describe these landscapes as without forests, but it is certainly not so: there are new "forests" that are being born, albeit with a different structure. The trees often extend as far as the eye can see, like a new forest, whose expansion is proportional to the growth of an agriculture that cannot do without it; this phenomenon assumes particular importance in the Po Valley, even if the regions of central Italy also had trees in the fields. The tree species used in agriculture provide a wide range of products such as faggots, wicker, firewood, poles, construction and work timber, and above all the indispensable integration of the scarce forage base through leaves. This happens especially through the techniques of pollarding or coppicing. Mainly elms, maples, poplars, mulberries and willows are planted, which are imposed in rental contracts, while in sales contracts they constitute an essential part of the value of the fund, a heritage that we have seen to be highly considered even in Roman times. Only in Brescia, in the sixteenth century, more than six million mulberries were planted per year, for silk production, while a "wide forest of mulberries" existed along Lake Como (Niccoli 1902).

The density of trees in some areas reaches values that do not suggest tree plantations, but rather real forests. If in fact in some areas there can be 50–70 plants per hectare, in others the double rows made the number rise even to 200 plants per hectare (Cazzola 1996), a quantity that would make one think of chestnut groves for fruit almost as "degraded forests", since their density is around 100–130 plants, not to mention the oak groves for the production of curved pieces for shipbuilding. As Cazzola explains, in the dry and anciently cultivated lands of Emilia, Romagna, Polesine, Ferrara, the Mantuan beyond and the lower Venetian plain, the city property organises agricultural production by establishing on the land a family of farmers, with sharecropping or rental contracts. The system, diversified in terms of farm size, but always proportionate to the peasant family and the pulling power of animals, presupposes mixed farming and the food self-sufficiency of the farmer with a wide variety of products. Wheat and wine are predominant among these. The cultivation of the vine takes place by marrying it to a living support, usually elm, field maple, willow or poplar arranged in rows at the edges of the field (Fig. 8.1).

The density of the trees and the width assigned to the field by local traditions are variable, but on every hectare of used agricultural surface we can encounter from 90 to

Fig. 8.1 Mixed cultivations of vines, poplars and wheat near Naples. The grapes were harvested using long wooden stairs. These types of agricultural systems are documented at least since Etruscan times in Italy (ninth–first century B.C.)

180 plants. The arable land with trees and vines is therefore the organisational model of a mixed but intensive agricultural system, capable of expressing the maximum efficiency from an energy point of view. Hundreds of trees and vines line up neatly at the edges of the field. Under the rows, grassy spaces extend that serve to drain rainwater towards the inevitable drainage ditches. Even their modest forage production is used. The woody production of the plantation is not negligible, placing itself between 5 and 10% of the gross marketable production. It serves the energy needs of both the peasant family and the master in the city. All agronomic literature, from the 1500s onwards, indeed dictates rules and precepts for a good tree endowment of the farm and for the associated viticulture. In an agriculture based on organic and with few external energy inputs, the plantation confers the maximum productive force compatible with pre-industrial technique and will reach the maximum commercial value in the estimates of the experts. In this system, animal power and trees are increasingly united by functional and complementary ties that the Padana plantation fully expresses. This is an intuition that Emilio Sereni had already developed in the mid-1950s about the agricultural landscape of the Emilian countryside (Sereni 1957), but that Henri Desplanques also proposed in the same years for other agricultural realities of sharecropping Italy. These mixed cultivations, therefore, constitute for the Italian agriculture, always lacking in forage base, also a real "suspended meadow", as Desplanques writes (Desplanques 1959).

For sure the forest, understood as a formation at full density, in the plain, is now very reduced. In the countryside of Pavia, between Ticino and Olona, the cadastre of Charles V shows that it is present only in 3.2% of the territory, while in Cremona the cultivated areas represent 80%, the meadows 14.7% and the forest remains a quota of 1.52%. We do not have data that allow an estimate of the percentage of the various tree species present. As a purely indicative and without the necessary relationship with the unit of surface, we report what written by Cazzola for the Duchy of Modena and Reggio and derived from a census of the nineteenth century (Cazzola 1999). Here there are about 1,500,000 elms married to the vines, 456,000 poplars, 262,000 willows, 79,000 mulberries, 36,900 maples, 12,600 walnuts and 6392 ashes, and 12,600 specimens of oak. This confirms the tradition originated in the pre-Roman period of marrying the elm to the vine above all, an agroforestry system about 2500 years old (Fig. 8.2).

From the 1920s onwards, however, there is a progressive decline in mixed farming, which goes hand in hand with the mechanisation and specialisation of agricultural crops that evolve towards monoculture for industrial purposes. Between 1936 and 1981, mixed farming in Bologna went from 124,000 to 15,000 ha, in Modena from 103,000 to just over 5000, in Forlì from 115,000 to 19,000 ha. In Emilia Romagna as a whole, where it occupied as much as 695,000 ha in 1941, it had already lost about 140,000 ha by 1961. In the following twenty years, the collapse would have been total: in 1981 only 81,000 ha of planted land remained to testify to the landscape that man's work had slowly created. The transition from a solar energy agriculture, i.e. largely self-sufficient for energy consumption but with relatively low productivity, to an agriculture with energy from external sources, largely of fossil origin, can be considered the beginning of the decline of the planted and mixed farming of the Po

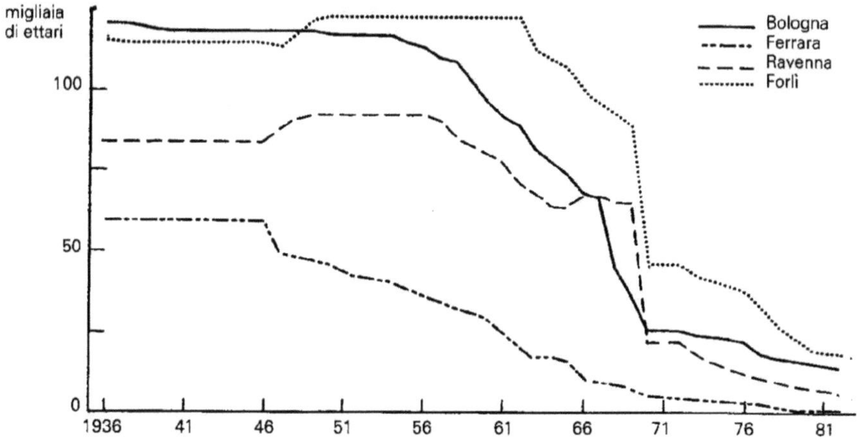

Fig. 8.2 The decline of mixed cultivations in Western Emilia and Romagna (1000 × ha)

Valley, as in central Italy. A new radical deforestation of the Po Valley lands thus took place after the Second World War in conjunction with the phase of rapid motorisation and mechanisation of agriculture. The widespread diffusion in the countryside of the liquid gas cylinder for domestic use began to free women and the elderly of the family from the millennial task of preserving and feeding the fire with wood and bundles obtained from the farm's trees. Rows of trees therefore became useless obstacles for a highly productive, mechanised cereal cultivation with prodigiously increasing yields thanks to the massive input of chemical fertilisers into the soil. The rapid replacement of working oxen with the company tractor made the production of elm leaves superfluous. These, if not periodically pruned, produced a harmful shadow on the crops. Specialised viticulture with dead cement support or steel poles, moreover, became more convenient for viticulture.

Nevertheless, we must consider that until about 1980, more than the half of the fuel wood in Italy was not produced from forests, but from trees in the field. This suggests that a different kind of forest, not the dense, homogeneous, forest cover we usually think of, but rather an agroforestry system, can play both an economic and environmental role. In the last decades the idea of putting trees in the agricultural areas has been proposed as a solution for improving the environmental quality of farmed land. However, there are still many areas in the word where we find a high percentage of agricultural land with greater than 10% tree cover: 96% of land area in Central America, 79% in Southeast Asia, 45% in Europe, among others. These are all examples of an agricultural heritage still surviving despite the industrialization occurred in agriculture at world level. The use of sustainable and ecological practices is the key feature distinguishing traditional agriculture, which was developed over decades or centuries, based on traditional experience and proven traditions. This kind of farming does not provide enormous production, but ensures sustainable yield over time, thanks to time-tested technologies and traditional know-hows, being more sustainable and less polluting than industrial techniques. This type of practices can be found most often in small scale farms and under the form of subsistence agriculture. Based on this idea, in 2002, FAO launched the idea of Globally Important Agricultural Heritage Systems (GIAHS), to identify and safeguard agricultural sites which have survived using traditional techniques and are still providing many services to the ecosystem, while maintaining magnificent landscapes, wild and agricultural biodiversity, ancestral knowledge transmitted through generations, and strong cultural and social values. The background of creation and maintenance of these systems are farmers' efforts to overcome disadvantageous geographic and harsh climate conditions and to increase and stabilize crop yields in a sustainable manner. These systems have been adapted to ever changing environment and climate conditions which finally have acquired resilience and robustness to ensure food and livelihood security in the local communities and reduce risks. The current efforts to adapt to and mitigate climate change are looking at traditional techniques as models of resilience and sustainability, thus, this form of agriculture may still play an important role (Agnoletti et al. 2023).

8.2 Mountain Reforestations

The work of reforestation and hydraulic arrangement of mountain basins represents the most important and problematic commitment undertaken by the Italian forestry administration from the Unification of Italy to the period between the two wars, with the aim of increasing the country's timber resources, but also of reducing the hydrogeological instability of the mountains. It was a work subject to contrasting pressures, linked to the evolution of political trends, socio-economic factors and technical orientations in the execution of the works, the effects of which have had a significant influence on the Italian landscape. Attention to the problem is affected by a debate that had already begun in the previous century and that has affected all of Italy and continues today. It is also interesting to reflect on this history considering the current scientific suggestions and political initiatives promoting extended afforestation for climate purposes. Sometime these plans lack of a careful considerations of the lessons learnt from the past, that, if not taken into careful consideration, may lead to the failure of these projects.

The problem of erosion control and the importance of preserving and expanding the forest were already present in the legislation of the he Italian states of the pre-unification period was necessary, but it took several years for them to be recognised as important enough to lead to a tool to regulate the matter at a national level (Trifone 1953). Attempts to unify forestry legislation began in 1862 with a first project by Minister Pepoli, followed by others over the next ten years, including that of Minister Finali in 1873. In the latter, for the first time, mention was made of the "chestnut tree limit", the altitude above which the chestnut tree does not grow, to allow agricultural cultivation in mountain territories, considered one of the main causes of deforestation. These were initiatives that, however, did not propose much for the use of the forest as a means of protection against hydrogeological instability (Trifone 1957). The 1874 law on communal uncultivated goods, which obliged municipalities to reforest or to cede uncultivated lands to third parties with the same obligation, was also of little relevance, because the municipalities had no funds, and few were interested in acquiring those lands (Borghesani 1909). Even the direct intervention of the State, albeit with initiatives that envisaged a limited financial commitment, was modest: the first free distribution of seedlings for reforestation in 1873 delivered only 148,000 seedlings from just ten forest nurseries present throughout the national territory (Sansone 1915).

The first noteworthy measures were recorded in 1867, when the Provincial Council of Cuneo allocated 2000 lire of the time for reforestation works, to which an equivalent sum was added by the Ministry of Agriculture. This represented the premise for the establishment of the first continuous consortium between the State and the province, formalised with a decree of 1872, which established a Forestry Committee with the aim of designating deforested and disused lands that "influence the disorder of the course of the waters and produce damage" and to provide for reforestation. The reforestation consortia constituted the most important bodies delegated to this activity and most of the works carried out until the early 1900s are due to their initiatives. At least until the 1877 law, they represented the result of special agreements between the State and the interested provinces, while the new measure explicitly promoted their establishment.

A fundamental turning point was therefore had with the 1877 law, which established in its principles the importance of the forest for hydraulic regulation and imposed the forest constraint, presenting itself as the moment of establishment of what after the 1923 law was defined as "hydrogeological constraint", in which the forests were protected in relation to their protective function (Raffaelli 1997). The law accepted the upper limit of the chestnut for the imposition of the constraint, effectively dividing the mountain into two zones with different regimes in this regard, introducing various modifications and exceptions compared to the Finali project which was more restrictive, admitting the constraint only with regard to the conservation of the territory and the water regime.

Both contemporary sources and recent historiographical literature interpret the law as the result of a policy inspired by liberalism, which tended to favour private property and to limit state intervention in matters of forest conservation (Trifone 1970; Osti 1913). A trend that appeared clear already in 1869 when, despite the foundation of the first Italian forestry school in Vallombrosa, there was an attempt to alienate all the state forests incorporated by the State in Tuscany, which was opposed by the Secretary General of the Ministry Luzzatti, who then obtained the establishment of the first State forest properties (Muzzi 1970). Therefore, the heated debates about the legitimacy of the constraint that preceded and followed the enactment of the law, including those distinguished by the Tuscan ultra-liberals, should not be misleading, because in reality the law did not reduce but extended the freedom of private individuals (Raffaelli 1997), following the idea of "removing laws inspired by principles of protectionism of other times", stated in Parliament as early as 1874 (Trifone 1953).

The measure limited interventions to only constrained lands, allowing private individuals great freedom to deforest, especially in the areas below the chestnut limit, where the imposition of the constraint was exceptional. Moreover, the choice of the chestnut obviously gave rise to discordant interpretations and significant differences, since this species in Italy can vegetate from sea level up to about 1500 m in altitude, with substantial differences from north to south of the country. Even though the regulation addressed this problem trying to limit its application difficulties, it remained a continuous source of uncertainty (Pampaioni 1878).

Against the new law immediately lined up the director of the Vallombrosa forestry institute Adolfo Di Bérenger, who had worked on a draft law on behalf of the Minister of Agriculture Minghetti since 1869. Judging that his opinions were not taken into due account, Di Bérenger openly expressed his dissent and for this he was prematurely put to rest, with a provision of Minister Majorana Calatabiano that caused a strong impression in the forestry world (Muzzi 1970). However, such a measure could certainly not strengthen the contents of the law. The constraints imposed on both forests and bare lands soon proved ineffective, nor did they allow a real development of reforestation activity, since both expropriations and the creation of consortia between the State, provinces and private individuals had been very slow. Similarly, the arrangement works, which could be financed for an amount equal to 65% of the total, had limited development, as they left the majority of interested parties the choice of the constitution of the consortium. The inadequacy of the law is confirmed by subsequent parliamentary initiatives aimed at modifying it, such as the bill of Minister Berti in 1882 with the significant title "Provisions intended to promote reforestation" (Trifone 1957). However, it was necessary to wait until 1888 for a legislative measure to be approved that recognized the authority of the Ministry of Agriculture to decide which lands were to be reforested and arranged, following specific projects prepared by the forestry administration.

This measure was greeted with many hopes by some foresters who saw in the restriction to only constrained lands a strong limit to the development of such activity, and believed that the new law put Italy on a par with nations like France that had passed a law on the matter as early as 1860 (Herzen 1889). But such hopes were soon disappointed, and there were as many as six reform projects presented in Parliament until 1910, to eliminate the inconveniences of a legislative system judged in part excessive, for the rigid prohibitions imposed, and in part insufficient for its narrowness of views (Trifone 1953).

In the period between 1867 and 1888 the effectiveness of the 1877 law was almost nil as far as the execution of large-scale works is concerned, and the 12 consortia established in this period reforested just 7383 ha. The most important reforestations were those made in the province of Genoa mainly with maritime pine, Aleppo pine and larch pine, which extended for 3522 ha, and in that of L'Aquila with 1480 ha, while the most important arrangement works were carried out in the province of Sondrio. The preference given to conifers was due to the need to use species with low requirements that could vegetate on poor, superficial soils, often compacted by the action of grazing and degraded by the passage of fire, which would not have allowed the immediate use of broadleaves. In addition to this, there was the hope of creating forests useful for the production of timber for work and construction, which Italy was very lacking in, a secondary objective of the reforestation work, but which responded to precise directives of the forest administration in terms of silviculture, aware of Italy's growing dependence on timber imports from abroad (Piccioli 1885; Agnoletti 2003).

Despite the limited results, the 1888 law showed a change in the state's policy and the guiding concepts of its intervention, because it acknowledged the substantial ineffectiveness of an attitude that limited the initiative of the public administration in this matter. In this light, the subsequent measures of 1893, the land reclamation law of 1900, which also concerned reforestation, and the subsequent special measures for the southern regions should be interpreted. This change in the state's philosophy culminated with the law of 2 June 1910, no. 277, which, by establishing the State Forest Domain, favoured reforestation and arrangement work by entrusting it with funds and competences also for the organisation of the reforestation consortia's activities (Sansone 1915). This was followed by the law of 13 July 1911, no. 774, in which the first two titles report measures for the hydraulic arrangement of mountain basins, then gathered in a separate text promoted with R.D. of 21 March 1912, no. 442. Here it was clearly stated both the link between the forest management of mountain basins and the hydraulic arrangement of river courses, and the need for a strong initiative of the State for the realisation of the works, adopting for the first time the term "hydraulic-forest" which did not appear in the 1877 law (Cantilo 1951). It was established that the arrangement and reforestation works were to be carried out "at the care and expense of the State" and the temporary takeover of private funds for the execution of reforestation. A measure different from expropriation for reasons of public utility, because the owner had the right to take back the fund but had the obligation to manage it according to a conservation plan, otherwise he could renounce and the land be purchased by the State (Trifone 1957).

The analysis of the interventions carried out between 1888 and 1912 shows a slightly better situation compared to the previous period. 16,826 ha were reforested, while the number of consortia in 1910–1911 had risen to 26. In 1886 the forest nurseries had reached 32, but as a result of the law of 2 June 1910 they reached the number of 168 distributed in 35 provinces of the Kingdom, on state-owned, municipal, private land and also on rented land, but which showed many design defects. This increase boosted the free distribution of forest seedlings to private individuals and municipalities, which went from a number less than a million in 1877–1878 to about 30 million in 1912–1913, with an increase of almost 100% between 1910 and 1914, but rising to almost 130 million between 1912 and 1922 (Pavari 1970; Sansone 1915).

8.3 First Fifty Years: The Reasons for a Defeat

Compared to the estimates of the forestry administration, which deemed it urgent to reforest at least 400,000 ha, the results obtained in this initial period were quite modest (Borghesani 1909). The judgement on the reduced effectiveness of the state's intervention must, however, consider a series of social and economic factors that constitute the context in which the action to restore the Italian mountain had to be measured. First among these is undoubtedly the demographic increase, which saw the Italian population double between 1861 and 1925, but which gave rise to a real "assault on the mountain", greater than what was described even for other European areas (Guichonnet 1991). A phenomenon that had no equal in modern times, affecting the entire national territory and bringing the mountain population from about 5,000,000 to 8,500,000 inhabitants in 1925 (Saraceno 1993).

This increase did not occur uniformly in the various parts of the country, assuming different growth rates, but had significant consequences at the forestry level, which the foresters themselves did not fail to notice (Caldart 1930). The need for new arable land was not solved with an increase in production per unit through technological development, but rather with the extension of the cultivated area, also affecting marginal areas through the cultivation of very steep areas and reclamation. The growth of foreign demand for agricultural goods that was recorded in those years did not make it convenient for landowners to introduce new technologies, as they had abundant and very low-cost labour available. When, at the end of the nineteenth century, American cereals entered the market, even the export of agricultural products ceased and this high labour-intensive agriculture functioned exclusively as a social buffer for the excess of labour force, even if the tendency to reduce forest areas remained unchanged. The industrial development recorded at the end of the nineteenth century also resulted, directly and indirectly, in an increase in the consumption of wood and timber. Italy did not have fossil fuel resources, and initially the energy needs of the industry were focused on wood and charcoal, which in 1861 provided more than 85% of energy needs. The use of solid and liquid fossil fuels that had to be totally imported grew slowly until it reached 60% by 1910, while electricity made its timid appearance towards the end of the century, contributing only 9% to the national energy balance on the eve of the first world war (Bardini 1991). All this, combined with the per capita consumption in constant growth, resulted in a huge consumption of wood, which according to the estimates of the period far exceeded the production of the forests. In fact, the need had to be met with an increasing recourse to wood collected outside the forest, such as trees, hedges and other agricultural plants, a production that increased by 50% between 1861 and 1912. The same growth was had for the needs of timber for the building industry, then in strong expansion (Romanelli 1990), which was imported from abroad in increasing quantities and values that increased fourfold in the same period (Agnoletti 2003). Timber imports constituted the third passive item of the Italian trade balance at the turn of the century, after fossil fuels and cereals.

The combined effect of the three factors: agricultural expansion, energy needs and timber production, influenced the Italian forests in various ways. The expansion of agriculture and pastures triggered a rapid process of reduction of the forested area, to which the effects determined by the 1877 law were added. As a result of the release, according to available statistics and the comments of some experts, at least 1,000,000 ha of forests were eliminated. Between 1870 and 1912, there was therefore a reduction in the forested area estimated between 15 and 30%, a trend that remained unchanged at least until the 1920s, when it reversed (Lunardoni 1904; Patrone 1970a, b). Where there were no deforestation processes, temporary or permanent, there was still an evolution of silviculture more suited to the needs of the rural economy, which however preserved the forest resource. It is in this period that a profound change in the mountain landscape occurs, in that territory with a predominantly forestry vocation in which 60% of the national forest surface was concentrated. The intense and multifunctional exploitation of the forest, mainly coppiced, and its transformation into agricultural crops, led to the consolidation of a mountain landscape in which the forest was a minority portion, especially compared to pastures.

8.4 The Technical Reflection of the First Decades of the Century

Along with a reduced effectiveness in quantitative terms, the implementation of reforestation and arrangement interventions also proved problematic from a technical point of view, leading to a reflection certainly not detached from the socio-economic context. It is beyond question that compared to the activity of other state bodies, the consortia are the main actors of the works of restoration of the Italian mountain until the 1912 law. However, the reports on the activity of the consortia highlight a problem that has been debated in the forestry sector for some time, with which they tried to explain some of the reasons for the modest effectiveness shown in achieving the set objectives. The works carried out were indeed criticised, accusing the consortia of having strayed from institutional tasks and original objectives, favouring hydraulic arrangement works of the riverbeds, basically the building of dikes to regulate waterflow along rivers, compared to reforestation. This trend appears documented by the data of the period in which it is observed that from 1888 to 1893 there is a certain disproportion between the expenses for the hydraulic arrangements and the reforestations. A difference that is maintained in subsequent periods, with a peak between 1911 and 1914, and can be summarised in a percentage value of 59% committed to arrangement expenses.

The criticism made to the consortia and the operational choices can be interpreted by briefly reconsidering the events of the Italian forestry school. At the time of its foundation, which took place in 1869, it was looking for its own technical-scientific identity for which it decidedly looked at models coming from the countries of Central Europe. From a general point of view, Germany was the main reference point, although this judgement seems to specifically concern economic aspects, while the development of silvicultural methods had also benefited from the experience of the French school (Patrone 1970a; b). The same Di Bérenger, first director of the Vallombrosa school, was of German origin and had carried out part of his education in Austria, before moving to Italy to work for the Vienna government and then move to the Italian forestry administration (Agnoletti 2001).

In the field of hydraulic structures, however, France was taken as the main reference point, at least in the period following the 1877 forestry law (Pavari 1970). French authors had given a great boost to the scientific development of this sector, thanks to the experiences made in the Alps which constituted a point of reference especially for what concerns works to prevent overflowing and flooding. In truth, in Italy there was a long tradition of studies on the subject. Already Bacci in 1576 had described the necessity of transverse structures for riverbed arrangements in relation to the Tiber river (Bacci 1576), a topic taken up by Manni in 1618 (Manni 1618), while their construction was addressed in detail by Viviani for the Arno in 1681 (Viviani 1681). The studies continued in the next century with Frisi (Frisi 1762), up to the fundamental nineteenth-century texts of Mengotti and Bergolli (Mengotti 1816; Bergolli 1847), when the problem was already the subject of considerable attention also from the public. Despite this, the Italian school was substantially unprepared to face with scientific rigour the techniques of hydraulic arrangement. The teaching of disciplines related to this topic was partially in the third year of the first courses at Vallombrosa, under the name of "Road and hydraulic architecture in relation to forestry art", while a course on the riverbed arrangements will come later. Only with the promulgation of the law of 1888 was a greater impulse given to reforestation, within a general reorganisation of the teaching courses related to the evolution of forestry science in an international context (Muzzi 1970).

For these reasons, the debate developed at the beginning of the 1900s and focused on the evaluation of the greater or lesser effectiveness of the masonry works compared to reforestation did not take into consideration what had already been said by Italian authors, but started from a reflection on the work carried out up to that point and on what was reported by the French forest engineers, especially Surell, Demontzey and Thiéry. Surell's work was the basis for the French law of 1860, but the results of its application were not known before the universal exhibition in Paris in 1878, where they were illustrated with drawings, photographs and scale models the effect of the works carried out and also the previous situation (Nico 1879; Agnoletti 1989). However, it was the work of Demontzey, whose technical and theoretical foundations appear in his work of 1882, to have a greater influence in Italy and to spark a heated debate (Demontzey 1882; Thiéry 1891). The text was reviewed in the "New Forestry Review" by Giuseppe Segala, who recognised its merit in having strongly affirmed the necessity of a good technical preparation for the construction of transverse structures for the stabilization of riverbeds, which in many cases must precede or at least accompany reforestation: "which, becoming its complement, does not lose its great effectiveness" (Segala 1883).

The publication of Demontzey's book was of extraordinary relevance for those times, considering the floods that occurred in north-eastern Italy in the late nineteenth century. The measures suggested to prevent the recurrence of such events are listed in a long article by engineer Alessandro Arnaud in the "New Forestry Review", where explicit reference is made to the need for the construction of transverses, emphasising their importance for the stability of the slopes. The need to maintain the forest is also clearly indicated, but without any reference to reforestation, while the rest of the article focuses on the construction technique of the artefacts (Arnaud 1884).

Perhaps for these reasons, more emphasis was given to the works of arrangement compared to reforestation, but it is certain that the beginning of the new century sees the comparison of very contrasting opinions on the matter. In 1910 the same Segala, criticising a previous article by Bandi, invited the "young" not to chant the woods "exaggerating the benefits" of forests in terms of their ability to prevent landslides and floods, because the causes are only partly related to the presence of forests, but mainly depend on rainfall patterns and rain. According to the author, once certain damage had occurred, the only recourse was to construct transverses, from which the forestry administration had often been excluded, but which Bandi considered useless. The article concluded by arguing that it was beneficial to be good "forest engineers" to solve all the problems of the mountain economy, affirming an orientation of study courses in the forestry field that would be maintained until the present day (Segala 1910).

Although Segala's criticism appears rather measured, the reactions were vehement. In the brief analysis of the measures taken by the government in the first decade of the century, the sub-inspector Bandi recognised a decisive negative judgement of the executive on the effectiveness of consolidation works (Bandi 1910). In support of this position, extensive references are made to France, where there had been deep reflection on the work done for the arrangement of mountain torrents. In the French Alps, most of the expenses incurred for the numerous large masonry works carried out had been useless, as many of them had gone to ruin. The blame for this failure was attributed to Demontzey, the main advocate of the need for these artefacts, whose work had significantly deviated from the work of Surell who in 1841 had suggested a rather sober use of such interventions, considered as exceptional interventions and not the rule, denouncing their defects (Surell 1842).

In truth, Surell's approach had been negatively judged by Lechalas, who in his introduction to *Restauration des montagnes* of Thiéry criticised his interpretation of torrential phenomena (Thiéry 1891). It is important to note that Demontzey himself had changed his mind about his initial beliefs, stating in a work of 1896, considered his "technical testament", that the most effective action could only be carried out by large forests, and that the philosophy of intensive arrangement works should be abandoned in favour of reforestation of large areas (Demontzey 1896).

The controversy between the two forest inspectors did not go unnoticed, as in the following issue of "Alpe" an additional article by inspector Canciani, who was involved in some arrangement works, tried to distance himself from the polemical tones taken by the discussion. He considered both types of intervention equally important, but still believed that reforestation was mandatory, and that arrangement works should only be done where necessary, calling for greater flexibility in the choice of method (Canciani 1910).

The judgement on the first French experience and its negative influence at the national level was reinforced a few years later by an article by Di Tella. In criticising the approach that seemed to favour masonry works over reforestation, he observed that initially the legislator had rightly identified the latter as the element that should have absolute prevalence, and indeed the work of the forestry consortia initially favoured this aspect. With the publication of Demontzey's first work, and because of the visit made by the students of the Vallombrosa school together with some representatives of the Ministry of Agriculture in the French Alps, a reversal occurred instead, that led to favouring masonry works (Di Tella 1914) (Fig. 8.3).

Di Tella's opinion was undoubtedly authoritative. Not only as the holder of the course on hydraulic arrangement works at the Royal Forestry Institute of Vallombrosa, but also as the author of the famous pamphlet "The forest against the torrent",

Fig. 8.3 Arrangement works in the Prebec river at the end of the nineteenth century (Piemonte region)

published by the Propaganda Commission for the forest and for the pasture of the Italian Touring Club in 1912, to highlight the urgency of reforestation and arrangement of mountain basins. Both in this text with a popular slant, and in the notes of the lessons given at Vallombrosa, it is affirmed that "the forest is the most effective means of consolidation and protection for lands exposed to the fury of meteoric erosion". However, cases are clearly contemplated in which the construction of transverses appears necessary and there is therefore no denial of the need for such works (Di Tella 1912; AAVV 1911). The caution expressed on the effectiveness of these works was undoubtedly the result of an evaluation of the frequent failures that the use of these artefacts had also encountered in Italy, but Di Tella himself observed that the judgement on the work of the Demontzey school necessarily had to take into account other aspects related to the economic and social environment of the mountain (Di Tella 1914). The preference given to the arrangement of riverbeds was also linked to the open hostility of the mountain people against the reforestations undertaken with the law of 1860, which took away lands from grazing, a problem that also accompanies the entire history of reforestations and the application of forestry legislation in Italy (Pavari 1970). It was therefore thought that the problem could be solved more quickly with an engineering approach, compared to the long waiting period necessary for the beneficial effects of reforestation, which was not well seen by the mountain populations. For this reason, the intensification of measures in favour of reforestation and of any kind of constraints in mountain areas developed an open hostility towards the foresters and the forest, which characterises a long-time span and which is also reflected in the number of sanctions applied to transgressors (Jandolo 1929). Such reactions paved the way for a rethink on mountain policy, which necessarily had to favour socio-economic development and not just impose constraints, however necessary to limit human pressure.

Beyond purely technical opinions, it is possible to glimpse other motivations that influenced positions on the matter. From a governmental point of view, the opinion expressed during the discussion of the bill for the establishment of the Forest Domain—by the Minister of Agriculture Raineri, who expressed scepticism about arrangement works and declared that more emphasis should be placed on reforestation, lends itself to a double interpretation. It may indeed be that this judgement was influenced more by spending considerations than technical motivations, given the high cost of transverses to control erosion and the increasingly frequent references to the use of wooden artifacts or small dry walls made with materials found on site (Valentini 1910). An opinion supported by the interventions of other parliamentarians who also criticised the excessive cost of the engineers' per diems for field work (Bandi 1910), but which introduces a further element to evaluate the ongoing controversy, if one reflects on the experts involved in the discussion.

Not coincidentally, in favour of hydraulic arrangements works was in fact Maganzini, president of the technical committee for the important works in the Sele river basin, who criticising the effectiveness of the laws of 1877 and 1888 noted that having considered reforestation exclusively was the cause of the aversion against the laws and their failure, and that the forester had to be associated with the engineer, As has already happened in France, Germany, Switzerland, Austria and other "civilised" countries (Maganzini 1911). Maganzini's words are not surprising, given that the article reproduced the text of the press conference held at the headquarters of the Society of Italian Engineers in Rome on 19 April 1911. Considering that the previous year Perona, director of the Vallombrosa Institute, had also firmly opposed masonry works, it can be assumed that there was probably a conflict of competencies, if not a contrast between foresters and engineers, to secure the execution assignments (Di Tella 1934). If only for the fact that most of the works were carried out by the Civil Engineering Corps with funds charged to the Ministry of Agriculture, rather than the Ministry of Public Works (Cantilo 1950). The role of engineers was nevertheless significant in the forestry school itself, given that the course specifically dedicated to hydraulic-forestry arrangements, proposed with the new study plan approved after the institute's move to Florence in 1914, was taught by Manfredi De Horatiis, a graduate of the Vallombrosa school, but also a civil engineering graduate from the University of Palermo.

Observations on an existing imbalance between masonry works and reforestation works continue until the 1930s, a period in which Di Tella publishes an article with the same title as the one published 16 years earlier and almost the same content, renewing his appeal for a different line of intervention (Cappuccini 1928; Di Tella 1931). However, an evolution of the concept of arrangement works is observed from the engineering point of view, which tries to overcome the inconveniences that occurred previously. Initially, interventions were mainly concentrated on watercourses, while later it is proposed to intervene on the entire basin to influence the water regime by reducing the torrentiality of the tributaries. As early as 1916, it was suggested to create, in morphologically suitable areas, large water collection basins that served to decant them from the alluvial material transported, stopping the moving mass that would thus be deposited with large dams. In addition to regulating floods, the larger basins could also be used for the production of electricity, then in strong growth, thus providing an additional service. A few years later a new concept was introduced, stating that it was not always necessary to build high dams, but it was possible to resort to flat areas to let the waters expand, decant them of solid materials and slow their race, thus referring to the modern notion of "detention basins" (Andreocci 1932).

However, it is not in the techniques of making the transverses that innovative concepts could develop in Italy compared to other European countries, but rather in those of actual reforestation that the country offered a significant contribution. The characteristics of poverty of our soils, the climate of particular dryness of the areas subject to reforestation and the steepness of the mountain slopes forced not only to identify species suitable for such conditions, but also to develop specific planting techniques, which assumed a certain importance also in the rest of the Mediterranean area. It is worth mentioning the terracing, necessary to collect the little existing soil and prevent the surface runoff of water, but very costly due to the wide use of labour, often forced to obtain them by chipping the bedrock with a pickaxe. Special mention must be made of the Allegretti method, tested in Sardinia in 1919 under very difficult conditions, which soon proved to be one of the most suitable for hot-arid climates (Morelli 1961).

Fig. 8.4 1906. Arrangement works made with wood and stones. Vesuvio mountain (Campania region)

These years were crucial as they allowed for experiences that made technically possible the expansion of the large reforestation programmes of the following decades. Once again, the conditions of the country, very different from those of Central Europe, required the development of specific techniques, which could not use solutions tested in very different geological and climatic conditions. A theme that presents a strong continuity throughout the history of the Italian and Mediterranean forest management, compared to the north of Europe (Fig. 8.4).

8.5 The New Philosophy of State Intervention

The establishment of the Special Company of the State Forestry Domain, which took place in 1910, overturned the philosophy of non-intervention previously followed, asserting the greater ability of the State to pursue public interests. The government's action was essentially to encourage private owners, but in cases where there were protective forests and reforestation of mountain basins and private individuals were unable to intervene, it had the duty to replace them in the protection of public interests and this represented the principle of the Luzzatti law shared by other economists of that period (Bolla 1919). The first decades of the company's activity, aimed primarily at the creation of a state forest ownership and its management, are characterised by a significant profit recorded by the budgets, but the expense for reforestation works appear rather limited (Camaiti 1959).

A further fundamental legislative measure in favour of reforestations and arrangement works was the Royal Decree Law of 30 December 1923, n. 3267, also known as the Serpieri law, named after the then Undersecretary of State, the main architect of this project. It was to represent the basic structure of the Italian forest legislative system, on which all its subsequent development would be based (Caruso 1977). The principle was affirmed that in unstable mountain basins it was necessary first of all to intervene with hydraulic and forest works, integrated with agricultural arrangement and land improvement works, hence the "organic" character of the measure compared to the previous ones.

In Title II the matter was distinguished into hydraulic-forest arrangement of mountain basins and reforestation and reinforcement of restricted lands. For the first point it was established that the related works were to be carried out at the expense of the State: in particular, reforestations, reinforcements and related construction works would be the responsibility of the Ministry of Agriculture and Forests, while the immediately connected hydraulic works were the responsibility of the Ministry of Public Works. For the second point, the law provided that both the forest administration and the provinces and municipalities could, alone or united in a consortium, promote the reforestation of restricted lands. The imposition of the hydrogeological constraint operated by the measure extended to all lands of any nature and destination, which due to their forms of use could with public damage undergo denudations, lose stability or disturb the water regime, bringing the extension of the bound lands to about 7,000,000 ha compared to the 3,825,548 calculated in 1879 (Trifone 1953). In addition to these initiatives, the Special Company of the State Forestry was reorganised, with the aim of providing more decisively for improving timber resources, given the strategic role of this resource and the increasing dependence on import from foreign countries (Miliani 1926), while the management of the forests of municipalities and other entities were to be subject to economic plans.

The 1923 law not only reorganised and standardised the 51 legislative measures operating in the Kingdom, to which the 27 of Austrian origin and linked to the new provinces had been added, but considered forestry policy as an element of a broader economic and hydraulic policy (Romano 1986), even though it was later objected that the purely economic aspects had been almost entirely ignored (Panegrossi 1955). Therefore, the mountain was considered as a whole, not only thinking about imposing constraints, but about achieving a generalised progress of the mountain economy in all its sectors, considering this an indispensable condition to guarantee the protection of the forest and the control of erosion (Serpieri 1926). This philosophy was reflected in the consolidated text no. 3256 of 30 December 1923 on hydraulic reclamation—which asserted more strongly than in the past the concept of the unity of the hydrographic basins, providing that a single directive should define an organic arrangement—, followed by the law of 18 May 1924, no. 753, on land transformations of public interest. These two laws were already associated in 1929 with the common name of integral reclamation, to which fascism gave much emphasis (Jandolo 1929), but for a measure that organically formulated this concept, it was necessary to wait for the Royal Decree of 13 February 1933, n. 215, on integral reclamation (Serpieri 1931). The unitary vision of the problem and the intent to reduce the problems of competences ignited further hopes, even speaking of a "battle won" and the crowning of decades-long efforts in the field (Di Tella 1934): a vision perhaps adequate to the crowning of a legislative effort, but very optimistic in relation to practical results.

8.6 From Unification to the Second World War: A Balance

Despite the organic nature of the measures mentioned, reforestation still yielded modest results, which is not surprising if we observe that between 1870 and 1938 only 7% of financial resources were allocated to reforestation and arrangement works, while 93% of the funds went to reclamation works in the plains (Cantilo 1950). The analysis of the effects of the practical measures, more than other considerations, gives a measure of the results achieved: it is enough to remember that in about 80 years just 194,720 ha were reforested. However, it is useful to specify the responsibility of the various entities involved and some aspects related to the temporal and spatial scale of the interventions carried out. From what has been said previously, we can clearly identify the reforestation consortia, the forestry administration, understood as the Ministry of Agriculture Industry and Commerce, and the state forestry estate company, as entities involved in the financing and implementation of reforestation and arrangement works. The intervention of these entities in terms of expenditure was not equal, and the sums spent also varied from one region to another; in addition to this, a small part of the works was carried out with funds from the Ministry of Public Works. It would be too long to analyse in detail the quantitative data, which are however extensively illustrated in a long article by Generoso Patrone (Patrone 1952); however, it is useful to make some summary references referring to this work for any further study.

In terms of expenditure, for the entire national territory and for the entire period analyzed, the forestry administration contributed for the 73%, the consortia for 21% and the company for 6%. The expenditure borne by the reforestation consortia is highest in the Alps at 33% and non-existent in Sardinia, gradually decreasing from the Northern Apennines (26%) to the Central Apennines (20%), to Sicily (13%), and to the Southern Apennines (5%). Conversely, the sums invested by the forestry administration decrease as you move up from the Southern Apennines (90%) to the Alps (63%). There are noticeable differences in the results achieved in the first 60 years of state intervention, i.e. up to the law of comprehensive reclamation, compared to the subsequent period. In the period between 1867 and 1914, there is a clear prevalence of the intervention of the reforestation consortia with 65% of the sums spent, followed by the forestry administration with 30%, and 5% borne by the company. In the subsequent period, i.e. from 1914 to 1924, the intervention of the forestry administration rises to 74%, with 8% borne by the company and 18% carried out by the consortia. This trend remains constant even in subsequent periods since from 1924 to 1950 the intervention of the administration remains around 80%, with a reduced percentage carried out by the company and an ever-decreasing intervention by the consortia, which reach 4% in 1950. The figures confirm in fact the change in policy implemented in this sector from 1910 onwards, which sees the state's initiative grow decisively.

From a geographical point of view, the Alps absorb a total of 32% of the resources committed over the entire period considered, 13% the Northern Apennines, 20% the Central Apennines, 16% the Southern Apennines, 14% Sicily and 5% Sardinia (Patrone 1952). For a more accurate evaluation of these parameters, however, the different extent of the individual mountain regions must be taken into account, which is maximum for the Alps, followed by the Central Apennines, the Southern Apennines and the Northern Apennines, and also the different conditions of hydrogeological disorder. In general, the expenditure per hectare decreases from the islands, to the South and to the Alps, and Patrone explains it with the greater conditions of degradation of the southern regions. The very high per hectare expenses of the Sicily region are remarkable in this context, almost three times higher than those of the Alps, the Southern and Central Apennines; on the contrary, as far as total expenses are concerned, the highest amounts concern the Alps, with an amount double or in any case much higher than the other regions (Fig. 8.5).

Observing the interventions carried out in the various regions, it is noted that in the first period of 47 years no intervention took place in Lazio, while the Alpine regions are very represented (Veneto 13.7%, Lombardy 12.2%, Piedmont 10.4%) also thanks to the better conditions in which the consortia could operate, due to the fewer conflicts with the local populations regarding the limitations imposed by reforestation on grazing. Some provinces attracted significant shares of the sums spent at the regional level. In Piedmont, 68% of the interventions were carried out

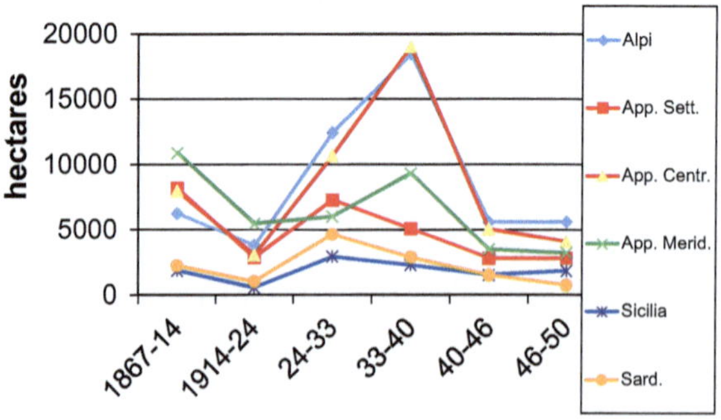

Fig. 8.5 Reforested area in the various Italian geographical areas from 1867 to 1950

in the province of Cuneo, where some imposing masonry traverses were built in the Prebec torrent basin, later destroyed by flood events that also questioned the effectiveness of such interventions (Agnoletti 1989). The same happened in Calabria for Cosenza (60%), and in Campania for Naples (65%).

Moving on to the actual reforested areas, in the period from 1861 to 1950, only 194,720 ha were reforested, with an average of 2346 ha per year. 26.9% of the new forests were created in the Alps, 15% in the northern Apennines, 25.7% in the central Apennines, 19.9% in the southern Apennines, 5.7% in Sicily and 6.8% in Sardinia. This shows an increase in forested area more than proportional to the forests already present in southern Italy and less than proportional in northern Italy, a result justifiable by the need for greater commitment in areas where bare surfaces were much more extensive. An interesting observation is that the forestry administration probably gave less development to reforestation compared to hydraulic arrangements, than did the consortia and the company. In fact, looking at the expenses made by the three entities and the areas reforested by each, there is a smaller proportion of reforestations due to the Administration, which could suggest that a certain quota of expenses went into arrangement works, rather than in reforestation. On the other hand, as already stated previously, most of the works were carried out by the civil engineering department with funds from the Ministry of Agriculture. However, the positive role of the new laws issued in the first decades of the century is evident, as confirmed by the speed of growth of the forest area, which is minimal (800 ha per year) during the period 1867–1914 and maximum from 1933 to 1940 with 8191 ha. As the sums spent increase, the extent of reforestation increases but to a lesser extent, and this means once again that arrangement works prevail over reforestations.

As already seen for expenses, from 1867 to 1914 it is the consortia that reforest, while after 1914 it is the forestry administration that increasingly takes on an active role. From 1867 to 1950 the expansion of the forest area is ensured for 57% by the forestry administration, for 18% by the company and for 25% by the consortia. The contribution of the three entities is different in the various geographical regions since in the southern Apennines the administration contributes 83%, but only 43% in the northern Apennines. The company intervenes significantly in Sardinia (47%) and very little in Sicily (6%), while the consortia have done the most in the northern Apennines and in the Alps with 41% and 39% but intervene minimally in Sardinia (0.4%) and in the southern Apennines (4%).

It is interesting to evaluate the reforestation action also through some important cases like that of the works for the Sele basin, in Campania, carried out at the beginning of the 1900s for the future Apulian Aqueduct, the costs of which were entirely borne by the State. This is one of the most important public works carried out in Italy whose balance is still fully in the black today. The study carried out by a commission of experts, established in 1901, led to the technical provisions contained in the law of 26 June 1902, no. 245, and to the R.D. 1 January 1903, no. 46, where the municipalities interested in the project (located in the provinces of Avellino and Salerno) were indicated and the technical provisions for the interventions. The handwritten report on the work carried out throughout 1910 confirms the negative judgement on the effects of the 1877 forestry law and the decisive action of fire

Fig. 8.6 Reforestation carried out in the Sele basin—Vallone delle Brecce—between 1903 and 1910 (Library of the Forestry Management Section, School of Agriculture, University of Florence)

used to extend the pastures, with the consequent development of erosive phenomena (Fig. 8.6).

The area of the Sele basin was made up of 826 ha of bare land, 563 ha of restricted woodland, 140 ha of unrestricted woodland and 755 ha of unrestricted and cultivated for agricultural production. The work carried out the stabilization of river beds and reforestation to stop landslides. The intention was to slow down the degradation of the banks, provide absorbing power for rainwater and reduce their speed, even if sometimes it was enough to suppress grazing and build dry stone terraces and wooden structures to achieve this result. Mainly black pine, larch pine and Scots pine were planted, with reforestation through sowing or hole planting, confirming the predominant role of these species, as also confirmed by their majority presence in all Italian forest nurseries and the indications for the most suitable species for southern Italy already in 1893 (Savastano 1893). The report notes that at the end of 1910 there were still 920 ha to be reforested, but there were significant problems due to the refusal of municipalities to restrict grazing areas, and in fact the seedlings were often protected from livestock even with wire fences. Table 8.1, which shows the breakdown of expenses in the period considered, shows that in reality there is no prevalence of expenses for the arrangement works, which indirectly goes back in favour of the opinions expressed at the time by Maganzini as president of the technical committee. Both here and in other interventions in the South, a wide use of trellises and fences built with plant material is observed. In the case of the works on the Bagni torrent in the province of Catanzaro, the expenses for wooden works accounted for more than 50% of the total, while reforestation works absorbed only

Table 8.1 Overview of the expenses incurred for the reforestation works and arrangements of the Seie basin from 1903 to 1910 for the construction of the Pugliese Aqueduct

	hectares	Expenses (lire of 1910)		
	Afforestation	Afforestation	Arrangement works	Other
1903	11	3274	1244	
1904	52	9970	17,498	
1905	53	11,031	7490	
1906	59	6908	8000	
1907	42	8649		
1908	80	11,001		
1909	107	11,776	350	
1910	110	16,496		
Total	515	79,140	34,585	21,413

18% of the expenses. In this area, 8 km of trellises were also planted with material. I live, which together with those made of wood have today been rediscovered by environmental engineering.

8.7 The Assertion of a "State Landscape" in the Post-war Period

The decade following the end of the Second World War finally marked a real reversal of trend, which was not only realised through legislative measures, but rather with the profound change in socio-economic conditions. Without a doubt, the legislative measures of 1949 and 1950 on worksites and reforestation sites, on the Fund for the South, on extraordinary interventions in the depressed areas of the centre-north, and above all the law of 25 July 1952, n. 991, enacted on the proposal of Amintore Fanfani (Muzzi 1955), were effective. The main purpose of the measures was not so much to create new forests but a social one: in fact, especially in the centre-south, hundreds of reforestation sites were created which had the merit of reducing unemployment. However, only once the mountain was freed from human pressure, thanks to industrial development capable of subtracting labour from agriculture and the technological progress of the agricultural sector, did forests and mountain territories separate their destinies from national demographic growth. This triggered a process that would lead in 50 years to the plantation of about 800,000 ha and the achievement of the objectives defined more than a century earlier (Merendi 1926),[1] to which was added a generalised growth of forests that would have brought them to an extent almost 50% higher than the 1800s.[2] The general push to increase the extent of coniferous forests compared to broad-leaved trees, to cope with the chronic shortage of construction timber, had achieved the result of increasing these forest formations from 15% in

1870 to about 60%, to the detriment of broad-leaved trees. Such results came at a completely different historical moment for mountain areas, now affected by a process of population reduction and natural reforestation of abandoned lands, and in an international timber market where Italy was now a transforming country, which bought almost all the timber for work abroad. Therefore, the calls made in 1961, at the national congress on reforestation, on the need to extend Italian forests to meet the needs for timber, appear illusory. Looking closely, there was no economic planning for the reforestation work, what was available in the nurseries was used without thinking about the economic results in relation to a possible timber market. In addition to this, the economic conditions in which forest uses took place in the mountains and the lack of high economic value forests would always have penalised national production compared to foreign timber (Camaiti 1961).

The new reforestations did not contribute to any improvement in productivity due to the lack of silvicultural care in totally artificial formations, often of poor quality, characterised by too high densities, mechanical instability and various pathologies (Agnoletti and Mercurio 1996). The improving function that conifers were supposed to perform in relation to the quality of the soil and the role of preparatory species for a subsequent entry of broad-leaved trees was not followed by the planting of new forests and these formations today represent a low-quality element of our landscape. In this way, even if belatedly, that "state landscape" was asserted which, in a sense, had been sought to be imposed from the previous century, contrasted with a "social landscape" characterised by the prevalence of pastures and agricultural areas on mountain slopes. From a landscape point of view, these reforestations are characterised by very rigid planting schemes, often at very high density (even 2×2 m), made for the vast majority with Austrian pine, created in areas with completely different characteristics of the landscape, in which such plantations stand out for a certain foreignness to the local context. Despite this, some facilities have over time taken on the role of witnesses to a historical period that reflects a scientific thought and a state action, then followed also by some regions and the European Union, although not with the same quantitative results. European agricultural policies have indeed favored, with contributions to agriculture, the planting of artificial forests on abandoned fields and pastures, which, like the previous ones, have not achieved any significant result from a productive point of view, but have left their mark on the landscape. It is worth noting that the doubling of the forest area due to agricultural abandonment occurred after the second world war does not seem to have solved the problem of the hydrogeological instability of the mountains. Looking at the current indications of EU policies (e.g. the Green Deal), with new millions of hectares of reforestations planned for climate purposes, it seems there will be more opportunities to reflect.

Notes

1. Even in 1926 it was stated that a forest extension of 7,250,000 ha would have been enough to cover the national timber need; today the forests are calculated at more than 11,000,000 ha and they cover only a minimal part of such need.
2. The uncertainty about the real extent of the increase in forests in Italy is due to the different detection criteria, a problem linked not only to nineteenth-century statistics, but also to the current data of the National Forestry Inventory of 1985 and ISTAT, which diverge significantly regarding what is meant by forested area. Some values could indicate an increase even greater than that reported here, which is however reliable for a rough estimate.

References

AAVV (1911) Appunti sulle lezioni di Correzione dei Torrenti, tenute dal Prof. G. Di Telia nel R. Istituto Forestale di Vallombrosa. Biblioteca della Sezione di Assestamento forestale del Dipartimento di Scienze e Tecnologie Ambientali Forestali, Università di Firenze

Agnoletti M (1989) Le raccolte fotografiche della Collezione tecnologico-forestale dell'istituto di Selvicoltura. Museologia Scientifica 5(3–4):287–292

Agnoletti M (2001) Fra storia e tecnica. Sviluppi e tendenze della storia forestale. In: Agnoletti M (ed) Storia e Risorse Forestali. Accademia Italiana di Scienze Forestali, Firenze

Agnoletti M (2003) Bosques e industria de la madera en Italia, de la unificaciòn al fascismo (1861–1940). Historia y economia del bosque en la Europa del Sur (siglos XVIII–XX). Prensas Universitaria de Zaragoza, Zaragoza

Agnoletti M, Mercurio R (1996) Rimboschimenti in terreni ex agricoli: il caso della fattoria di Gargonza. Atti e Memorie della Accademia Petrarca di Lettere Arti e Scienze 57:341–361

Agnoletti M, Santoro A, Fiore B, Piras F, Romano F, Bazzurro A (2023) Agricultural heritage systems in Europe, Asia, Africa, Central and South America, Springer, Dordrecht-New York

Andreocci A (1932) Criteri di sistemazione montana. L'alpe 2–3:52–56

Arnaud A (1884) Triglie, traverse e serbatoi artificiali. Nuova Rivista Forestale 7:134–161

Bacci A (1576) Del Tevere lib. 3. All'Ill.mo Senato e Popolo Romano. Venezia

Bandi V (1910) La sistemazione dei bacini montani ovvero parere e non essere. L'alpe 6–7:161–171

Bardini C (1991) L'economia energetica italiana (1863–1913): una prospettiva inconsueta per lo studio del processo di industrializzazione. Rivista di Storia Economica 8:81–114

Bergolli G (1847) Delle serre o chiuse nei torrenti fiumi. Tip. di Carlo Vincenzi, Modena

Bolla GG (1919) Gli scritti di politica e legislazione forestale di Luigi Luzzatti. Annali Dell'accademia Dei Georgofili 97:3

Borghesani G (1909) Le origini della questione forestale in Italia e l'opera del Marchese Luigi Tanari, senatore del Regno. L'alpe 4:97–111

Caldart F (1930) Sistemazioni montane, demografia e bonifica integrale. L'alpe 17(9):417–423

Camaiti A (1959) L'Azienda di Stato per le foreste demaniali. Soc. ABETE, Roma

Camaiti A (1961) La politica dei rimboschimenti e della ricostituzione dei boschi degradati. Accademia italiana di scienze forestali, Atti del congresso nazionale sui rimboschimenti e la ricostituzione dei boschi degradati, Tip. Coppini, Firenze

Canciani G (1910) La sistemazione dei bacini montani. L'alpe 8–9:229–233

Cantilo F (1950) Politica forestale e tecnica delle sistemazioni idrauliche. L'italia Forestale e Montana 5(3):94

Cantilo F (1951) La sistemazione idraulico-forestale dei bacini montani e leggi 10 agosto 1950, nn. 646 e 647. L'Italia Forestale e Montana 5(5):217–231

Cappuccini G (1928) Come e perché si sistemano i bacini montani. L'alpe 6:157–165

Caruso C (1977) Evoluzione della legislazione forestale nazionale in materia di rimboschimenti. L'italia Forestale e Montana 32(4):141–146
Cazzola F (1996) Disboscamento e riforestazione ordinata nella pianura del Po: la piantata di alberi nell'economia agraria padana, secoli XV–XIX. Storia Urbana 20(76–77):35–64
Cazzola F (1999) Clima e produzione agricola nell'Italia del Seicento. Qualche ipotesi per l'area padana. Società italiana di Demografia Storica, La popolazione italiana nel Seicento. Bologna, Clueb
Demontzey P (1882) Traitépratique du reboisement et du gazonnement des montagnes. Rothschild, Paris
Demontzey P (1896) Les retenues d'eau et le reboisement dans le bassin de la Durance. Remondet-Aubin, Aix
Desplanques H (1959) Contribution à l'étude des paysages ruraux en Italie Centrale: l'arbre fourrager. In: Géographie et histoire agraires. Actes du colloque International organise par la Faculté de Lettres de l'Université de Nancy. Université de Nancy, Nancy, 2–7 septembre 1957
Di Tella G (1912) Il bosco contro il torrente. Commissione di Propaganda per il bosco e per il pascolo, Touring Club Italiano, Milano
Di Tella G (1914) Rimboschimenti ed opere murarie nella restaurazione dei bacini montani. L'alpe 4:97–106
Di Tella G (1931) Rimboschimenti ed opere murarie nella restaurazione dei bacini montani. L'alpe 5:241–249
Di Tella G (1934) Per la montagna italiana. Dalla legge forestale del 20 giugno 1877 alla legge sulla bonifica integrale del 13 febbraio 1933, XI. L'Alpe 1:1–3
Frisi P (1762) Del modo di regolare i fiumi, e i torrenti, principalmente del bolognese, e della Romagna. Per Vincenzo Giuntini, Lucca
Guichonnet P (1991) Le popolazioni alpine fra età moderna e contemporanea. Lazzarini A, Vendramini E (eds) La montagna veneta in età contemporanea. Storia e ambiente. Uomini e risorse. Edizioni di Storia e Letteratura, Roma
Herzen A (1889) Rimboscamenti e Inondazioni. Nuova Rivista Forestale 12:101–118
Jandolo E (1929) La montagna e le leggi per la bonifica integrale. L'alpe 1:7
Lunardoni A (1904) Vini, uve e legnami nei trattati di commercio. Cooperativa tipografica editrice, Roma
Maganzini I (1911) Sulla sistemazione dei bacini montani. L'alpe 5–6:252–253
Manni D (1618) Trattato della direzione dei torrenti, Pistoia
Mengotti C (1816) Idraulica fisica e sperimentale. Tip. Andreola, Venezia
Merendi A (1926) Politica forestale e realtà. L'alpe 8:245
Miliani GB (1926) Linee direttrici della politica forestale. L'alpe 7:220
Morelli A (1961) La tecnica dei rimboschimenti in Sardegna. Accademia italiana di scienze forestali, Atti del congresso nazionale sui rimboschimenti e la ricostituzione dei boschi degradati, Tip. Coppini, Firenze
Muzzi S (1955) Dalla legge forestale del 1877 alle odierne direttive di economia montana. L'italia Forestale e Montana 10(6):262–270
Muzzi S (1970) Vicende storiche della scuola forestale italiana. Accademia italiana di scienze forestali, L'Italia forestale nel centenario della fondazione della scuola di Vallombrosa. Tip. Coppini, Firenze
Niccoli V (1902) Saggio storico e bibliografico dell'agricoltura italiana dalle origini al 1900. Nuova enciclopedia agraria. Utet, Torino
Nico P (1879) Sulle condizioni forestali della Francia, relazione letta al club scientifico di Vienna il 27 marzo 1879 dal barone A. Seckendorff. Nuova Rivista Forestale 2:191
Osti G (1913) Appunti per uno studio sistematico della legislazione forestale. L'alpe 3–5:83–84
Pampaioni T (1878) La nuova legge forestale italiana. Nuova Rivista Forestale 1:97–118
Panegrossi T (1955) La situazione della popolazione alpina in Italia e la legislazione a favore dell'economia montana. L'italia Forestale e Montana 10(5):225–239

Patrone G (1952) Il contributo dello Stato e degli Enti alle sistemazioni montane e al miglioramento e all'ampliamento dei boschi e dei pascoli dal 1867 al 1970. L'italia Forestale e Montana 7(6):321–349

Patrone G (1970a) Cenni di geografia economica forestale. Accademia italiana di scienze forestali, Firenze

Patrone G (1970b) Il contributo tecnico scientifico della scuola forestale italiana dal 1869 al 1924. Accademia italiana di scienze forestali, L'Italia forestale nel centenario della fondazione della scuola di Vallombrosa. Tip. Coppini, Firenze

Pavari A (1970) Le sistemazioni montane e i rimboschimenti. Accademia italiana di scienze forestali, L'Italia forestale nel centenario della fondazione della scuola di Vallombrosa. Tip. Coppini, Firenze

Piccioli F (1885) Le principali foreste d'Italia. Nuova Rivista Forestale

Raffaelli T (1997) Dalla repressione del «danno pubblico» alla produzione del «bene pubblico»: cultura economica e politiche forestali (1861–1915). Il Pensiero Economico Italiano 5(2):41–79

Romanelli R (1990) L'Italia liberale. Il Mulino, Bologna

Romano D (1986) I rimboschimenti nella politica forestale italiana. Monti e Boschi 6:7–12

Sansone A (1915) Relazione sulla Azienda del Demanio Forestale di Stato (1° luglio 1910–30 luglio 1914). Ministero Agricoltura Industria e Commercio. Stab. F.lli Capaccini, Roma

Saraceno E (1993) Il problema della montagna. Franco Angeli, Milano

Savastano L (1893) Il rimboschimento nell'Appennino meridionale. Tip. Giannini e Figli, Napoli

Segala G (1883) A proposito del trattato di rimboscamento e di inerbimento delle montagne. Nuova Rivista Forestale 6:159–171

Segala G (1910) La sistemazione dei bacini montani ed i criteri per compierla. L'Alpe 5:129–133

Sereni E (1957) Note per una storia del paesaggio agrario emiliano. In: Zangheri R (ed) Le campagne emiliane nell'epoca moderna. Saggi e testimonianze. Feltrinelli, Milano

Serpieri A (1926) L'ambiente economico e sociale. L'Italia forestale

Serpieri A (1931) La legge sulla bonifica integrale nel primo anno di applicazione. Ministero dell'Agricoltura e delle Foreste, Roma

Surell A (1842) Études sur les torrents des Hautes-Alpes. Carilion-Goeury & V. Dalmont, Paris

Thiéry E (1891) Restauration des montagnes, correction des torrents, reboisement. Librairie Polytechnique, Paris

Trifone R (1953) Il vincolo forestale prima e dopo la legge del 1877. Annali Dell'accademia Italiana di Scienze Forestali 1:61–76

Trifone R (1957) Storia del diritto forestale in Italia. Accademia italiana di scienze forestali, Tip. Coppini, Firenze

Trifone R (1970) Direttive della legislazione forestale italiana. Accademia italiana di scienze forestali, L'Italia forestale nel centenario della fondazione della scuola di Vallombrosa. Tip. Coppini, Firenze

Valentini C (1910) Gli Atti del Congresso Forestale Italiano di Bologna (1909). L'alpe 10–11:313

Vecchio B (1974) Il bosco negli scrittori italiani del Settecento e dell'età napoleonica. Einaudi, Torino

Chapter 9
The Civilisation of the Chestnut

Abstract The chestnut is perhaps the most representative species of the historical Italian forest landscape, given that this species, unlike all others, has played a decisive role in the life of rural populations. Historically, the chestnut has had a range of functions incomparable to other species, both as main products, wood and chestnuts, and as by-products, such as leaves and bark, as well as the multifunctionality of the chestnut grove, integrating perfectly with grazing. Paleobotanical studies have shown that in central Italy around 1000 BC there was a presence of chestnut pollen equal to 8% of the total tree flora. This percentage increased strongly during the expansion period of the Roman Empire, reaching 48% at the beginning of the Christian era. The landscape of the fruit chestnut grove is one of the most fascinating forest landscapes for its ability to combine aesthetic values and cultural aspects. The density of the chestnut grove, generally between 90 and 120 plants per hectare, makes it a very bright forest, while the cleanliness of the undergrowth, almost always a meadow kept low by grazing, makes it clean and pleasant also for recreational purposes. The fruit chestnut grove has often been the subject of land arrangements, so we find it both on terraces earth and on dry stone terraces, but also with the classic "lunettes", small semicircular stone terraces built around individual plants. Such attention and expenditure of labour were justified by the profitability of the chestnut grove, superior to any other forest species. The study of a specific case in the Apuan Alps, in Tuscany, helps to better understand the role of the chestnut grove in the mountain landscape and the evolution that has taken place in the last century. This case study also highlight the peculiar protection applied in the European Nature 2000 network site identified in this area, that indicates the mixed forest with chestnuts trees, deriving from the abandonment of the previous chestnut grove, a natural habitat deserving strict protection.

9.1 Origin and Distribution

The chestnut is perhaps the most representative species of the historical Italian forest landscape, given that this species, unlike all others, has played a decisive role in the life of rural populations. Jean-Robert Pitte, in the handwritten note to the monograph

M. Agnoletti, *History of the Italian Forest Landscape*, Environmental History 19,
https://doi.org/10.1007/978-3-031-81631-4_9

on this species, defines it as a tribute to the most beautiful tree of creation. Therefore, if the triad "wheat, olive, vine" is, according to the French historian Fernand Braudel, the symbol of the Mediterranean rural landscape, at least for Italy the chestnut is so from a forestry point of view. Historically, the chestnut has had a range of functions incomparable to other species, both as main products, wood and chestnuts, and as by-products, such as leaves and bark, as well as the multifunctionality of the chestnut grove, integrating perfectly with grazing. For this reason, we believe it deserves a separate discussion from all other types of forest, also because compared to coniferous or broad-leaved forests—more often composed of mixed species—the fruit chestnut grove is a monospecific forest, that is, composed of a single tree species (Bounous and De Guarda 2002). All the above justify the term "chestnut civilization" that we used for this chapter.

Pollen information tells us that the chestnut as a component of the Pleistocene Italian flora is a tertiary relic whose natural area has gradually shrunk to southern Europe following successive glaciations (Paganelli 1998). In Europe, Asia, America and in the tertiary flora of Greenland, Alaska and Scandinavia, traces of chestnut leaves and fruits have been found, signs of the expansion of the Castanea genus when the climate of those regions was temperate. The chestnuts known to prehistoric man were probably very small and were perhaps also consumed cooked, as evidenced by the remains of fruits, together with acorns, in terracotta pots, found in Liguria and Savoy and dating back to the Iron Age (Pitte 1986). Human intervention has expanded the geographical distribution of the chestnut. Asia Minor, and more specifically the Transcaucasian region, widely considered the area of origin of chestnut cultivation, is to be considered the centre that first domesticated, cultivated and spread the species. Its path therefore seems to start from Lydia to reach Greece first and then Italy (Camus 1929) (Fig. 9.1).

Paleobotanical studies have shown that in central Italy around 1000 BC there was a presence of chestnut pollen equal to 8% of the total tree flora. This percentage increased strongly during the expansion period of the Roman Empire, reaching 48% at the beginning of the Christian era. This testifies to the role of Roman civilisation in the progression of areas cultivated with chestnut in Italy, and in fact the chestnut is currently the most widely distributed forest species in Italy. It is indeed present in all regions, in the Islands, in the Apennines and in the basal areas of the Alps and Prealps, with a very wide altitudinal distribution, ranging from sea level in the centre to 1500 m in Sicily.

Its importance in the Italian landscape naturally did not escape the observations of the Grand Tour travellers. Michel de Montaigne described it in his journey to Italy in 1580, crossing Tuscany, which he admired for its mountains "all well cultivated and green to the top, rich in chestnuts, olives and also vines, which are planted here in circular patterns like concentric steps" (De Montaigne 2003). In the early 1800s, Stendhal and Shelley were struck by the splendour of the chestnut groves that descended compactly from the slopes of the Lario (Como lake) mountains almost to its shores, highlighting the beauty of one of the most interesting woods in Lombardy (Farneti et al 1972). Similarly, Edward Lear described it in his travel diary in Calabria in 1847, as one of the many magnificent woods in this part of Italy (Lear 2003).

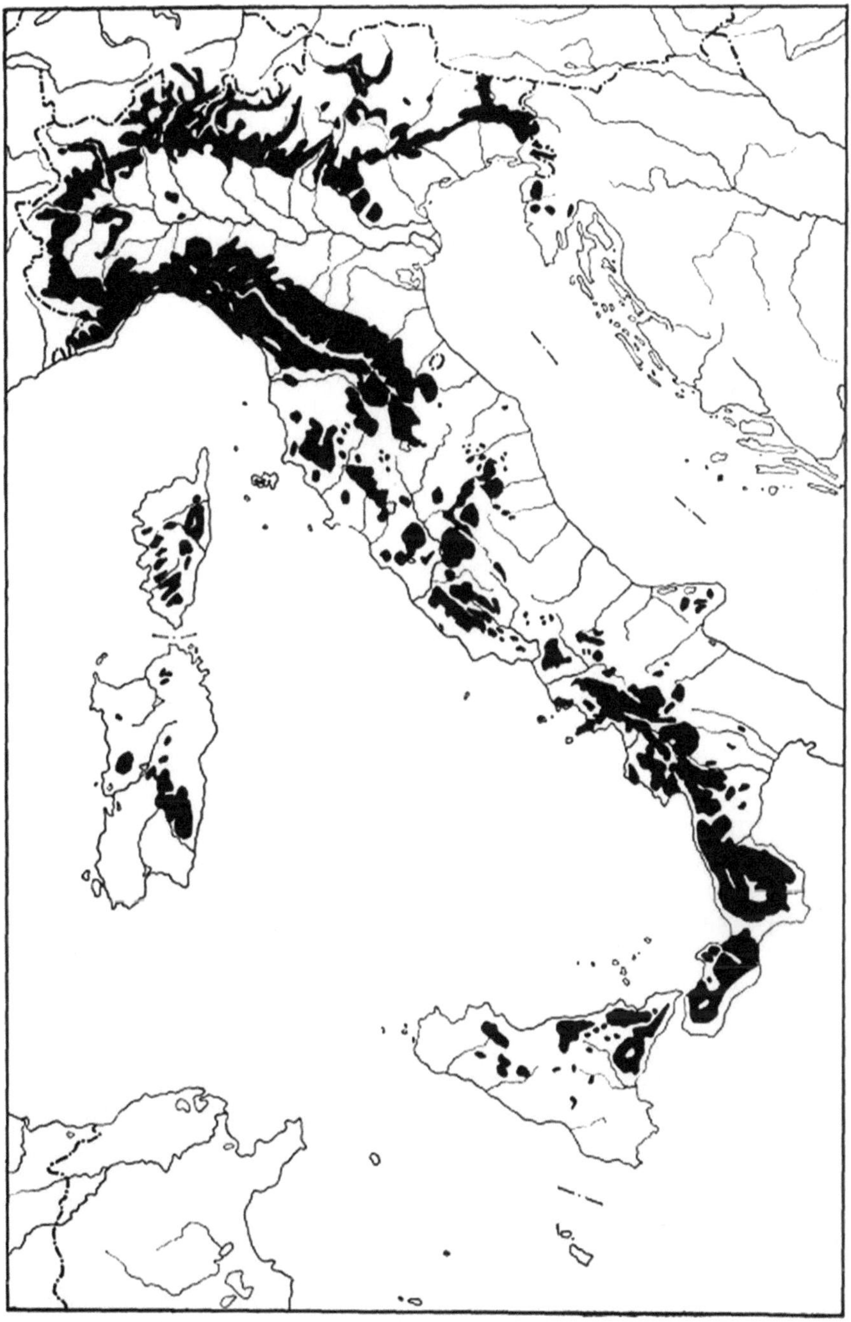

Fig. 9.1 Distribution area of the chestnut in Italy. The species is present from the north to the south of the peninsula and from sea level up to 1500 m altitude on Etna. This distribution is entirely due to human activity (Pitte 1986)

9.2 From the Roman Period to the Middle Ages[1]

In the agricultural manual De re rustica by Marcus Terentius Varro (first century BC), the chestnut is identified with the term castanea. In the Bucolics (first century BC), the poet Publius Virgil Maro, under the guise of the shepherd Titiro, describes the Padanian landscape as rich in tender chestnuts, as well as cheese and ripe apples (I eclogue), and in the II eclogue he states: "I myself will gather the whitish apples, the chestnuts and the walnuts, which Amarilli loved so much". Chestnuts, for sale in the fruit markets of the Via Sacra in Rome, along with grapes are rustic gifts for the beloved, according to the gallant precepts addressed by Ovid, in his Ars amatoria (first century BC), to young lovers. In the Naturalis Historia (first century AD) Pliny the Elder attributes the origin of chestnut cultivation to Asia Minor and confirms the spread of the species in Italy. He considers chestnuts similar to acorns and wonders why nature has hidden with so much zeal in a "dome bristling with thorns" a fruit of "little value", especially if it is "Sardianos balanos", that is wild chestnuts, unlike the acorns of Zeus, object of cultivation. Describing the practice of grafting, Pliny describes the valuable varieties, cultivated especially near Taranto and Naples, but also dwells, like Martial later, on the methods of use: the large fruits are tasty if roasted slowly, the small ones should be boiled; dried and ground, they can replace common bread, especially during the period of abstinence from cereals in the rites for the worship of Ceres. In the treatise De re rustica (first century AD), the agronomist Lucius Junius Moderatus Columella provides detailed teachings on the sowing and cultivation of this tree which, in addition to offering precious fruits, can be considered, in agreement with Pliny, the supplier of the best wood for stakes to be used in viticulture, as we described in the second chapter. With the conquest of new territories by Rome, the culture extends well beyond the peninsula, with a very rapid expansion, involving all Central and Southern Europe especially Portugal, Spain, France and Switzerland—but also Southern England (Fig. 9.2).

At the beginning of the ninth century, in the *Capituiare de villis vel de curtis Imperatoris,* compiled by Charlemagne to revitalise the crops of his vast domains in different European countries, it is recommended that the broli, or gardens, be enclosed by hedges and walls and contain, in addition to vegetables, flowers and aromatic plants, fruit trees such as chestnuts as well as apples, pears, plums, cherries, loquats, figs, walnuts and laurels. The same emperor has the habit of consuming chestnuts cooked in wine (Roche 1935). During the Middle Ages, starting from the thirteenth century, the chestnut tree seems to be the protagonist of the woodland landscape of many regions. Documents relating to census yields, statutes, charters of Carthusian monasteries and monasteries attest to the process of agricultural valorisation and the advance of the chestnut tree especially in Piedmont, Liguria, Tuscany, Campania and Calabria and, to a lesser extent, in Lombardy, Veneto, Emilia-Romagna, Umbria and Lazio, which took place between the 10th and the beginning of the fourteenth century (Comba 1983; Pitte 1986).

---

[1] For the Sects. 9.2 to 9.4 I have to thank Prof. Giancarlo Bounous for allowing me to use most of the text included in his work of 2002.

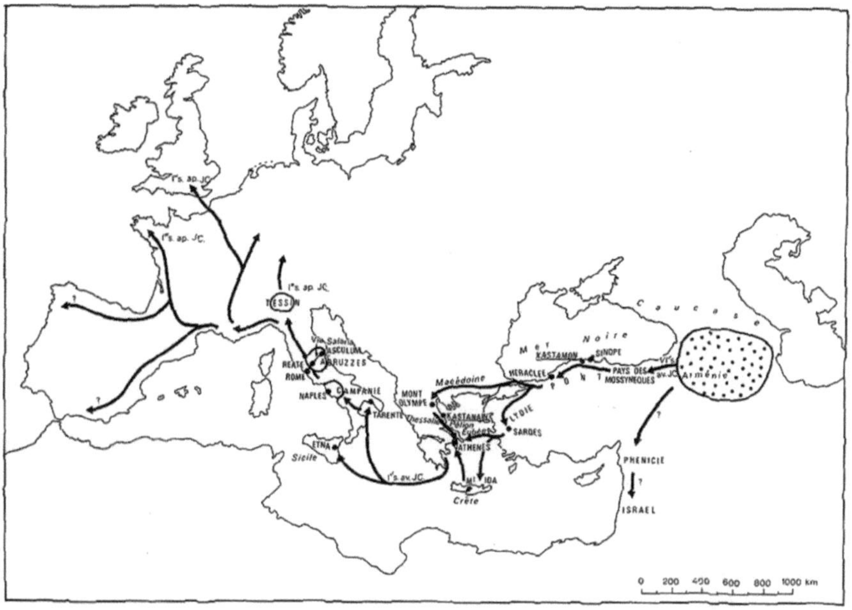

Fig. 9.2 The area of origin and the spread of the chestnut in Europe (Pitte, Terres de Castanide, cit.)

The demographic increase already described for the period following the year 1000 increases the demand for new spaces to be ploughed, cleared and cultivated and the woodland areas undergo a strong contraction, but the chestnut grove, being able to produce flour and fruits, represents a species that is still worth cultivating especially in mountain environments. It therefore becomes a concern of every community, which lives in a regime mostly autarkic, to regulate, through laws and statutes, also the management of woods and chestnut groves in particular. The medieval statutory texts contain severe rules that protect chestnut groves by punishing damage, regulate the traditional rights of use exercised by virtue of centuries-old customs and regulate civic uses. In the chestnut groves, access to animals is prohibited, especially at the time of fruit fall, and the prohibition applies to all animals. Damages to the woods, whether they are common or private, are punished. Some communities that boast prestigious woods for the large dimensions of the trees, to safeguard them from the dangers of ploughing or collective exploitation, bind them with particularly restrictive rules. In general, the penalties provided for damage to chestnut trees far exceed those for other fruit trees (Barelli and Gay di Quarti 1966; Naso 1987) and are doubled if the perpetrators are foreigners (Poli 1999). These latter are explicitly forbidden to procure wood while the local inhabitants, who by custom used it for heating and construction (residential, sacred, defensive, civil), are prevented from excessive withdrawals.

The cuts are regulated and also coppicing is organised by parcels so that the cutting of each one takes place at intervals of 5–7 years. In some cases, there is an obligation to graft every year, a certain number of chestnut trees are planted on both private and common lands. The forest rapidly changes its characteristics, the composition of other species decreases, while fruit chestnut groves and crops advance. The expansion rates of the species are not the same everywhere, but the forest landscape records an expansion of the chestnut tree mostly at the expense of the oak grove but also of other species such as alder, beech, elder, ash, hornbeam, birch, poplar (Nisbet 2000; Comba 2000). This is also testified by Bishop Ottone of Freising (Bavaria), who, visiting Italy in 1154, reports in "Gesta Frederici seu rectius Cronica-", "The country is so fertile that you can find forests of fruit trees, mainly chestnuts, figs and olives" (Pitte 1986).

If the chestnut grove is spontaneous it is "boscus silvaticus", if it is the object of care it is "boscus allevatus" or "domestic chestnut grove" or "alevamina castanearum or silvae castaneae". The constant threat of fires, particularly feared by medieval man who has only his own strength to contain it, imposes exceptional legislative measures such as the obligation to rush to extinguish fires, the prohibition of the practice of debt near a forest and the prohibition or limitation of the construction of charcoal kilns. Heavy fines ("solvat prò hanno solidos centum… vel plus") are charged to those who, by deceit or not, "conbuxerit nemora", or even the cutting off of the hand ("si sol vere non poterit amittat manum"). Strict statutory rules prohibit any illicit or fraudulent collection. Cases of permitted withdrawal are very small quantities "prò una cena vel plandeo" or a "faudata (Codex Demontis, 1305–1509) and gleaning. In all communities, free collection is publicly granted by municipal authorities (Barelli 1966; Botteri 1892; Gabotto 1901; Tamagnone 1969) and no one can access to spontaneous or cultivated chestnut groves, until after the harvest. In the statutes of Mombasiglio (CN) of 1331, for example, once the harvest is completed, access is granted to gleaners for one day, then for two days to pigs to eat rotten or wormy fruits and finally to sheep (Barelli 1936). The chestnut grove becomes an increasingly valuable asset and it is common for wealthy owners to entrust its management to leaseholders. For small owners, on the other hand, it becomes the bread tree that, in addition to providing timber for building, viticulture and heating, is able to meet the food needs of the family and, depending on the progress of the harvests, to sell on the market.

Overabundant productions are marketed in neighbouring countries or in the plain, as shown by the accounts of the tolls, and the "gabella castanearum" affects not only exports but also transports from municipality to municipality (Daviso di Charvensod 1961). In the years 1320–1321, in all the accounts of the Castellarne Sabaude, even in those places where the species is no longer widespread today, the castellans noted whether the product of the chestnut groves was good and abundant or if the fruits were small or rotten due to atmospheric calamities.

Chestnuts became a well-known and appreciated fruit during the Middle Ages. The marrons of Brianza are praised in the manuscript *Pacuinum sanitatis,* illuminated by Giovannino de' Grassi in the last decades of the 1300s, and those from Milan by the Bolognese agronomist Pietro de' Crescenzi in the treatise *Liber ruralium commodorum,* written around 1305, where it is reiterated that the "chestnut is a well-known tree. It produces very large fruits, which the Milanese call marrons". The *Cris de Paris,* edition of Paris 1486, attributes to the chestnuts coming from Lombardy the prestige of being the best available on the markets of that city. The 15th-century Savoyard doctor Pantaleone da Confienza praises the mountain diet mainly consisting of chestnuts, milk and dairy products which—he claims—is able to offer complete nutrition. Bartolomeo Sacchi, known as Platina from the Latin name of his native town Piadena (CR) and arrived in Rome in 1465, considers the chestnuts from Taranto, Naples and Lombardy to be the best (Fig. 9.3).

Pietro de' Crescenzi also focuses on the methods of collecting chestnuts. "Chestnuts are harvested when their maturity causes their husks to fall to the ground, or when they begin to show in the husks, they are shaken with poles". The shaking, a form of collection sometimes also carried out in recent times, was practiced on a large scale during the Middle Ages especially in relation to the preservation system using husk storage. With long poles, the husks, still green and closed, were made to fall, then collected with wooden tongs and gathered in husk storage (ericerio). In fact, the author continues: "and collected with the husks, they are reduced to a heap within a hedge for fear of pigs. And when they have been so enclosed for several days in the husk storage, they open: and these are the best to keep them fresh, as I

Fig. 9.3 A seventeenth-century land register representing Monte Pisano where the only tree species drawn are the olives and the chestnuts, testifying to the value of these two crops (Agnoletti [ed], *The Tuscan agro-forestry landscape,* cit.)

have said, or green: because they can be kept green until March". The piles of still closed husks are covered with a layer of well-pressed leaves and soil, and watered, at regular intervals, to promote the fermentation process of the fruits, which can be kept healthy and fresh until spring. As for the husk storage, it is mentioned in many medieval contracts: in those of the Carthusians in Val Susa (TO), the leaseholders had to bring the product due to the monks to their husk storages in the chestnut groves cared for by the lay brothers (Patria 2000).

Michelangelo Tanaglia in his *De agricultura* also proposes the shaking and the constitution of the husk storage but only with the purpose of promoting the opening of the husks; the fruits should indeed be freed from the husks after about two weeks and stored in sand to be preserved until May (De Angelis 1975). While the shaking is carried out by men, the collection of chestnuts is entrusted mainly to women and children. This operation is followed by the division of the product by size: the first choice goes on the markets or constitutes the "beautiful and acceptable chestnuts" of the lordly censuses, while the smaller pieces are intended for domestic consumption or drying. Regarding this preservation system, the medieval Bolognese agronomist always adds: "But those, which ripened by the heat, fall from the tree, can hardly be kept up to fifteen days. The green ones, placed in the smoke so that they dry, can be kept for a long time, and it is said, that they are tastier than the others. Keep me green in the sand, as said, or placed in the smoke on grates for almost two months, so that when they are dry, they are cleaned and can be kept for a long time". As you can see, the author hints at drying but does not mention the "curing", a method that dates back to the 1700s.

In his treatise, de' Crescenzi invites the use of chestnut wood for the staking of vineyards because of its particular quality, as already indicated by Latin authors. The statement is supported by the case study: although since ancient times the vineyard was preferably planted near woods with chestnuts, in the Middle Ages the phenomenon expanded, as evidenced in the documents of abbeys and monasteries (Gullino 2000). The vineyard, in fact, especially if conducted on a trellis, required a considerable amount of poles, generously offered by the chestnuts present in spontaneous woods or coppices, in a period of about 8–10 years, equal to that of deterioration due to wear. Chestnut wood, still sought after for its qualities, was in fact praised especially for the regularity and compactness of the trunk as well as for resistance to rot, due to the tannin it contains. In Tuscany, the accounting of sharecropping companies recorded separately the chestnut wood, in particular the assortments for the staking of vineyards.

The chestnut can be *virida,* that is fresh, not ripe, and perhaps not even freed from the husk; *munda,* if devoid of husk and sifted; *sicca,* if dried and peeled; or *pista* if, after drying and peeling, it is ground and crushed for the preparation of soups or porridge (Nada Patrone 1981). White or dry chestnuts could be boiled in water or wine and on the tables of the rich were always served accompanied by good wine, as reported by Giovannino de' Grassi at the end of the 1300s. To make dry chestnuts more digestible it was suggested to cook them in two stages and not to use in the second cooking, which had to be particularly slow, the boiling water. To enhance their flavour it was recommended to combine them with legumes or dried fruit such as walnuts, almonds, raisins, pine nuts or plums: chestnuts, from poor people's food, turn into a delicacy "if cooked with prunes" (Guainerio 1481). Chestnuts can be roasted in a perforated pan with cooking over a high flame, or under the embers, over a slow fire. The Milanese writer Bonvesin de la Riva, in his panegyric of 1288 *De magnalibus urbis Mediolani,* argues that this second method of cooking improves the taste and nutritional quality of chestnuts. This opinion is confirmed also by de' Crescenzi and Ugo Benzi, a doctor at the Casa Estense in the fourteenth century.

The use of chestnut flour is affirmed especially towards the end of the 1400s, a period of wars and frequent moments of crisis, and spreads quite rapidly among mountain populations, especially in periods of shortage of cereals. It should also be borne in mind that, like drying, the coarse grinding of chestnuts it could be done domestically, avoiding the payment of taxes on milling (Nada Patrone 2000). On the poorer tables, flour is used to bake bread and prepare thick polenta, chestnut cakes, pancakes, nutritious soups or patties. The good nutritional value of the chestnut, already affirmed by Galen and Avicenna, is emphasised by de' Crescenzi who indicates it as "more nutritious than all the grains as it is close to the grains of bread". Due to its composition, similar to that of wheat, the chestnut is defined as "tree bread" (Braudel 1977). Despite its good nutritional value and the wide diffusion that makes it an increasingly important, if not prevalent, food for increasingly large sections of mountain dwellers, the fruit remains the food of the poor and the name often takes on a derogatory value, as we read in a statute of Castel del Piano (in the province of Grosseto, Tuscany) from the 1500s: "chestnuts are the bread of the poor people, and they have no other support. The men are so poor that they do not make riots of other grains and fodder than chestnuts and in those they are occupied from September until all December, to be able to live and sustain themselves all year round". This is reiterated by the expressions "worth less than a chestnut" used by authors of the 1300 s such as Giordano da Rivalto and Antonio Pucci or the verses of the poet Pietro de' Faitinelli who, in exile, would prefer the "bitter chestnut cake" at home to the "bread of tender wheat" outside his Lucca city.

The chestnut is not exclusive food of the mountain dwellers but also of the peasants of the plain or the hill and, as a seasonal fruit or as transformed, it is also found on the markets of the cities or on the tables of the rich. On the markets of Naples, Messina, Venice, Genoa, Rome, Siena, Cuneo, Milan, fresh or dry chestnuts are sold. Already from the 1300s those produced in the various Italian regions, object of export, are found in the markets of Paris, Tunis, Constantinople, Acre in Syria, Alexandria of Egypt. Chestnuts appear on the noble tables, in fact Michele Savonarola, a doctor Paduan at the service of Niccolò d'Este, in a bookl, remembers the bread of chestnuts. Also Platina, in a treatise of the fifteenth century, addressed mainly to the Roman world that gravitates around the Vatican, dedicates space to chestnuts and suggests to propose them as the conclusion of the meal after having reduced them in the past and stirred with a sugary syrup.

For the wealthier classes, the chestnut integrates and makes a diet often hypercaloric. An example is the recipe for chestnut that we find in the "Libro de arte coquinaria" of the master Martino da Como, personal cook of the patriarch of Aquileiai from the mid-1400s. To particularly valuable ingredients (saffron, spices, rose water), he combines abundant quantities of cheeses, meat and eggs, with the result of an elaborate and particularly nutritious dish. In the fifteenth century, chestnut groves are now widespread everywhere in the hills and mountains, and chestnuts, the basic diet of a large part of the population, effectively become a commodity of exchange and payment like wheat. If the productions are abundant and exceed the needs of self-consumption, they are dried and sold. On the markets of Cuneo, in March 1428, dried chestnuts from the possessions of the House of Savoy (Razè 1976) are found. In most cases, however, the fruit, despite having a high content of carbohydrates, vitamins, mineral salts and a high caloric and protective value of the organism, fetches increasingly lower prices compared to other floury and nuts (1/2 compared to oats, 1/3 compared to rye, 2/3 compared to walnuts) (Nada Patrone 1981).

In Liguria, the species has been widespread from the tenth-twelfth century even in the lower areas classified today as Lauretum. Towards the end of the Middle Ages, the range begins to shrink to the shady and humid slopes, while near the coasts the mixed cultivation of vine and olive progresses. The consumption of chestnuts remains mostly limited to the populations of the hinterland. The wood, in addition to being used in construction and for the manufacture of tools and furniture, is used in large quantities as fuel, for example in the foundries of the bell factories of Sarzana in 1000–1100 and in the Genoese glassworks of the 1300s. The species begins to spread in Tuscany around the eleventh century and extends from Lunigiana and Alta Versilia to Garfagnana, from the Pistoia and Prato mountains to Mugello, from Casentino to Amiata.

Within each community, agricultural and forestry activities are regulated and, as for other regions, the numerous statutory texts confirm the importance of the species in the various provinces. In Garfagnana, for example, the first general law on the chestnut tree from the end of the fifteenth century is of particular interest. It prohibited the cutting of chestnut woods larger than 2000 m^2 without explicit permission which, if granted, was kept in the archives of the municipality of Lucca. Where cuts were made, grafting from 50 to 100 chestnut feet per coltra (about 4000 m^2) and cleaning of the land were imposed, with the privilege of being able to collect the fruits of the grafted plants for 8 years, even if the forest was communal. In case of fire, it was mandatory for all men from 18 to 60 years old to help extinguish it, under penalty of 20 shields. The concession to the Pisans to extract from Lucca 800 chestnut poles (chestnut poles) seasoned for works related to a bridge over the Era river dates back to 1420. The production in Casentino valley, in addition to chestnuts, of raggiolane chestnuts, coming from grafted plants, is documented from the second half of the fifteenth century. As elsewhere, the work of promoting the monks has had a significant weight in the spread of the culture in the region. Tuscan agricultural techniques and chestnut cultivation are, with great probability, exported to Corsica during the dominion of the Republic of Pisa.

In Campania regione, the chestnut tree, established since ancient times as a fruit species, is widespread during the Middle Ages and, in some areas, is already cultivated in the eleventh century, in association with the vine, in a similar way to the hazelnut (Lizier 1907). There are numerous agricultural contracts dating back to the Lombard and Norman dominations that mention it. From the documents of the Archive of the Abbey of Cava or of the Holy Trinity of Cava de 'Tirreni (SA), built by the Cluniac monk Alferio in 1011, the commitment and strength of the monks in propagating the culture and improving it with the spread of grafting are evident. In the possessions of the Abbey, in addition to the chestnut groves, those wild ones, several varieties were produced: zenzale, robiole, granacee (Vitolo 1987). The region, which offers environments particularly suited to cultivation, already records quality productions in the Middle Ages and, among the best varieties, those of Serino and Montella excel (Marciarli 1952). In the twelfth century, the cultivars Ensetazze and Palumbelle are reported for the Avellinese (Vitolo 1987). In the 1300s, the chestnuts of Naples are known and marketed even on the markets of Tunis (Cherubini 1984). The production of timber is also particularly important: the shipowners of the Republic of Amalfi, who export chestnut wood everywhere, become, like the medieval monks, promoters of cultivation.

Since the early Middle Ages, the chestnuts of Lombardy have been renowned, and the records, dating back to the eleventh century, of the monastery of Reichenau (Germany), with possessions in Gravedona, Tremezzo, Limonta and Lecco, mention "12 bushels of chestnuts as a contribution from the Lombard possessions". In a deed of 10 May 1258, drawn up in Como, the chestnuts of the territory of Vercana appear as tithes. The statutes of Gravedona of 1403 regulate the taxes related to chestnuts and dried or green marrons. In the region, the fruits were consumed both by wealthy families, who had them through land income, and by the poorer ones, who cultivated them directly. The difference in the two consumptions was mainly quantitative: in the diets of the poor it was almost the only item (Comalini Fumagalli 1997). In Veneto, "Castagnedum" is mentioned in a document from 1192 and rental contracts from the Montello area mention the species as early as 1291. However, there is little mention of chestnuts in the inventories and accounts of monasteries and abbeys. The oldest statutes relating to the Marca Trevigiana, which concern the collection and food exploitation of the chestnut, date back to the fourteenth century, perhaps induced by the famine of 1347 and the plague of 1348 (Cagnin 1997).

9.3 Chestnut Groves in the Modern Age

As the centuries passed, the species, now known throughout Europe, took on a significant role in Italian agricultural history. Chestnuts became a constant item in the statistics carried out by various states, especially in times when they were threatened by wars and famines, when the need to know the mouths to feed and the consistency of available agricultural goods loomed. The deliveries, mandatory for all subjects, "of grains and the main foodstuffs" as well as the "number of mouths" (people over the age of 2–3 years), were noted by each municipality in a special notebook. In Piedmont, in the "General Summary of mouths and victuals delivered by order of H.A. of 4 November 1607", the chestnuts "all reduced to white" are 36,894 sacks out of a production of 1,096,477 total sacks of "ferment, barbariato [mixture of wheat and rye], jarina, segla, legumes, millet, rice and oats", for a population of 622,887 people. In the "Summary of the number of people, grains, and victuals extracted by me, the undersigned secretary, from the reports of the deliveries made from n. 576 lands and places of the State of Her Majesty from this side of the mountains by virtue of the order of 28 February 1612", resulting from the deliveries made only on part of the territory of the Savoy state, "the white chestnuts" are 20,402 sacks, those grogliate (with shell) are 976, on a production of 533,675 total sacks of "ferment, barbariato, sigla [rye], flour, legumes, millet, oats, rice, barley, milica [very likely sorghum]", for a population of 650,875 people (Gribaudi 1955). These summaries, although succinct and imprecise, offer an approximate idea of the type of agricultural production and the predominant components in the diet of the population at that time. Interesting information on the Piedmontese production of chestnuts related to that period, are Offered by Monsignor Francesco Agostino Della Chiesa (1593–1662), Bishop of Saluzzo and historian in the service of Duke Carlo Emanuele I, in his *Report on the Current State of Piedmont*—"The country... being almost entirely made up of hills and mountains, is not very abundant in wheat..., but where this is lacking, nature compensates by producing in the mountains and valleys... large quantities of all other types of grains... but above all, a very large quantity of chestnuts".

Equally interesting, but debatable, are the observations of Lodovico Bertaldi, physician to Carlo Emanuele I, in *Rules of Health and Nature of Foods by Ugo Benzo Senese enriched with annotations and copious natural and moral discourses by Lodovico Bertaldi, physician to the Most Serene Highnesses of Savoy and newly added in this second impression to the same subjects the treatises of Baldassar Pisanelli and his Natural Histories and annotations of the physician Gaina* (Turin 1620). Referring to chestnuts, he states: "Our Alpine people are very prolific because many of them feed on these, almost all year round, hence their women are prolific". Among the best chestnuts, he says, "are the large ones called by the French *marrons*, and of these, biscuits are made, which are preserved all year round, dried first in smoke... The wild and small ones called by us Piedmontese pognenghe or prickly ones are criticised: the most praised are the Astigiane ones although there are few of them, and similarly those from the Ligurian Alps and, among those that come from the western Alps, these from Envie, a place not far from Saluzzo, and Bargie". Bertaldi also advises adding wild fennel seeds to make fresh chestnuts cooked in water with their skin more digestible and tasty, which are considered astringent but "full of wind and harmful". Another way to cook them is "under the embers, or on the iron grill, and then recooked with butter, salt and pepper, and lemon juice" to serve them "at the second table", that is, as a second course. The Savoyard doctor also offers a detailed overview of the distribution area of the crop: "In the States of Savoy many, for most of the year, live on nothing but chestnuts, and they are robust and healthy, as are those who inhabit the Ligurian Alps: such as the Marquisate of Ceva, in the district of Mondovì towards Frabosa, valleys of Oneglia, Maro, Tenda, Cunio, and towards the Alps Code, the Marquisate of Busca, that of Saluzzo; valleys of Lucerna, San Martino, Perosa, Pinerolo, Giaveno, valleys of Susa; Marquisate of Lanzo, Valperga, valleys of Pont, Brosso, Aosta, Ivrea, Biella up to the valleys of

Sesia, in the Duchy of Milan, which valleys are very abundant in chestnuts, dairy products and other things necessary for human living".

The General Statistics of 1752 for Piedmont reports the quantities of products from each community, derived from the average of a three-year or five-year period, and the prices practiced in individual places (Prato 1908). According to these Statistics, the chestnut tree covers 4% of the total area of Piedmont and is particularly important for the provinces of Mondovì (36,800 ha) and Cuneo (11,500 ha) which, together, represent 54% of the Piedmontese chestnut-growing area (about 90,000 ha). In the province of Cuneo, the municipalities of Chiusa Pesio, Peveragno, Robilante, Boves, Bemezzo, Roccavione, Demonte and Roccabruna account for 74.3% of the Cuneo chestnut-growing area. For some in these municipalities, chestnut cultivation is a monoculture: in fact, if we exclude the areas occupied by high-altitude woods and pastures, almost all the cultivable areas are dedicated to chestnut trees. Significant production is also recorded in the provinces of Biella and Ivrea. The production of the Savoy State in 1752 was about 740,000 quintals, significantly higher than that reported by the 1607 Statistics, which, although referring to white chestnuts, i.e. dry ones (where it should be remembered that in general to obtain 1 quintal of white chestnuts, 3 quintals of green or fresh chestnuts are needed), was about 32,000 quintals.

In the 1752 Statistics, it is evident that the price of chestnuts is very close to that of oats but always significantly lower than that of wheat or rye. The best prices are for the products of Pinerolo, while those of Cuneo and Acqui have medium–high prices. As for the productivity of the plants, according to the 1752 Statistics, that of Cuneo is high (about 11 quintals/hectare) while it is quite low in the province of Mondovì (5 quintals/hectare), against a Piedmontese average of about 8 quintals/hectare. The data from Cuneo are also confirmed by the Report that the Count of Brandizzo makes of every city and land located in the province of Cuneo that he administered as superintendent in the years 1750–1752 (Dadone 1982), where the production of the municipalities with a more pronounced chestnut tendency, such as Boves and Peveragno, is about 12 quintals/hectare.

Further confirmation of the strong increase in cultivation during the period in question are the numerous dryers built everywhere to transform an autumn fruit into precious food or exchange goods for all months of the year. If it was a small quantity of product, natural drying was used, spreading the chestnuts in the sun on the terraces. For larger quantities, depending on the financial resources of the owners, premises inside the same house or small isolated buildings were used, built near the houses or at the edge of the chestnut grove or in the middle of them, with different types depending on the regions.

The dryer (secou. or scàu in Piedmont, metato in Tuscany) is square, or rectangular in shape, building, with a roof initially made of straw, then of stone slabs or tiles. It consists of two floors and, sometimes, an attached room for tools or a porch where threshing operations can be carried out and firewood can be stored. The building has two openings: the one on the second floor allows the chestnuts to be poured and turned, the one below allows access to the fire. On the lower floor, which functions as a boiler, a fire is lit, which must be contained, fed three times a day with chestnut wood or waste forest products, such as stumps and chestnut husks. On the upper floor there is a single grate, made of wood or metal, and on it the chestnuts are arranged in an initial layer of 20 cm. More can be gradually added, up to 50 cm, taking care to turn them periodically and maintain a constant temperature. The drying time, variable depending on the quantity of fruit to be dried, is of about a month (Bounous 1996, 1999). Thus, alongside channels for the regimentation of water, terraces for the containment of slopes and the various improvement works carried out over the centuries to transform spontaneous woods into fruit chestnut groves, the idea of building dryers on the collection site emerged, reducing the weight of the goods to be transported to the pantries or markets. The hospital of Cuneo had them built in the Castagnetta region in 1516 (Camilla 1970) while as many as 22 are the dryers built for about 200 ha of chestnut grove in the Marro area (municipality of Chiusa Pesio). It should be noted that 15 of these dryers are located in the lower part of the plots to facilitate the transport of chestnuts and wood from the plot to the building (Pitte 1986). The dried chestnuts were shelled through the "beating", generally carried out by men with clubs or within long and narrow bags of heavy canvas. Women, on the other hand, were responsible for the operation of the sieving, carried out by moving large wicker baskets, in such a way as to favour the exit of the husks. With the sorting, the broken and smaller fruits were separated, to be transformed into flour, from the most beautiful ones to be used for oneself or to be marketed.

During the modern age, the chestnut continues to be essential food for mountain populations but begins to be known also outside this environment. Landò Ortensio, in fact, a doctor and writer in the service of princes and prelates in Italy and Europe, in addition to praising the Chiavennaschi chestnuts (province of Sondrio), mentions the sugar chestnuts that are found on the Neapolitan markets. Costanzo Felici, a doctor and naturalist from Marche, reaffirms this at the end of the sixteenth century, in his treatise "Dell'insalata e piante che in qualunque modo vengono per cibo all'homo," when he says that "it is then very common on tables and everywhere" and describes its most frequent uses: "it is then eaten dry by itself and in soups making good company to other dishes, like among others the chickpea, and of these dry ones where there is plenty and that scarcity of wheat, reduced to flour, very good bread is made". Also the Piedmontese doctor of the sixteenth century Francesco Gallina recognises that "chestnuts hold the principality among all forms of acorns and these alone among all the saved fruits give the body notable nourishment: but they are not praised in daily foods" (Nada Patrone 1991). Among the most traditional and simple uses, that of the necci is attested in Pistoia in 1591: "women… with certain irons named ronchetti make chestnut leaves to make the castagnacci that are cooked between two chestnut leaves" (Mantovani 1979).

The Brescian Agostino Gallo, in the work "Le dieci giornate della vera agricoltura e piaceri della villa," published in Venice in 1565, enlightens us on the preservation of marrons: "collected moderately mature for old moon, well dried, they are placed in a cool place or in the sand or in some pot also cool, which is so closed that in no way can anything be extracted from the air because in a short time they would be corrupted by it". To make them tastier, they can then be preserved with rose leaves, as also suggested by the Renaissance Modenese writer Giacomo Castelvetro who also advises: "having boiled them a little in excellent sweet white wine, they are drained and put to dry in the smoke; and so prepared they are exceedingly good, and are called biscotti, and they are preserved all year round". The writer's sympathies for the Protestant Reformation forced him to take refuge in London to escape ecclesiastical persecutions and Right in that city, he published in 1614 his treatise: *Brief account of all roots, all herbs and all fruits that raw or cooked in Italy are eaten. With many enjoyable secrets (not without purpose written within it) both regarding the health of human bodies and useful to necessary good farmers.* He directs his advice to the English aristocracy and recalls that "chestnuts can be eaten roasted with salt and pepper and instead of the sugar that is used here, we (Italians) use orange juice and some, in fact most... drink a little afterwards, everyone goes to sleep... If they are then cooked in water alone, these are called boiled". He also recalls the use of chestnuts to stuff "capon, geese and turkeys that want to roast, with dried plums, raisins and grated bread".

As these references show, the ways of using chestnuts increase in number and also extend to the bourgeois or upper classes, increasingly taking on the role of fruit or dessert, but not only. Chestnuts are now a fixed item in gastronomic treatises and their processing sometimes becomes elaborate. In a *book containing the way of cooking and various secrets and remedies for diseases and other,* composed by an anonymous Reggiano for the Counts Cassoli of Reggio Emilia in the second half of the eighteenth century, appears for the first time the recipe for candied chestnuts, while in *The Piedmontese confectioner trained in Naples who teaches the way to candy fruits in different ways. Make biscuits, marzipans, canestrelli, spirits, sorbets and many other things belonging to such art,* published in Turin in 1790, there are, in addition to the recipe for marron glacé, those for chestnut compote, candied chestnuts, chestnut biscuits, chestnut paste, caramel chestnuts, harlequin chestnuts, iced fruit chestnuts (Nada Patrone 2000). The chestnut has now assumed its rank in the kitchen and its use in quantity and mode is linked to tradition and the class of the consumer.

Over the centuries, in times of greater demographic pressure, man has faced the food problem by trying to extend, where possible, the cultivable areas to cereals, reducing the uncultivated, reclaiming the marshy areas, intensifying the exploitation of cultivated lands and, where cereal cultivation was not possible, spreading the fruit chestnut tree. Statistical data from the mid eighteenth century highlight a significant difference in population density between chestnut and non-chestnut areas: while the density for the most chestnut-rich Piedmontese provinces (Cuneo, Mondovì, Ivrea) is 94 inhabitants/km^2, a figure very close to that reported for other chestnut realities, the density varies from 10 to 40 inhabitants/km^2 in municipalities where the chestnut is present but not cultivated.

In Liguria region, according to a survey of the Republic of Genoa in 1531, the vast majority of Ligurian communities surveyed possessed cereals in insufficient quantities (1/3) compared to their own needs and made up for the missing part with chestnuts. Between the sixteenth and seventeenth centuries there is a crescendo of new settlements of populations coming from nearby valleys or from the coast, who derive from chestnut cultivation the means of their own subsistence. They plant new chestnuts or graft existing ones and benefit from concessions from the Republic or local feudatories. The spread of the culture is made evident by the multiplication of dryers: in the woods of Ovada (then Ligurian) between 1585 and 1650, 8 of them arise, dedicated exclusively to the operation. Around them also arise, in large numbers, new houses (cottages) consisting on the ground floor of the stable, and on the first floor of a central kitchen with a large fireplace that vents to the upper floor where chestnuts are arranged to dry in autumn (Moreno and De Maestri 1975). Ligurian chestnut cultivation thrives until the nineteenth century and, according to data from the Napoleonic survey of 1809, it covers a good 50% of the Apennine territory (Quaini 1973).

In Tuscany, the production of chestnut flour is considerable in relation to that of cereals. In the Report of Commissioner Gio. Battista Ledaldi on the city and captaincy of Pistoia in the year 1569, it is reported an average annual harvest of 350,000 staia of wheat and 180,000 staia of fodder, against 100,000 staia of dry chestnuts (therefore at least 300,000 staia of fresh chestnuts) without taking into account "those that are eaten fresh, which are also in a very good quantity". The same commissioner noted in 1553 that the Public Palace of Cutigliano (in the province of Pistoia) had the ceiling of the large hall made of chestnut wood and that the inhabitants of Cutigliano are "very poor, and seven eighths of them all year round eat nothing but chestnut cakes and drink water". Also, the prince of Massa and Carrara, referring to Alta Versilia, noted in 1577 that "a large part of his subjects fed on chestnuts most of the year". In 1676 in the territory of the amiatini municipalities of Arcidosso, Abbadia San Salvatore, Castel del Piano and Piancastagnaio, 3400 moggia of chestnut flour were obtained compared to 1370 moggia of cereals and beans. In Castel del Piano, in particular, the ratio was eleven to one.

At the legislative level, there is a continued interest in the chestnut, although sometimes contradictory: during the Medici period, for example, perhaps in an attempt to curb illegal cuts, Cosimo I (1565) allows the cutting of up to ten chestnuts without authorisation, while during the grand duchy of the Lorraine of Pietro Leopoldo (1765–1790) there is news of incentives for the planting of new chestnuts. The data from the 1762 census confirm the importance that the culture has in this period. The productions are copious and large quantities of fresh, dry and "green marroni" and "dry" chestnuts are exported to Romagna, then under the direct dominion of the Papal State (Bellina 1998).

Widespread use continues to be made of the wood of this cupuliferous tree as fuel to melt iron in the various foundries present in the Tuscan region. In Veneto, many of the chestnut woods of the low mountain and the hill disappear due to the continuous grazing, the intense production of charcoal and the use of large quantities of timber in the construction of the pile foundations of the houses of Venice. The agricultural law that from 1756 imposes the obligation to plant, for each "suitable field", four olive trees or, failing that, four chestnuts or other fruit trees, is an attempt by the government of Venice to face the persistent threat that hangs over the chestnut cultivation of the region, already contained in surface and production.

The Neapolitan political writer Giuseppe Maria Galanti in his Geographical and Political Description of the Sicilies, referring to the Campania of the late eighteenth century and more specifically to the provinces of Avellino and Benevento (Principato Ulteriore), states: "The products of which it abounds are chestnuts, walnuts and hazelnuts", and adds for the Cilento that "the mountains are covered with large trees, among which chestnuts abound which are of great profit to the Inhabitants for the wood that is consumed in the Kingdom and beyond, and for the fruit of which there is a great deal of trade, especially in the capital… In the mountainous regions, pigs are raised. Chestnuts and acorns, of which the region is abundant, make the industry of these animals easy. As already seen in the Middle Ages, on the Amalfi coast, the use of chestnut wood continues in the construction of boats as well as barrels, exported to Genoa and Sicily, for the preservation of wine, oil, salted fish" (Cherubini 1984) (Fig. 9.4).

In Molise, according to Galanti, "there are chestnuts, but perhaps due to a lack of cultivation they are wild, and their small and tasteless fruit serves as bait for pigs". Not so for Abruzzo, where especially in the province of L'Aquila, chestnuts are "many" and the fruits of the Tagliacozzo area have an important place in the diet. For Calabria, Galanti speaks of "excellent chestnuts" that occupy the slopes of the mountains of the provinces of Cosenza, Catanzaro and Reggio Calabria while already in the 1500s the productions of chestnuts from the territory to the north and around Cosenza and the Aspromonte area up to Catanzaro (Galasso 1975) were mentioned. Sicily, which had suffered a strong reduction in the forest cover in the Middle Ages, maintains on the north-eastern reliefs of the Madonie, the Peloritani and on the slopes of Etna large forests mainly made up of conifers, oaks and chestnuts. The chestnuts of Paternò are renowned, and also known outside the island is the processing of chestnut wood by the craftsmen of the Val Demone and the coopers of Messina.

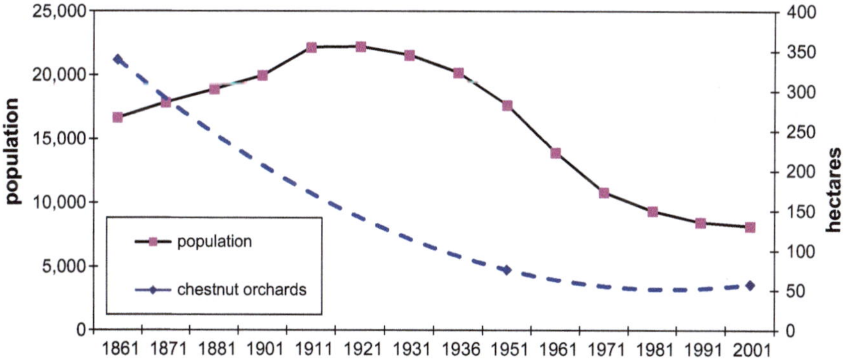

Fig. 9.4 In the graph with the dashed line, the trend of the population of Firenzuola (FI) in the northern Apennines, and with the other line, the historical dynamics of the area cultivated with chestnut trees. As can be seen, the decrease in chestnut trees can be considered an indicator of depopulation

9.4 The Contemporary Age

During the repeated and rapid political, social and economic transformations that characterise the twentieth century, Italian chestnut cultivation evolves in a troubled and, in some respects, contradictory way. On one hand, with its diversified productions, it continues to maintain, at least for the first half of the 1900s, the strategic role for the survival of a large section of the Italian mountain population. The productions remain important, especially compared to other fruit sectors. Italian chestnuts, in addition to satisfying a significant national consumption, are the subject of a lively trade both on European markets and overseas: quantities equal to about 1/10 of the total production are exported, especially between November and December, to the USA (24%), France (23%), Switzerland (15%), Germany (7%), Hungary (6%), Austria (5%) and other countries (Fenaroli 1945; Vigiani 1943). The average price achieved in the years 1921–1922 is about 1200 lire/t, close to that of wheat (1250 lire/t) (Bignami and Salsotto 1983). To this must be added that from the commercialisation of the timber, in 1922, about half of the gross value of the production of the fruits is realised.

On the other hand, the first signs of a crisis, already perceived in some cases during the 1800s, that afflicts, in different times and ways, the chestnut at a European level, are accentuated. In line with other authors, Vigiani, in 1908, lamented: "Although our country offers natural conditions as favourable as any for the cultivation of the chestnut, for several years the extent and production of chestnut groves has been steadily decreasing" (Vigiani 1908). According to a statistic made by G.C. Siemoni in 1872, the surface area of chestnut groves in Italy was 638,042 ha; today it amounts to about 508,140 ha. Despite conflicting data, drastic reductions in chestnut-growing areas are confirmed throughout Italy: in Piedmont from 1815 to 1937 millions of trees are felled and it is reported that already "between 1823 and 1832, 43,043 trees are felled in 8 municipalities of the Saluzzo prefecture…". The situation is similar in Tuscany where, between 1834 and 1934, there is a 50% decrease in the chestnut area on the Pistoia Apennines.

The studies of the first half of the 1900s by Piccioli (1922), Vigiani (1908, 1943), Remondino (1926), Fenaroli (1945) and the investigations on Italian chestnut cultivation by Polacco (1938) and Merendi (1928, 1938) all lament the state of crisis of the culture and offer extensive discussions on the pedoclimatic needs of the species, on the most appropriate cultural practices, on the existing varieties and on the defence measures to be adopted especially against ink disease. From 24 to 29 October 1922, the Chamber of Commerce and Industry of Cuneo, a province among the most interested in chestnuts both for tradition and for the extent of the culture, organises the "Chestnut Week", with the aim of focusing on and addressing the main technical, varietal, pathological and commercial problems, which have become pressing. The situation of gradual decay of the culture and reduction of the surfaces lasts approximately for the whole of the 1900s. The decrease in surface area and productive trend is evident, even if, especially for the first half of the 1900s, it is preferable to give indicative meanings to the data rather than actual statistical measurements (Fig. 9.5).

Fig. 9.5 Farmers transporting the leaves of the chestnut trees to the stable using a special wooden frame suited to hold the leaves. Every single part of this tree was used for a specific purpose

There are many factors contributing to the crisis of the culture, but among them the advent of the industrial era and the accentuation of the differences in living conditions between the plain and the mountain, which induces the mountain population to abandon the more difficult crops or to decrease the care and agroforestry practices to the chestnut groves. The harsh winters of the "little ice age" (Flohn 1968), the replacement, especially in the most favourable areas, of the chestnut with shorter cycle and more profitable crops (cereals, potatoes, fodder), the improvement and expansion of the road network that facilitates the transport of all kinds of food and makes the economy of self-sufficiency less and less significant, contribute to the constant decrease in the surface area cultivated with fruit chestnut and to the crisis of the traditional role that the species has played for the chestnut areas. Already in the 1800s there is a gradual abandonment of large chestnut areas and the felling of large quantities of trees to meet the strong demand for timber as well as for construction, for the production of charcoal and for the equipment of the first industries. The demand becomes considerable in the 1900s when the industries extracting tannic acid from chestnut wood increase and increase production. The first establishments, founded in 1854 and 1856 in Cuneo, produced an acid used mostly for dyeing silk black, as the female fashion of the late 1800s required (Bignami and Salsotto 1983). Following technological innovations, numerous industrial complexes arise that produce an excellent tannin sought after by the leather tanning industry: while at the beginning of the 1900s there were 13 industries in Italy, during the following decades their number increases and in 1949 they work about 800,000 t/year of timber, with a production of 55,000 t/year of extract exported to Canada, Australia, South Africa, Japan (Pitte 1986; Bignami and Salsotto 1983).

Phytopathological problems, especially ink disease and cortical cancer, have massively contributed to the decrease in chestnut harvests and the reduction of the area dedicated to chestnut groves. The ink disease, which spread in the second half of the 1800s on European chestnut trees, first in Spain, Portugal and then in other countries including Italy, has been the subject of numerous studies, including those of Petri who, having isolated the pathogenic agent Phytophthora cambivora (Blepharospora cambivora) in 1917, described the pathology, the methods of fight and began to investigate the degree of resistance to the fungus of the Japanese chestnut (Petri 1918, 1924, 1931). The degradation and abandonment of the culture have favoured the spread of cortical cancer, imported from the United States of America in 1938 and spread in Italy especially in the second half of the 1900s (Biraghi 1947, 1950), causing serious damage in all chestnut regions. At the International Chestnut Conference, held in Cuneo from 12 to 14 October 1966 and promoted by the Chamber of Commerce, Industry, Crafts and Agriculture of Cuneo, the threat of the fungi is partly mitigated: the European chestnut has "opposed an unexpected resistance" to the ink disease, and the cortical cancer shows a decreased virulence (Carlone and Paglietta 1967).

Even in 1954, it was estimated that at the Italian level, of the 826,000 ha of chestnut groves, 446,000 were high forests and 293,000 were coppice woods (De Philippis 1958). According to the national forest inventory of 2005, the consistency of the chestnut groves was about 788,406 ha with the minimum extension in Molise (390 ha) and the maximum in Tuscany (156,869 ha); this would place Italy in first place in Europe for the extension of these woods and for the production of chestnuts, but these are estimates that include also the woods where the chestnut has a strong presence derived from the abandonment of the chestnut groves. The analytical data of the previous National Forest Inventory, in 1985, indicated just 90,000 ha the extension of the still cultivated fruit chestnut groves, with a reduction of about 80% in just 30 years, and a very strong reduction of a forest of great value for the Italian landscape. A regional analytical data confirms this situation. According to the Tuscan forest inventory, the total surface of the woods with a prevalence of chestnut, referred to 1991, was 176,928 ha but with only 15,520 ha of managed fruit chestnut groves, 16,816 ha of "abandoned chestnut groves" and, for the remainder, coppices of which a modest percentage managed to become high forest. The surface of the woods in which the chestnut is simply "present" is however 266,096 ha (Hofmann et al 1998).

9.5 The Chestnut Tree in the Apuan Alps: Cardoso

The landscape of the fruit chestnut grove is one of the most fascinating forest landscapes for its ability to combine aesthetic values and cultural aspects. The density of the chestnut grove, generally between 90 and 120 plants per hectare, makes it a very bright forest, while the cleanliness of the undergrowth, almost always a meadow kept low by grazing, makes it clean and pleasant also for recreational purposes. The fruit chestnut grove has often been the subject of land arrangements, so we find it both on terraces earth and on dry stone terraces, but also with the classic "lunettes", small semicircular stone terraces built around individual plants. Such attention and expenditure of labour were justified by the profitability of the chestnut grove, superior to any other forest species. The study of a specific case in the Apuan Alps, in Tuscany, the subject of an investigation carried out in 2002, helps us to better understand the role of the chestnut grove in the mountain landscape and the evolution that has taken place in the last century (Agnoletti 2005).

The area studied is located within the regional park of the Apuan Alps, from the peak of Monte Pania della Croce (1858 m above sea level) to the northeast, passing, in a southerly direction, through Monte Forato (1230 m), to the Fosso di Casalina and then to the Cardoso stream and going up, to the border with the municipality of Retignano, to Poggio del Castello up to the mouth of Mosceta, partially including two protected areas, the Site of Community Interest (SCI) "Monte Corchia—le Panie" and the Special Protection Area (SPA) "Primary and secondary meadows of the Apuan Alps", both belonging to EU NATURA 2000 network of protected areas. The potential vegetation band typical of the area is that of the oak-hornbeam forests, generally between 300 and 1000 m in altitude, characterised in particular by the dominance of the black hornbeam *(Ostrya carpinifolia.)* and the downy oak *(Quercus pubescens),* enriched by an undergrowth composed of heather *(Erica carnea),* by *Amelanchier ovalis* and by *Viburnum lantana* (Ferrarini 1972). However, this vegetation band has been profoundly modified by human activity, which from the Middle Ages onwards planted the fruit chestnut grove, on which the economy of the entire area depended until the 1950s.

Most of the terracing and embankment works on the mountain, both in the chestnut forests and in the arable lands present in this area, are dated back to the 1500s. In the latter, the terraces were not supported by dry stone walls, but by grassy embankments, from which a certain amount of hay could be obtained (Bartelletti and Corfini 1986). The start of the terracing works, typical of this century, as well as the full consolidation of an economy based on the exploitation of natural resources, suggest the existence, already in the 1500s, of a system of temporary dwellings, arable lands and pastures. These seasonal settlements seem to have originated from ancient leasehold conductions on communal lands for collective use for the purpose of improving soil productivity. Before the administrative reorganisation interventions promoted in 1776 by Grand Duke Pietro Leopoldo, in fact, public goods were rented, and with the new conductors, works were started to cultivate lands originally uncultivated and unproductive (Bartelletti and Corfini 1986). By the end of the eighteenth century, the use of the territory for agricultural and pastoral activities had already saturated every possibility of development (Nice 1952). With an act of 1784, in fact, the Grand Duchy of Tuscany favoured the construction of new agricultural residential buildings (Paolicchi et al 1987).

The cartographic processing, using geographic information system (GIS), of the information contained in the Leopoldine cadastre makes it clear how the nineteenth-century landscape is divided into four main elements, which in their interaction determine the structure of the landscape mosaic: the areas for grazing, chestnut woods, bare arable land and vine cultivation. In the richness of the forms of land use proposed by the nineteenth-century survey—a good 67 classes of land use—, "pasture" alone constitutes 36.8% of the entire area, followed by "chestnut grove" with 23.52%, "arable" with 11.07% and "vineyard arable" with 9.75%, representing together more than 80% of the total surface, while other types of forest are little widespread. The morphological structure, together with human activity, is the main discriminating element in regulating the balances between the many pieces of the landscape mosaic, more than six hundred small plots surveyed by the cadastre. The simple structure of the steepest slopes of the reliefs, characterised by the dominance of bare pastures, contrasts with the greater fragmentation of the softer morphologies and lower altitudes, interrupted only by the significant extension of the chestnut groves, mainly distributed in the medium–high band of the slopes (Fig. 9.6).

The organisation of the nineteenth-century settlement and economic model is partly perceptible from an almost regular altimetric distribution of the main elements. At higher altitudes, in addition to the summit pastures, there are wooded pastures, bare arable land and the forage meadows of the mountain pastures followed, at lower altitudes, by chestnut forests and the mixed cultivation of the vine, a factor shaping the agricultural landscape of the valley floor, also characterised by the sporadic presence of hemp fields. However, it must be considered that vine cultivation is also present at higher altitudes (about 650 m above sea level), a clear sign of how the spread of crops was subordinate to deforestation and the modelling of the steepest slopes, carried out through a constant work of terracing the mountain slopes. The multiplicity of crops in the nineteenth-century agricultural fabric emerges above all considering the variety of the tree component not only of the wooded pastures, but also of the mixed crops. Among the wooded arable lands, the cadastral categories report the presence of mulberries and fruit trees, as well as chestnut and walnut. As far as the forest landscape is concerned, it is clear how the spread of the chestnut tree represented the main species of the forest community, distributing itself on the territory in pure formations, and together with the walnut as a valuable tree component in the wooded arable lands. It is therefore clear how the chestnut tree represented a central element of the local agro-silvo-pastoral economy used both as firewood, but above all as a fruit tree for the production not only of chestnuts but also of marrons, although to a much lesser extent, confirming the fundamental role of these formations in the Tuscan landscape.

Fig. 9.6 The chestnut tree of the hundred horses (Sicilia region). According to some written sources of the 18th century the name is due to the large number of horses that could be accommodated under the crown of the tree

9.6 Working in the Chestnut Grove

The most prized type of chestnuts, namely the marrons, were the object of the greatest attention from the farmers, while the company accounts usually reported two separate estimates given the higher commercial value of the latter. In the chestnut grove, about every fifty domestic plants, a wild chestnut tree was interspersed for pollination. The old felled trees were replaced with young ones, born spontaneously and allowed to grow in the clearings. At the right time, the wild chestnuts were grafted with "scions" taken from the most productive plants, and the grafts were performed by specialised personnel, farmers or craftsmen.

Except on holidays, the chestnut grove was usually dedicated to the moments of the farmer's life not linked to agricultural production: tasks such as "pruning", i.e., the cutting of branches for the formation of the crown and the cutting of shoots that damaged production, were carried out, and the old trunks were also cleaned of rot and wood accumulated inside, as chestnuts empty out as they age, thus producing excellent fertiliser that was entirely reused. The plants were also cleaned of dry branches, an operation that was carried out every year during the chestnut harvest. A few weeks before the harvest, the ground of the chestnut grove was cleaned, eliminating ferns, leaves or any other shrub. Everything that could serve as fuel was taken home. In October, when the marrons began to fall, women and children went to the woods in the morning to collect them and ensure that no one else stole them. When the husks opened, the whole family moved to the woods to collect the husks they were beaten from the trees with long poles. The elderly, women and children collected all the chestnuts that had come out of the husks, while those still closed were put in baskets and transported near the paths, then transferred home and the best marrons selected and packed separately, often to be sold to a middleman. The second choice was used for domestic consumption, while the third choice was for the animals. The widespread and pervasive lack of fodder, especially where sheep farming was prevalent, made it necessary to also collect leaves and branches to feed the livestock. Therefore, in the chestnut groves and especially in the autumn period, leaves were collected and taken home. From pollarding, a normal practice in the management of the chestnut grove, young branches were often obtained for the winter feeding of the sheep. In addition to the production of chestnuts, documents of the time confirm a wide use of the chestnut tree also as a construction and work material. Young chestnut trees were used as beams for roofs, the branches, as well as the shoots of the coppiced chestnut trees, for barrel hoops, as well as staves for barrels (Savi 1811). As can be understood from these brief notes, the chestnut grove was a place that involved the whole rural society in its management, and this was conducted at the level of the individual plant (Fig. 9.7).

9.7 The Twentieth Century

As a result of the abandonment of chestnut cultivation that began after the Second World War, the landscape structure of this part of the Apuan Alps is heading towards rapid and profound transformations. Comparing the landscape in the 2000s with that of the nineteenth century, we observe the dominance of a vast forest cover that extends from the valley floor up to the higher mountain slopes, where summit grassland formations remain, mainly distributed near the peak of Pania della Croce. The monotony of the forest landscape of the lower and intermediate bands of the mountain slopes, now covering 74% of the entire study area, is interrupted only by the small inhabited centres and the residual agricultural areas of the terraced cultivations. Pasture areas, including surfaces classified as rocky outcrops, occupy the highest altitudes, covering 20% of the total surface.

The extremely simplified structure of the landscape mosaic corresponds to a diversification of forest formations compared to the original monospecific chestnut grove. The most widespread types of land use within the forest canopy are evidently those characterised by the presence of chestnut (64%), followed by woods dominated by black hornbeam (about 20%) mainly distributed on limestone matrix soils in the north-east part of the study area. Other forest types are mixed woodland (5.4%), distributed almost exclusively in the south-western part of the study area, high scrub (5%), beech forests located at higher altitudes (2.4%), shrublands (0.4%) and riparian

Fig. 9.7 On the left, the village of Cardoso (Apuane alps - Tuscany) at the beginning of the 1900s (archive Mauro Agnoletti); on the right, Cardoso a few years ago (photo Mauro Agnoletti). A secondary mixed forest has now covered the drystone terraces existing on the slope above the village (left). A civilisation has disappeared

formations of valley floor (0.2%). The measure of the simplification of the landscape that occurred in this period can be understood by observing the variations of the landscape mosaic, which goes from a number of patches greater than six hundred in 1832 to only 84 in 2002, with a reduction in the number of land uses from 86 to only 16 in the same period.

Summarising and synthesising the categories of land use based on three general classes defined as forest, pastures and cultivations, for three different dates, 1832, 1981 (date of establishment of the Apuane Regional Park) and 2002, significant transformations become appreciable. The strong increase in forested area corresponded to a drastic contraction of pastures and cultivations; but while the reduction of pasture areas has subsequently slowed down sharply after 1981, cultivated areas and the forest canopy show, in the transition from 1981 to 2002, still evident changes.

However, considering that the last period is characterised by a shorter time span, it is significant to underline how the landscape presents an evolution towards a lesser human impact and a growing importance of natural evolution, the which tends to create a continuous forest cover that reduces the diversity of the pre-existing landscape mosaic. This phenomenon confirms for the Apuane Alps the trend highlighted in most studies done in Tuscany (Agnoletti 2002) and previous studies for more limited periods such as that of Vos and Stortelder (1992), but finds similar results with what has been observed for other European areas (Mander and Jongman 1999), especially in marginal areas where the Traditional economic systems have been abandoned (Hanousková et al., 1999).

As already indicated, the study area is entirely within the Apuan Park, while a portion of its territory is included in two protected natural areas, the Site of Community Interest (SCI) "Monte Corchia—Le Panie" and the Special Protection Area (SPA) "Primary and secondary prairies of the Apuan", established for the protection of birds. It is therefore an area subject to three types of protection linked to the regional park of the Apuan Alps, the SCI and the SPA, to which the specifics of the Tuscan regional law n. 56 of 2000 fully apply, which implements the European "Habitat" directive, identifying the list of natural and semi-natural habitats of regional interest to be protected. Scrolling through this list, it often seems difficult to find a direct relationship with the environments that have characterised the landscape of the study area and, especially in the tree component, elements that are the result of transformation phenomena are much more represented. The prairies, with various plant associations, are present in the list and certainly represent a habitat in danger of extinction, having been reduced by 97% compared to 1832. In the same way, all the wooded pastures—which constituted 27% of the landscape of 1832 and today have decreased by 93%—as well as mixed crops with vines, decreased by 89%, are absent.

It is therefore difficult to justify the absence of the chestnut grove from the habitats to be protected, while instead the forest with a majority of chestnut trees, which is not in any danger and represents a phase of an ecological dynamic subject to further evolution, is protected. This is clearly the result of conceptions linked to a negative vision of the chestnut grove from an environmental point of view, which in the past, namely by the former director of the Park, have even questioned the aesthetic value of these formations, as well as presenting data not exempt from criticism regarding aspects related to the alleged reduced specific diversity of these formations compared to other types of forests, as demonstrated by other authors (Romane and Valerino 1997). The chestnut grove in pure formation has therefore almost disappeared and, within an extensive reforestation process underway in the last two centuries, motivated by the depopulation of the Apuan mountain, today a mixed forest which can hardly be defined semi-natural, but rather semi-cultural dominates the landscape.

The protection established through the various environmental restrictions does not protect or recover the historical chestnut landscape, but rather "safeguards" its transformation into something different. In 1996 the area was the subject of an environmental disaster, concerning very intense rainfall events that caused landslides leading also to the destruction of part of the town of Cardoso. A good part of the causes of such events were directed at the chestnut groves, which had become too old with large and heavy trees, often being monumental plants. In fact, the lack of maintenance of terraces with chestnut trees and cultivated is a contributing cause of such phenomena, but the solution proposed by the authorities, instead of the recovery and maintenance of a historical and landscape heritage, is renaturalisation, therefore the evolution towards other types of forest, to definitively sanction the disappearance of the chestnut civilization.

9.8 Trees as a Historical Monument

The presence of forests with a low density of trees per hectare has favoured the historical persistence of specimens that over the centuries have acquired characteristics of monumentality, given that the free space around the plant allows it to develop also in width and not just in height. Fruit chestnut groves, especially those abandoned, along with other types of forest, offer multiple examples of trees with these characteristics, already recognised in nineteenth-century treaties where it is mentioned that the chestnut tree "has a very long life and acquires prodigious thicknesses" (Savi 1811). In recent environmental literature, they are often presented as remnants of primary forests or "old growth", but in reality, at least in Italy, they are trees related to past traditional management practices, e.g. pollard trees, or trees belonging to ancient wood pastures or pasture woods.

Monumental specimens, especially for their dimensional aspects, have always struck observers since the most remote times. Observing a monumental tree, it is easy to imagine the profound changes that have occurred over time, of which it has been a witness, and it is their longevity that represents one of the fundamental reasons that have led many peoples in the past to attribute a sacred value to them, while their scarcity in the current landscape makes them the object of renewed attention. Large trees express the extreme manifestation of resistance and adaptability, conditions often indispensable for long-term preservation. Other values associated with them concern their architectural and landscape significance.

In the global context, there are specimens that naturally reach considerable dimensions, just think of the sequoias of the United States or the African baobabs; at our latitudes and in our cultural context, however, there are neither species with such characteristics, nor have the conditions of intense forest utilizations allowed the conservation of many monumental specimens. It is therefore normal that the oldest trees are often those resulting from human action, such as the trees present in former wooded pastures or in grazed woods or in any case with low densities that have allowed a great development of the crown or trunk, compared to denser woods. Not to forget that trees such as the beech can reach ages of several centuries without presenting aspects of dimensional monumentality. The search and protection of monumental trees, now implemented by law in Italy, is therefore a commendable initiative, but also the monumental specimens that the scientific literature and the media periodically report refer to woods or pastures widely used at the time of the birth of the monumental plant (Fig. 9.8).

The monumental tree, with its single presence or arranged in a row or group, constitutes an element that contributes to fully perceive a landscape, helping to remember the past socio-economic context. This helps to identify that type of landscape, making it unique in the current territorial context, where uniformity and

Fig. 9.8 A monumental chestnut tree in the Scesta Valley (Apennine mountains Toscana region)

simplification of cultivation often occur, characterising vast Italian areas. The fact that monumental trees, whether close to or far from civil structures, are cut down or die is perceived in most cases as almost a mutilation, diminishing their historical, architectural and landscape value. In other cases, the monumental tree holds a prominence that descends from its botanical rarity: for example, plants outside their typical climatic band and poorly numerically represented. This type of tree can also include species not typical of the Italian territory that, following the custom developed particularly during the eighteenth century, were planted in gardens and parks of historic villas and palaces.

The classification of monumentality can therefore concern: *dimensional and chronological monumentality-*, linked to the circumference and height of the plant; *monumentality related to the shape and bearing of the plant-*, which concerns the external and aesthetic aspect of the plant, understood as shape and/or bearing. Environmental and anthropic pressure conditions the morphological aspect of the tree, such as the shape that the crown assumes in the presence of prevailing winds, or obstacles that determine twisted aspects of the trunk or finally the type of cultivation due to grafts or particular prunings that have determined the current shape of the tree; *monumentality linked to botanical rarity-*, refers to species not typical of the environment in which they grow or vegetating in extreme conditions; *architectural monumentality-*, refers to specimens that are located near rural artefacts, rural buildings or structures of high historical-cultural value. In this case, the monumental tree appears as an element that enhances the adjacent anthropic elements also creating visual effects; *historical-cultural monumentality-*, the importance of the plant is linked to particular events of local history, traditions, legends, etc.; *aesthetic monumentality-*, plants located in a context of high aesthetic value or whose presence characterises a certain place. This is not an intrinsic characteristic of the plant, but concerns the context in which it is inserted and the role it plays on the aesthetics of the places.

Monumental plants can to some extent be restored and managed continuously, but avoiding the felling of trees, or their replacement with new seedlings of chestnut, regardless of the productive capacity or the phytosanitary state, unless there are serious risks of collapses and crashes with damage to things or people. This often happens when visions of a productive or phytosanitary type guide the restoration of an ancient chestnut grove. In senescent or sick plants that often present themselves as monumental specimens, vegetative development is slowed down and the size of the fruits reduced. On these trees it is possible to obtain the reversibility of ageing by revitalising them through more or less drastic pruning of the crown. The recovery exploits the marked regenerative potential of the chestnut and, with extraordinary pruning interventions comparable to a real "restoration", allows the still sufficiently vital plants to regain over the years the lost functionality. We intervene with renewal cuts and more or less energetic prunings, as well as pollarding. These operations, preceded by interventions to clean the undergrowth from intrusive species, must be completed over the years for the successful outcome of the "restoration". Subsequent interventions consist of: cultivation and selection of new shoots to regulate the conformation of the crown, restoration of the network of farm roads, maintenance and pruning prunings and possible fertilisations.

References

Agnoletti M (2002) Le dinamiche del paesaggio fra XIX e XX secolo. In: Agnoletti M (ed) Il paesaggio agro-forestale toscano, strumenti per l'analisi la gestione e la conservazione. ARSIA, research report, Firenze

Agnoletti M (2005) Dinamiche del paesaggio, biodiversità e rischio idrogeologico nella zona della Pania di Cardoso fra 1832 e 2002. Edizioni Regione Toscana, Firenze

Barelli G (1936) Statuti e carte di franchigia di Mombasiglio. Biblioteca Storica Subalpina, Torino

Barelli G (1966) Statuti di Lesegno, collezionati sui codici da Guido Gay di Quarti. Deputazione Subalpina di Storia Patria, Torino

Barelli G, Gay di Quarti G (1966) Statuti di Lesegno, Parte III. Deputazione subalpina di storia patria, Torino

Bartelletti A, Corfini L (1986) Indagini sugli usi civici. La situazione in Alta Versilia avanti le Riforme Leopoldine. Studi Versiliesi 4:19–26

Bellina A (1998) Il castagno nella Comunità Montana Mugello, Alto Mugello, Val di Sieve. Atti del Convegno nazionale sul castagno, Castello Brandolini, Cison di Vaimarino (TV), 23–23 ottobre 1997, Comunità montana delle Prealpi trevigiane, Vittorio Veneto

Bignami GR, Salsotto A (1983) La civiltà del castagno. L'Arciere, Cuneo

Biraghi A (1947) Una gravissima minaccia per i nostri castagneti: il cancro della corteccia. L'italia Forestale e Montana 2(1):8–16

Biraghi A (1950) Caratteri di resistenza in Castanea satina nei confronti di Endothia parasitica. Bollettino Stazione di Patologia Vegetale 3:5

Botteri G (1892) Memorie storiche e statuti antichi di Chiusa Pesto. Tip. A. Fina, Torino
Bounous G (1996) Rilancio della castagna «Siria» della Valle Grana (CN) per la produzione di «castagne bianche». Monti e Boschi 3:30–34
Bounous G (1999) Tra i castagni del Cuneese. Ed. metafore, Cuneo
Bounous G, De Guarda A (2002) Origine e cenni storici. In: Bounous G (ed) Il castagno. Coltura, ambiente ed utilizzazioni in Italia e nel mondo. Il Sole 24 Ore Edagricole, Milano
Braudel F (1977) Capitalismo e civiltà materiale (secoli XV–XVIII). Einaudi, Torino
Cagnin G (1997) Il castagno nel paesaggio agrario e nell'economia a Treviso nel medioevo. Studi Trevisani, Bollettino Degli Istituti di Cultura Del Comune di Treviso 8:5–49
Camilla P (1970) Archivio Storico dell'Ospedale Civile di S. Croce in Cuneo. Biblioteca della Società per gli Studi Storici, Archeologici e Artistici della provincia di Cuneo, Cuneo
Camus A (1929) Les chàtaigniers. Monographie des genres Castanea e Castanopsis. Paul Lechevalier, Paris
Carlone R, Paglietta R (1967) Possibilità di trasformazione del castagneto in frutteto di castagno o eventuale sua sostituzione con altre colture agrarie. Atti del Convegno internazionale sul castagno, Cuneo
Cherubini G (1984) La civiltà del castagno in Italia alla fine del Medioevo. In: Cherubini G (ed) L'Italia rurale del basso Medioevo. Laterza, Roma-Bari
Comalini Fumagalli A (1997) La «Castanea sativa» nelle Tre Pievi. Notiziario Banca Popolare di Sondrio 74:156–165
Comba R (1983) Fra XIII e XV secolo: uomini e risorse agricolo-forestali nel variare delle congiunture. In: Comba R (ed) Metamorfosi di un paesaggio rurale. Domini e luoghi del Piemonte sud-occidentale dal X al XVI secolo. CELID, Torino
Comba R (2000) Testimonianze sull'uso dell'incolto e sul dissodamento e sul popolamento nel Piemonte meridionale (XIII–XIV secolo). Bollettino Storico-Bibliografico Subalpino, Torino
Dadone P (1982) Un aspetto del problema dei rapporti tra società e ambiente nel medioevo: la diffusione del castagno nel Cuneese medioevale. Tesi di laurea, Università di Torino, Facoltà di Lettere e Filosofia, a.a. 1981–1982
Daviso di Charvensod MC (1961) I pedaggi delle Alpi occidentali. Deputazione subalpina di storia patria, Torino
De Angelis L (1975) Tradizione classica e pratica agraria nel «De agricultura» di Michelangelo Tanaglia. Tesi di laurea, Università degli Studi di Firenze, Facoltà di Lettere e Filosofia, a.a. 1974–1975
De Montaigne M (2003) Viaggio in Italia. Rizzoli, Milano
De Philippis A (1958) Lezioni di selvicoltura speciale. Università di Firenze, Firenze
Farneti G, Malatesta S, Pedrotti F (1972) Guida alla natura della Lombardia e del Trentino Alto Adige. Mondadori, Milano
Fenaroli L (1945) Il castagno. Tip. Ramo editoriale degli agricoltori, Roma
Ferrarini E (1972) Carta della vegetazione delle Alpi Apuane e zone limitrofe. Webbia 27:551–582
Flohn H (1968) Clima e tempo. Il Saggiatore, Milano
Gabotto F (1901) L'agricoltura nella regione saluzzese dal secolo XI al secolo XV. Tip. Chiantore e Mascarelli, Pinerolo
Galasso A (1975) Economia e società nella Calabria del Cinquecento. Guida, Napoli
Gribaudi P (1955) Scritti di varia geografia. Giappichelli, Torino
Guainerio A (1481) Pratica medicinae, Antonius de Carcano, Pavia
Gullino G (2000) La vite e il castagno in Piemonte: aspetti di un connubio (secoli XIII–XV). In: Comba R, Naso I (eds) Uomini boschi castagne. Incontri nella storia del Piemonte. Società per gli studi storici, archeologici ed artistici della provincia di Cuneo, Cuneo
Hanousková I, O'sullivan PE, Witkowski Z (1999) Perspective in sustainable land use of marginai areas, land abandonment and restoration In: Farina A (ed) Perspectives in ecology. Backhuys Publishers, Leiden
Hofmann A, Goretti D, Merendi GA, Tabacchi G, Vignoli M, Bernetti G (1998) L'Inventario Forestale. Edizioni Regione Toscana, Firenze

Lear E (2003) Diario di un viaggio a piedi. Reggio Calabria e la sua provincia. Laruffa, Reggio Calabria

Lizier A (1907) L'economia rurale dell'età prenormanna nell'Italia meridionale. Palermo

Mander U, Jongman RHG (1999) Ecological and socio-economic consequences of land use changes. In: Farina A (ed) Perspectives in ecology. Backhuys Publishers, Leiden

Mantovani M (1979) Appunti sulla vita di un territorio montano nell'epica moderna. La montagna pistoiese nel primo secolo del principato Mediceo (1530–1630). Tesi di laurea, Università degli Studi di Firenze, Facoltà di Lettere e Filosofia, a.a. 1978–1979

Marciarli F (1952) Alcuni aspetti storici e folcloristici dalla castanicotura irpino- salernitana. Monti e Boschi 2:79–84

Merendi A (1928) I castagneti da frutto in Liguria. L'alpe 15:34–40

Merendi A (1938) Il problema dei castagneti da frutto in Toscana. Accademia dei Georgofili, Firenze

Moreno D, De Maestri S (1975) Casa rurale e cultura materiale nella colonizzazione dell'Appennino genovese tra XVI e XVII secolo. I paesaggi rurali europei, Deputazione di Storia Patria per l'Umbria, Perugia

Nada Patrone AM (1981) Il cibo del ricco ed il cibo del povero. Contributo alla storia qualitativa dell'alimentazione. L'area pedemontana negli ultimi secoli del medioevo. Centro Studi Piemontesi, Torino

Nada Patrone AM (1991) L'enologia nelle considerazioni di un medico piemontese del Cinquecento: Francesco Gallina. In: Comba R (ed) Vigne e vini del Piemonte moderno. L'Arciere, Cuneo

Nada Patrone AM (2000) Le castagne: gastronomia e medicina fra tardo medioevo e prima età moderna. In: Comba R, Naso I (eds) Uomini boschi castagne. Incontri nella storia del Piemonte. Società per gli studi storici, archeologici ed artistici della provincia di Cuneo, Cuneo

Naso I (1987) Uso e tutela del bosco nel Piemonte medievale. La legislazione statutaria dei secoli XIII–XV. Per un museo dell'agricoltura del Piemonte, Il bosco e il legno. Associazione Museo dell'Agricoltura del Piemonte, Torino

Nice B (1952) Le alpi apuane. Studio antropogeografico. Camera di Commercio Industria e Agricoltura di Lucca, Lucca

Nisbet R (2000) Alcuni aspetti di storia naturale del castagno. In: Comba R, Naso I (eds) Uomini boschi castagne. Incontri nella storia del Piemonte. Società per gli studi storici, archeologici ed artistici della provincia di Cuneo, Cuneo

Paganelli A (1998) Evoluzione storica del castagno nell'Italia nord-orientale dal pleistocene superiore attraverso l'indagine palinologica. Atti del Convegno nazionale sul castagno, Castello Brandolini, Cison di Vaimarino (TV), 23–23 ottobre 1997, Comunità montana delle Prealpi trevigiane, Vittorio Veneto

Paolicchi SD, Bellatalla GM, Pierotti M (1987) Storia e struttura territoriale del Comune di Stazzema (Lucca). Tesi di laurea, Università degli Studi di Firenze, Facoltà di Architettura, a.a. 1986–1987

Patria L (2000) De castaneis se pascant et alimentent: produzione e consumo delle castagne in Val di Susa nel tardo Medioevo. In: Comba R, Naso I (eds) Uomini boschi castagne. Incontri nella storia del Piemonte. Società per gli studi storici, archeologici ed artistici della provincia di Cuneo, Cuneo

Petri L (1918) Studi sulla malattia del castagno detta dell'inchiostro. Morfologia e biologia del micelio parassita. Annali del R. Istituto Superiore Forestale Naz 3:3–35

Petri L (1924) Esperienze sul grado di resistenza del castagno giapponese alla Blepharospora cambivora. Annali del R. Istituto Superiore Forestale Naz 9:9

Petri L (1931) La coltivazione del castagno giapponese in Italia. Bollettino della R. Stazione di Patologia Vegetale 11:275–284

Piccioli L (1922) Monografia del castagno. Stabilimento Tipo Litografico G. Spinelli, Firenze

Pitte JR (1986) Terres de Castanide. Fayard, Paris

Polacco F (1938) Indagine sulla coltivazione del castagno da frutto in Italia. Boll. mensile di Statistica agraria e forestale

Poli I (1999) Del castagno in Garfagnana. Storia coltura poesia. M. Pacini Fazzi Editore, Lucca

Prato G (1908) La vita economica in Piemonte a mezzo il secolo XVIII. Documenti finanziari degli Stati della monarchia piemontese, Officine Grafiche della Soc. Tip. Ed. Naz., Torino
Quaini M (1973) Per la storia del paesaggio agrario in Liguria. Camera di Commercio, Industria, Artigianato e Agricoltura, Savona
Razè E (1976) La proprietà fondiaria dei Savoia e il loro controllo delle attività commerciali a Cuneo attorno al 1430, sulla base di due resoconti inediti della castellala omonima. Facoltà di Lettere e Filosofia, Università degli Studi di Torino
Remondino C (1926) Il castagno. Paravia, Torino
Roche D (1935) Le chàtaignier dans l'Ardèche. In: Congrès méditerranéen du chàtaignier, Aubenas, 25–27 octobre 1935
Romane F, Valerino L (1997) Changements du paysage et biodiversité dans les chàtaigneraies cévenoles (sud de la France). Ecologia Mediterranea 23(1–2):121–129
Savi G (1811) Trattato degli alberi della Toscana. Piatti, Firenze
Tamagnone A (1969) Gli statuti di Rifreddo e Gambasca. Deputazione Subalpina di Storia Patria, Torino
Vigiani D (1908) Il castagno. Casa Ed. F.lli Ottavi, Casale Monferrato (CN)
Vigiani D (1943) La coltivazione del castagno. Soc. Anonima Editrice Dante Alighieri, Genova
Vitolo G (1987) I prodotti della terra: orti e frutteti. In: Musca G (ed) Terra e uomini nel mezzogiorno normanno-svevo. Dedalo, Bari
Vos W, Stortelder A (1992) Vanishing Tuscan landscapes. Pudoc Scientific Publisher, Wageningen

Chapter 10
Forests and Charcoal

Abstract The use of charcoal dates back to ancient times and can be considered to have progressed hand in hand with metalworking. In Italy, the consumption of plant fuels remained widespread and extremely high until the second post-war period due to the scarcity of fossil fuels and the maintenance of traditional social and economic structures. This marks a significant difference from other European countries where charcoal was already in decline in the 1700s due to the use of coke in metallurgy and a different model of industrialization. There are evidence of the production of charcoal as early as the Etruscan era, until 1970 it was still a quite common activity. Charcoal was produced in almost all Italian forests from the Alps to Sicily and at all altitudes. However, it is not possible to define a type of "charcoal forests", as this was made almost regardless of the type of forest, even though some forests had more favourable characteristics than others, especially coppiced forests. Charcoal is less bulky than firewood, which facilitates its transport, but it also ignites more easily and has a higher calorific value, being about double that of wood. Therefore, considering the calorific power per unit of weight, it was often much more convenient to produce charcoal in the forest and then transport it to the roads than to simply produce wood. Charcoal burners usually worked together in "companies" that were formed at the beginning of the season and based on friendships and acquaintances. Life in the forest could have not a few difficulties of cohabitation, but generally each company remained united for several years, often being made up of a family nucleus with or without external components. Charcoal burning has left many traces in the forest landscape. Charcoal remains from former charcoal kilns provide information at fine spatial scales, capturing local woody-species composition and revealing the history of landscape management. In Italy is quite common to find remains of charcoal kilns platforms in areas protected for the naturalness of the vegetation, suggesting a previous history of forest utilizations. Charcoal burning platforms could reach a density up to 10 per hectare.

M. Agnoletti, *History of the Italian Forest Landscape*, Environmental History 19,
https://doi.org/10.1007/978-3-031-81631-4_10

10.1 Eight Thousand Years of History

The use of charcoal dates back to ancient times and can be considered to have progressed hand in hand with metalworking. In Italy, the consumption of plant fuels remained widespread and extremely high until the second post-war period due to the scarcity of fossil fuels and the maintenance of traditional social and economic structures. This marks a significant difference from other European countries where charcoal was already in decline in the 1700s due to the use of coke in metallurgy and a different model of industrialisation. In Italy, wood and charcoal have maintained their importance for millennia, almost up to the present day, especially wood, highlighting a substantial immutability of work in the forests in terms of techniques and practices, even coexisting with the great transformations of the second post-war period and leaving clear traces in the forest landscape, since the civilisation associated with wood and charcoal quickly dissolved, not only in the economy but also in the country's culture. We have evidence of the production of charcoal as early as the Etruscan era, when it was used specifically for furnaces for metal smelting, but Pliny and Theophrastus also mention it, describing the carbonisation methods of the time, not so far from those used until the middle of the last century.

Charcoal was produced in almost all Italian forests from the Alps to Sicily and at all altitudes. However, it is not possible to define a type of "charcoal forests", as this was made almost regardless of the type of forest, even though some forests had more favourable characteristics than others. The reasons for the preference for charcoal over firewood lie in some of its peculiarities. Charcoal is less bulky than firewood, which facilitates its transport, but it also ignites more easily and has a higher calorific value, being about double that of wood. In firewood, given the higher water content, the calorific value instead decreases drastically, so much so that, even if the percentage of water present in the wood is around 20–25%, the calorific value reaches 3000 cal. Therefore, considering the calorific power per unit of weight, it was often much more convenient to produce charcoal in the forest and then transport it to the roads than to produce wood. Firewood was sold stacked and the measure of the stacks was the cubic metre, which corresponds to a cubic metre in which the voids between the various trunks of the stack are considered flat: it was calculated that the real volume of the wood was 2/3 of the apparent one. On the mountains of Mugello valley, from a cubic metre of beech wood, 70–72 kilos of charcoal were obtained and the yield of carbonisation ranged between 14 and 17%, although, according to scientific literature, it was 20%. The transformation of wood into charcoal therefore allows a lighter product to be obtained, even if, considering a yield of 20%, to obtain 7500 cal from a kilo of charcoal, 10,000 of the 17,500 cal of the five kilos of necessary wood are lost. Carbonisation was nevertheless functional for the following main reasons:

- only with charcoal was it possible to achieve the high temperatures necessary for various melting and metalworking processes, ceramics, etc.;

- The regularity of combustion, almost absence of smoke and flames, good calorific power and ease of use were valid reasons for preferring coal in food cooking and domestic heating;
- The costs of transport, handling and storage favour the use of coal due to its smaller volume and weight.

Charcoal is made up of almost pure carbon since the other components only make up 20% of the total. It is distinguished as strong and sweet, and each of the two types can be classified as "cannel" (made from whole branches and trunks), split charcoal (from split trunks and branches), and log charcoal (from roots and knotty woods). Chestnut charcoal (sweet charcoal) was the best for use in blacksmiths' forges, where it burned quickly, producing the heat necessary to heat the iron for processing.

Determining whether a forest was destined to produce charcoal depended, alongside its structural characteristics, on other elements such as the market prices of forest products, labour and transport, which varied in space and time. Thus, the same type of forest located in two different economic situations related to different historical moments and different social situations could in some cases be destined to produce charcoal and in others not. Naturally, there were more important characteristics to guide the transformation of forests into charcoal, such as the quality of the assortments that distinguished timber forests from those for firewood. Within this latter category, the position of the forest in relation to the market place, the cost of labour, transport and the prices of charcoal and wood, suggested whether to produce wood or to transform it into charcoal. For this reason, we find historical descriptions regarding the manufacture of charcoal from the Alps to Calabria, in coniferous forests and in coppices, even though the coppice was the preferred form of management to produce charcoal given the small assortments that facilitated the construction of the charcoal kiln and the carbonisation process. Serpieri and Segala, at the beginning of the twentieth century, explained that in the Alps it was mainly the coppices that produced charcoal in the Apennines, and in the islands in addition to the coppice also some high forests were transformed into charcoal (Serpieri and Segala 1917). In those years it was estimated that 2/3 of the firewood was produced in Italy, of which three million cubic metres came from coppices, 2.3 million cubic metres from deciduous forests and half a million from coniferous forests. As can be seen from these data, the percentages between deciduous and coniferous, and between coppices and high forests, are equivalent, so it is difficult to define forests specifically destined for charcoal, but it is certain that carbonisation appears to be an activity widespread throughout the national territory.

10.2 The Charcoal Kilns

Carbonisation, or the technique by which charcoal is produced, has been handed down over the centuries and has undergone, at least in Italy, one of the major peaks between the nineteenth and twentieth centuries, when, during the industrial revolution, given the lack of coal mines, the main source of energy used for both industries and transport was precisely wood charcoal. The carbonisation process was carried out using two main methodologies, but the most common one involved a process carried out with a scarcity of oxygen and direct heating. Carbonisation took place through the construction of specific structures: the charcoal kilns. These could have various shapes and sizes, but they were mainly distinguished into two categories: vertical charcoal kilns and horizontal ones (Piccioli 1923).

The vertical charcoal kiln, more widespread, took the form of an Apollonian paraboloid, the dimensions of which varied depending on various factors, first and foremost the space available for its setup. The construction of a vertical charcoal kiln first involved the choice of the square or *charcoal yard,* that is, the site where it would be erected; this had to meet many requirements, first of all being large enough to contain the entire structure and also have the necessary space for the charcoal burners to move around to manage the wood cooking. The type of soil was also taken into account, which had to be sufficiently permeable; soils that were too compact, for example, with a too high clay content, were not suitable, as they did not ensure a sufficient influx of air, thus preventing an adequate carbonisation process (Fig. 10.1).

Of fundamental importance was also the position of the charcoal kiln platform that had to be sheltered from the wind, a negative factor for the process as it indiscriminately increased the draught causing excessive combustion. The platform then had to be located in a place that facilitated both the transport of wood from the cutting area and the removal of coal at the end of the process; moreover, it had to be close to a place for the supply of water necessary in case of problems related to fire management. The location of the charcoal kiln also depended on the morphology of the area, the slope was particularly influential, which conditioned the width of the platform. Often to overcome situations on particularly steep slopes, terraces were created supported by dry stone walls or with timber. In some cases, terraces already used previously could also be exploited, but taking into account that this would have caused a decrease in yield of about 10–15%.

In the case where a new charcoal kiln platform was built, the process began by choosing the most suitable place for the creation of the square and, starting from the centre, using a rope, a circumference of area equal to the width to be given was traced. to the clearing; subsequently, the surface was levelled, leaving a rise of about 20–30 cm in the centre to facilitate the drawing of the charcoal kiln during the carbonisation process. The width of the threshing floor, as previously mentioned, had to be greater than the actual size of the kiln, to allow the movement of the charcoal burners who were tasked with monitoring and managing the entire process. After the creation of the platform, a small ditch was created at the edge of the clearing for the collection of rainwater, so that there would be no possible stagnation on the threshing floor

Fig. 10.1 Illustration of a traditional charcoal kiln in a forest setting. The kiln is dome-shaped, constructed from stacked logs and covered with earth. A ladder leans against the structure, and a pile of logs is nearby, indicating preparation for charcoal production. The background features trees and a hilly landscape (Landi and Piussi 1988)

that could affect the carbonisation of the wood. Once the clearing was completed and the collected wood was transported to the immediate vicinity of the platform, the actual construction of the charcoal kiln began. The first operation involved the construction, in the centre of the threshing floor, of the chimney, which constituted the duct through which the charcoal kiln was ignited and drawn during the carbonisation process. Various methods of chimney construction have been handed down, but the most common one involved planting three wooden poles in the ground, about 30–40 cm apart from each other and of a height equal to that expected for the charcoal kiln. These three poles, positioned in a triangle, were then tied together using one or more willow rings. The rings were positioned at a height slightly lower than that of the pieces to be carbonized (Fig. 10.2).

This created a framework on which the wood was supported, leaving an empty channel in the centre of the kiln. The construction of the charcoal pile continued by vertically placing the pieces of wood around the chimney, arranged in two or three tiers. The first pieces of wood leaning against the structure were placed vertically, while the others gradually assumed a more inclined position. When the first tier had reached a diameter of about one metre, the arrangement of the pieces for the second tier began. For the pieces used, a preliminary selection had to be made; in fact, to achieve the same carbonisation of all the timber used, the larger pieces and those of species with heavy wood were placed around the chimney, both at the base and in the middle where the temperature was higher. The smaller and lighter material

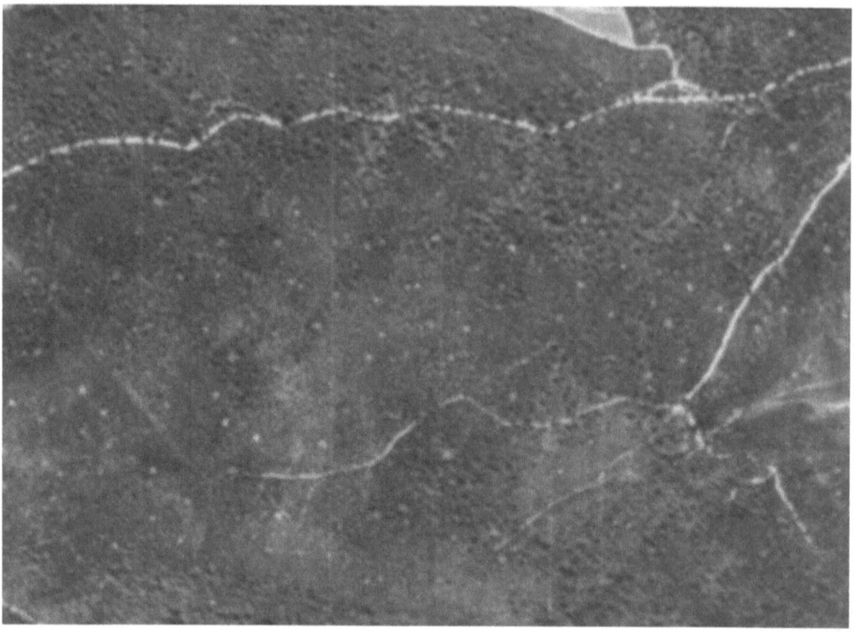

Fig. 10.2 In this aerial photo from 1954, the charcoal kiln platforms present in an area of the Montagnola Senese are observed as white dots in the middle of the dark grey area. The high density of these, which appear with almost geometric regularity in the landscape, and the dense service road network for the transport of coal are noticeable (Agnoletti [edited by], *The Tuscan agro-forestry landscape,* cit.)

was instead used for the outer part of the charcoal pile and constituted the so-called "coat", the layer in contact with the covering mantle, or was placed around the central structure to facilitate the ignition. To obtain a paraboloid shape at the end of the arrangement of all the tops, the first tier had to have an inclination of the pieces of wood not too pronounced, while the inclination of the tops of the second layer was greater, while the part at the top of the structure saw the presence of wood arranged almost horizontally. Throughout the construction process, particular attention was paid to avoid creating empty spaces between the wood so as not to have excessive oxygenation that would have negatively affected the carbonisation process. Once the construction of the stack was completed, it was covered at the base with grassy slabs or a layer of live plant material mixed with wet clay clods that went to build a mantle with the purpose of regulating the influx of air. The remaining part of the structure was instead covered with straw, hay, grass or branches to form a layer of about ten centimetres on which fine earth was pressed. The thickness of the mantle was not uniform for the entire structure but decreased towards the top of the charcoal pile. In order to ensure greater stability of the structure, a base armour was also made at the base of the kiln, for whose construction a lattice was created around the base or branches of equal height to the first stage were stuck into the ground.

At the end of the construction of the kiln, the ignition was carried out through the chimney. Lit charcoal was introduced into the hearth and immediately after small pieces of readily combustible wood, the *stubs*. Subsequently, to encourage the start of a uniform combustion process throughout the structure, holes were made at the base of the charcoal pile with the aim of promoting air circulation. Once it was ascertained that the fire had taken all parts of the structure, the mouth of the chimney was closed with a sod or a stone and, every three-four hours, additional material was inserted to prevent the fire from going out until all the hearth was full of glowing coals. Once the ignition process of the charcoal pile was completed, the phase of *governance,* began, during which for the first 24 h there was the so-called "carbonisation to the blind" that saw the rise of the fire along the chimney until it reached the top and started the actual process of carbonisation. At the end of this phase, which allowed the dehydration of the wood, the chimney was fed again with new material for a time that varied from two to four days. The control of the carbonisation process took place through the colour of the fumes, indicators of the level of carbonisation of the material of the pile. The progress of the entire process was never regular and was not only linked to the skill of the charcoal burners but also and above all to weather conditions: many difficulties were created, for example, by the presence of strong winds or rains (Italian Federation of Agricultural Consortia 1954) (Fig. 10.3).

Although the carbonisation process was almost the same in all areas where it was practiced, the construction techniques of the charcoal piles often changed, which could, for example, provide for a side ignition with a chimney dug underground. This type of construction was planned for small-sized charcoal piles that would not have withstood the weight of the charcoal burner who had to go to the top of the pile for the ignition and feeding of the charcoal pile. In addition to the differences in the construction of the chimney, which varied depending on the size of the charcoal pile and local customs and uses, there can also be differences in the material used for the creation of the *mantle,* often in fact the wall at the base was not created but the entire charcoal pile was covered with dry leaves and earth and to prevent it from sliding a hedge was built with intertwined branches that surrounded the structure.

However, carbonisation did not only take place in vertical structure charcoal piles but, depending on the type of material used and the place where this transformation was carried out, there could also be horizontal charcoal piles and carbonisation in pits. In the first case, we are faced with a type of structure carried out in places where the material used was mainly made up of large and regularly shaped resinous stumps. For the construction of these charcoal kilns, a slightly sloping terrain was chosen, on the flat area three poles were arranged longitudinally on the ground, one in the centre and two on the sides, and vertical poles were driven into the lower part of the threshing floor. The logs to be charred were then placed transversely to the poles that had been initially arranged on the ground. In the case of horizontal charcoal piles, there was no real rock pile, but a small opening was created in the stack from the lower part of the clearing, which constituted the entrance to a channel that ran through the entire structure and had another opening on the opposite side. The completed charcoal pile had a trapezoidal shape and was also covered with branches and earth. Charcoal production in pits, on the other hand, was less used due to the

Fig. 10.3 A detailed black and white illustration of a traditional earth-covered dwelling with a wooden frame and thatched roof. The entrance is supported by two poles, and a ladder made of branches leans against the structure. In the foreground, there are simple wooden tables and stools. The background features a rocky landscape with a small waterfall Mugello (Tuscany) (Landi and Piussi [edited by], *The work in the woods,* cit.)

low yield of charcoal and the low quality of the product, as well as differing greatly in the structure and technique used. This was carried out especially when the wood material was in surplus and of poor quality. In this case, square or circular pits were opened in the ground with a depth of one metre and a width of 1.5–2 m. Already lit branches were placed at the bottom of the pit, onto which the material to be charred was then thrown and partially burned, then more material was added until it reached the upper edge of the pit, and when the operation was finished, it was left to cool, covered with earth and grassy sods.

The charcoal production process was carried out at all times of the year, during the summer period in mountainous areas and in the winter period in coastal areas and in Sardinia. The species used are many, but, from species to species, not only the duration of the process and the difficulty of execution vary but also the quality of the product obtained. Generally, for example, coniferous wood requires a longer charring period than that of broad-leaved trees, and the use of more seasoned wood, given its lower water content, reduces the process times. From the news that reaches us, the best quality charcoal comes from wood from plants grown in infertile soils and south-facing, and of better quality is also the charcoal produced from coppiced plants rather than from tall trees. As for the species, the most valuable charcoal comes from oak species, such as holm oak and oak, from hornbeams, beeches and maples; lower quality charcoal, on the other hand, from species including firs, pines, larches, chestnuts, alders and poplars. Particular importance was given to the wood assortments used for the construction of the charcoal piles. The length of the material ranged between 80 and 120 cm, but the diameter varied greatly from 3 to 5 cm in Tuscany, while in Lombardy it even reached 25 cm. In Molise, up to the first decades of the twentieth century, charcoal piles were made up of 70% round pieces and 30% quarters, i.e. trunks divided into four parts with an axe to have a size that allowed good charring, since a trunk of a mature tree of 40–50 cm in diameter could not be charred whole.

10.3 The Work

Talking about the production of charcoal allows us to delve deeper into some aspects of work in the woods and its organisation, which involved complex phases in which various professional figures were involved. In this sense, the case of Tuscany is particularly significant, as the region had a long history regarding the production of charcoal and was almost self-sufficient for these forms of fuel. Before creating the charcoal kiln, the forest to be cut was purchased, which was usually sold "standing" in Tuscany for the larger forests, i.e., before cutting them. The forest was sold behind the payment of the so-called "macchiatico", that is, a monetary value corresponding to the difference between the price of the products obtainable (wood, charcoal, etc.) and the costs of cutting and transporting out of the forest. In the usual procedure of the sharecropping farms, the factor negotiated the sale with the merchant during a brief meeting. Before such a meeting, the merchant usually knew, either from direct knowledge or by having sent an expert woodcutter to view the forest, the value of the forest to be cut. Payment was made by depositing a down payment, usually half, and the rest on balance. Some large properties proceeded with sales through auctions, but these were procedures little used for this type of forest, much more used for timber forests.

Around the activity of the woodcutters and charcoal burners, a complex productive and commercial system moved, regulated mainly by customs applied in a context of personal relationships based on mutual esteem, friendship, kinship ties. It is a circle of interests largely based on trust where rarely were there contracts or written procedures, in which a given word was worth more than a signature on a contract (Landi and Piussi 1988). A system in which, from other points of view, the harshness of working conditions and the equally harsh law of business and the state of need of the workers, was accompanied by a strong component of paternalism, along with frequent prevarications and non-compliance of difficult claim. In the 1930s, in Mugello, cutters and charcoal burners worked as employees with piecework wages, they were usually hired for more or less long periods of time, and in the months of unemployment they carried out other activities; very few had fixed work relationships with a farm or a contractor with daily pay.

Charcoal burners usually worked together in "companies" that were formed at the beginning of the season and based on friendships and acquaintances. Life in the forest could have not a few difficulties of cohabitation, but generally each company remained united for several years, often being made up of a family nucleus with or without external components. Men generally carried out the jobs of greater responsibility, while women and boys participated in the work, even the heaviest ones, hoeing the grassy piote to cover the charcoal pits, arranging the wood, preparing meals, supplying water and guarding the hut where the family lived. In general, the company produced less than a charcoal burner who worked alone, in fact it had a per capita production of about 210–240 kilos of charcoal per day, while working alone a charcoal burner reached 300–310 kilos. Despite the lower yield, the company was almost indispensable for the complexity, difficulties, and hardness of work in the forest. It also met the need to give work to more people, involving 5–6 charcoal burners. A precise but not rigid hierarchy regulated the relationships within the company, with a boss "capopala" who negotiated the work with the merchant or with the owner of the forest, and was responsible for production, deliveries, calculation of the charcoal produced and respect for the rules. A working season usually lasted 180 days and in these six months a company of three men produced about 1300 quintals.

A very dense network of difficult and tortuous paths wound over the mountains where the charcoal burners cooked the charcoal. As understood from previous chapters, forests were indeed much more populated in past centuries compared to the absence of humans that characterises today's woods. The entire network of paths had to be kept clear of branches and shrubs to the width of a man with outstretched arms, to allow the transit of mule caravans carrying bundles in which charcoal was transported. This network of paths connected all the charcoal platforms scattered in the woods to the road and to the charcoal burners' huts. The containers for charcoal were almost exclusively the "bundles" with well-defined and homogeneous measurements. The Tuscan magonal bundle contained from 90 to 115 kilos of charcoal, the Roman magonal bundle contained instead from 70 to 95. The name derives from "magona", a term with which the iron-producing industries from the Alps to Calabria were indicated. Alongside the charcoal burner's profession, there was also that of the carters, i.e., the personnel who transported the charcoal out of the woods to the roads and then to the destination markets, generally with long caravans of mules and carts. Mules were the only means of transport capable of moving in the woods along winding and rugged paths to reach the places of sale served by roads. The carter usually owned mules and carts that were hitched to the mules once they reached the drivable roads, and each cart could carry up to 25 bundles of charcoal, about 17–18 quintals. Generally, along with the carter, there was also the figure of the packer, who was responsible for measuring the charcoal and preparing the bundles.

A special mention deserves the charcoal burners' huts, an essential element in the system of charcoal production and the civilisation connected to this activity. The construction of the hut intended for the life of the woodcutter absorbed about a week's work for two people, and it was an important work since it had to constitute the only dwelling for the charcoal burners for all the months of work duration. The hut was as functional as could be made within a forest, but it is difficult to realise the discomforts connected with the use of such a primitive dwelling, especially with families: often the testimonies of the charcoal burners report that they "lived like beasts". These are structures with a decent variety of typology on a local scale, but for the most part entirely built with materials of vegetable origin coming from the forest itself. In Mugello, a hut built for 3–4 charcoal burners was about 4–5 m long, 2 m wide and 2 m high (Landi and Piussi 1988). In Friuli, huts are reported built with a dry stone base up to 80 cm high, on which fir logs were placed, and with double-pitched roofs. The location of the hut was sought within the work area, possibly near a spring and in a shaded place, it had inside rudimentary beds, on which they slept dressed, seats and some minimal furnishings.

10.4 Charcoal and Landscape

Charcoal burning has left many traces in our forest landscape. Charcoal remains from former charcoal kilns provide information at fine spatial scales, capturing local woody-species composition and revealing the history of landscape management (Ludemann 2003). In Italy is quite common to find remains of charcoal kilns platforms in areas protected for the naturalness of the vegetation, suggesting a previous history of forest utilizations.

Given the prevalence of mountainous and hilly areas in our country, the charcoal platforms was generally a small semi-circular terrace more often of earth, which created a flat surface, often arranged along very steep slopes. Such platforms still dot many Italian woods with densities that can reach up to 10 per ha. To give a point of comparison in order to assess the intensity and role of charcoal burning in the Italian landscape compared to other countries, we can remember that in Slovakia same case studies shows a density of 0.75 per ha (Máliš et al. 2021), while in Sweden an average of 1 every 12 ha has been detected. This explains why our woods were not only strongly affected by charcoal production but also very populated, if we consider all the activities occurring in forest areas. The intensity of the Carbonisation is one of the explanation for the conversion of many high forests to coppice woodland, which is the main management form in Italy. This was not only because charcoal was preferably produced from coppice, given that the diameters of the shoots were smaller than a trunk of a mature tree and did not need to be split to build the charcoal pile, but also because the quantity of assorted timber produced was higher than that produced by the trunks of trees grown in tall forests.

The population's consumption occurred with different models for the city and the countryside. In the city, the consumption of wood mainly used wood and charcoal from forests, while for the rural population the situation was different. Before the second world war, it was much higher than that of the urban population (a ratio of about 5:1 in 1911), but its needs were more often met with the use of wood collected outside the forest, with the cutting of trees, hedges and pruning of agricultural plants,

Fig. 10.4 A charcoal burning platform in a beech forest (Moscheta, Toscana region). These structure can be found all over the mountain areas in Italy, most often they are not recognized for their function both by tourists and sometime even by managers

a production that seems to increase by almost 50% between 1861 and 1912, going from 9 to 13.5 million m^3. As we already said at the national level, firewood did not predominantly come from the forest, in fact, that coming from trees outside the forest has always been almost double compared to that obtained from proper forests. In Lombardy in the early decades of the century, it was estimated that about 2.5 m^3 per hectare per year were drawn from the trees in agricultural areas, divided between 1.5 m^3 of firewood and 1 m^3 of timber, almost all consumed internally to the farm and therefore escaped from the influence of the timber market (Fig. 10.4).

This is a value equal to the average annual increase of all national forests, but higher than the average increase of Lombardy forests (Federico 1946). These data combine well with what we have said about the spread of mixed crops and the planted Padana, since in traditional agriculture the density of trees planted in the fields was very high, higher than that of many forests, from which many products were drawn, including firewood. This rural "self-consumption", partially detached from the wood and charcoal market, is therefore an element to always keep in mind when linking the production of wood and charcoal to the issue of degradation of Italian forests (Agnoletti 2010). It is in fact quite common in the international scientific literature to find articles concerning forest degradation, or even the destruction of forests due to charcoal production. As some research findings suggest, coppicing taking place over centuries has often enhanced diversity of forest understorey vegetation, suggesting strong links between current biological processes and the historical human legacies

(Máliš et al. 2021). However, it always depends on whether this activity, as any other utilization of the forest, are carried out following proper management rules.

References

Agnoletti M (2010) Paesaggio rurale. Strumenti per la pianificazione strategica. Edagricole, Milano

Federazione Italiana dei Consorzi Agrari (1954) Enciclopedia agraria italiana. Tip. Ramo editoriale degli agricoltori, Roma

Federico S (1946) Problemi forestali in Lombardia. L'italia Forestale e Montana 1(5):191–214

Landi M, Piussi P (1988) Il lavoro nei boschi. Boscaioli e carbonai a Luco e Grezzano tra il 1930 e il 1930. Comunità montana Zona E, Gruppo d'Erci

Ludemann T (2003) Large-scale reconstruction of ancient forest vegetation by anthracology—a contribution from the Black Forest. Phytocoenologia 33:645–666. https://doi.org/10.1127/0340-269X/2003/0033-0645

Máliš F, Bobek P, Hédl R et al (2021) Historical charcoal burning and coppicing suppressed beech and increased forest vegetation heterogeneity. J Veg Sci 32:e12923. https://doi.org/10.1111/jvs.12923

Piccioli L (1923) Selvicoltura. Seconda edizione riveduta e ampliata. Utet, Torino

Serpieri A, Segala G (1917) Il legno greggio. Comitato nazionale per le tariffe doganali e per i trattati di commercio, Roma

Chapter 11
The Landscape in the Short Century

Abstract From unification of Italy, it is possible to follow the transformations of the surfaces, and some elements related to the governance and structure of the woods with a certain approximation, thanks to the availability of sources for the whole country. The analysis of the available data presents some difficulties, mainly due to uncertainties about the methods of detection and the lack of systematic detection. There are also difference in what is considered "forest" in terms of a surface covered by a certain number of trees that can be legally defined forest. The last data available comes from the third Italian national forest inventory of 2015 (CREA in Le Foreste Italiane, 2021) that estimated the total wooded area to be more than 11,054.458 ha which is 36.7% of the country area. It includes the high forest and temporarily unstocked areas, plantations for timber and wood production and other wooded land (short trees forests, sparse forests, scrubland and shrubs). The first three categories cover 30.2% of Italy and make 82.2% of the Total wooded area. According to the most reliable statistics, after a decrease occurred towards the end of the nineteenth century, between 1936 and 2015, the extension of the forest area has more than doubled. The spread of dense and homogeneous forests has contributed to decrease the diversity of the landscape mosaic. Due to the disproportionate increase of forests, negatively affecting the conservation of the landscape, the Regional Government of Tuscany has produced a law allowing the removal of forests on previously cultivated land. Despite this very significant increase of forests, entirely due to socioeconomic factors, they represent only about 0.21% of the forest of the world, thus, their contribution to carbon sink is quite limited. According to scientific evidence there are no natural or virgin forests in Italy as all the Italian forests have been influenced by human influence. In the inventory, there are no data on forests having historic and cultural value. Social and cultural values represent the third pillar of sustainable forest management in Europe and that forest are a part of the cultural landscape of the nation, a new law is going to come into force to detect and preserve them.

M. Agnoletti, *History of the Italian Forest Landscape*, Environmental History 19,
https://doi.org/10.1007/978-3-031-81631-4_11

11.1 The Dynamics of Forest Surfaces

In the previous chapters, we outlined the genesis of the forms of the forest landscape, demonstrating the cultural matrix of Italian woods. From the post-unification period, it is possible to follow the transformations of the surfaces, and some elements related to the governance and structure of the woods with a certain approximation, thanks to the availability of sources for the whole country not available for previous centuries. It is also a particularly interesting historical period, especially from the second post-war period onwards, when the events of the forest increasingly separate from those of the economy and the needs of the population. It is the period in which the forest issue, meaning also the set of laws and regulations concerning the woods, is now affected by the perception of an almost entirely urban population and cultural influences characterised by global scale issues, linked to environmental themes and proposed by a long series of United Nations initiatives that have to do with the concept of sustainability. These are initiatives that do not act directly on the woods, as happens with a cut for economic purposes, but rather indirectly, nevertheless determining a transformation of the landscape. They are therefore to be considered in all respects factors capable of acting on the dynamics of the forest landscape.

The analysis of the available data in forest surfaces presents some difficulties, mainly due to uncertainties about the methods of detection and the lack of systematic detection. This is a problem that deserves some in-depth study, especially for the period around the turn of the century, when forest and social dynamics seem particularly correlated, but the data are also more uncertain. As for the inconsistencies between the various surveys and a certain "unreliability" linked to some statistics, especially those of ISTAT (the Italian governmental statistical agency), it should be noted that even the most recent surveys present methodological settings with significant differences in the final data, which must be taken into account. For example, the first national forest inventory of 1985 indicated the extension of the woods at 8,302,000 ha, in the same period ISTAT indicates 6,822,000 ha, while the FAO calculated the Italian forest area to be more than 10,000,000 ha.[1] These are differences perhaps more important than the possible detection errors attributable to surveys carried out between the second half of the nineteenth century and the early twentieth century, despite the availability of certainly more sophisticated detection technologies. It also highlights the different meaning assign to the concept of "forest" across time from the technical and scientific point of view, that has caused uncertainties even from the legal point of view, without forgetting further difference due to the meaning of forest among local populations.

The last data available comes from the third Italian national forest inventory of 2015 (CREA 2021) that estimated the total wooded area to be more than 11,054.458 ha which is 36.7% of the country area. It includes the high forest and temporarily unstocked areas, Plantations for timber and wood production and Other wooded land (short trees forests, sparse forests, scrubland and shrubs). The first three categories cover 30.2% of Italy and make 82.2% of the Total wooded area. They meet to the FAO definition of Forest: the canopy cover of tree species must be able to reach 5

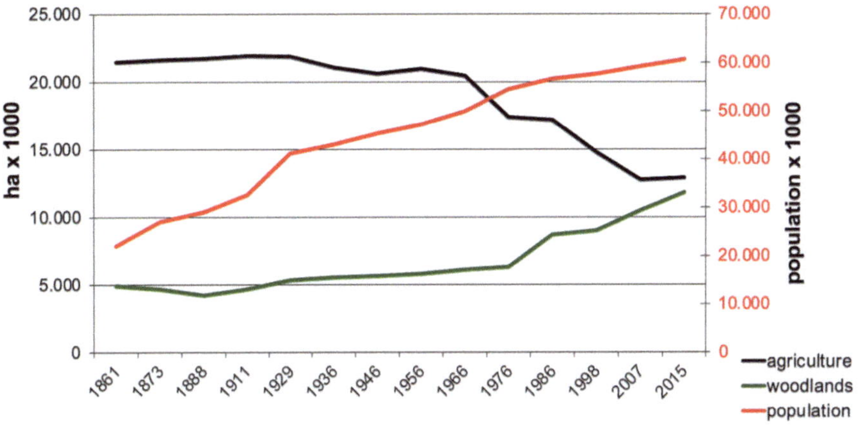

Fig. 11.1 Trend of forest, agricultural surface and Italian population from 1929 to 2015

m height in situ and must be more than 10% on an area larger than 5000 m^2 and wider than 20 m; the predominant use must not be agricultural or urban land. They include small open areas, rivers and small water bodies below the above-mentioned thresholds, forest roads and temporarily unstocked areas due to clear cuttings or natural disasters. According to the last national inventory if we consider only high forest Tuscany is the most wooded region, but if we consider all the categories, specially scrublands and shrubs, Sardinia is the most forested region (Fig. 11.1).

In order to understand the variations of the forested surfaces from the unity of Italy, various statistics were considered, among which the first national estimate of 1862 should be mentioned, which was heavily criticised by ISTAT, which questions its sources and also notes the absence of specific descriptions of what is meant by forest (ISTAT 1958); in the estimate, moreover, data for Veneto region and Rome (State of the Church) were missing, because they were still not included in the Italian Kingdom (Valenti 1911), in addition to chestnut groves considered as agricultural surfaces and not forests. This last aspect is particularly interesting to evaluate what was meant by forest in those years, and the absolute importance of the fruit chestnut grove in the rural economy. In addition to the statistical Annals, the Forest statistics ordered by Minister Castagnola in 1868 should be remembered, which however did not include the province of Rome, nor do we know if it included the chestnut groves. Although it was a work carried out using forest agents who examined land registers and archive maps, with results gathered by municipalities, districts and provinces, it was not immune to the criticisms of the subsequent 1874 statistics.

This concerns a survey that was part of a more general study on the agriculture of the Kingdom (Ministry of Agriculture, Industry and Commerce 1876), in which it was criticized the non-inclusion of chestnut groves in the previous one and the inclusion of surfaces that in reality were not forests (bushes, uncultivated lands, etc.), but which at the same time gives the Castagnola statistics a certain modernity, given today's conception of forest that also includes bushy surfaces. The 1874 statistics in turn lacked much of Tuscany (only the province of Lucca was included), which was finally included in a work of 1886 that integrated the previous one and where an estimate of the average extension of the forests between 1876 and 1881 was also produced (Ministry of Agriculture, Industry and Commerce 1886). Other information on forests is contained in a study of 1892 carried out by the Ministry of Agriculture, Industry and Commerce as part of a general agricultural statistics, for which there are however no detailed news on the methods followed (Ministry of Agriculture, Industry and Commerce 1894).

For the beginning of the new century, the work of Lunardoni of 1904 is fundamental. It presents a table concerning the forests existing until 1870, the state owned forests non existing any longer, the forests removed from constraints and forests turned into cultivated areas, thus providing an important estimate of the deforestation that occurred also for the disastrous outcomes of the 1877 law. This is an important work, which contrasts sharply with the data from the agrarian land register that will be developed in the following years, and on which many of the following statistics are based. This discrepancy also sparked a heated parliamentary debate on the calculation methods of the lands previously forested and then ploughed in the new land register, which indicated a total extension of the forests greater than that of 1870 but judged very unlikely (Lunardoni 1904). The first data of the new land register relating to forest surfaces were published by the Ministry of Agriculture in 1913, but only for some provinces were the new data used; for 9 of them, the data from the pre-unitary land registers were still used and for 16 the topographic maps of the Military Geographic Institute (Ministry of Agriculture, Industry and Commerce 1913). Other interesting data on forest surfaces come from the work of Carloni of 1925, proposed within a volume on the Italian forest situation, but also in this case the author does not cite the origin of the data (Carloni 1926).

From these years until the second post-war period, the most important estimates are those of the agrarian land register and the Forest Map of Italy of 1936, while from 1950 onwards the ISTAT series are available which present a continuity and methodological uniformity that at least allows the comparability of data (National Committee for the celebrations of the centenary of the Unity of Italy 1961; National Forest militia 1936; Agnoletti et al. 2022). Through these surveys it has been possible to build the forested surface series, an operation that partly reflects the work of Bardini who corrected discrepancies due to the uncertainties of some surveys, applying an index to interpolate the data for the period 1862–1913 (Bardini 1991). However, we must emphasise that, even when only reporting statistical data, the general trends remain unchanged. For the most recent years, the national forest inventories provide a series of data that have the merit of using a comparable methodology, so from 1985 onwards we can say that the statistics present characteristics of greater reliability even if our work mainly wants to highlight trends rather than the detail of the numbers (Fig. 11.2).[2]

The main data commonly considered for forest studies is that relating to the extent. It is observed that the period between the Unification of Italy and the years immediately preceding the First World War is the one in which an important contraction of the Italian forest surface. The data shows a continuous process of reduction from

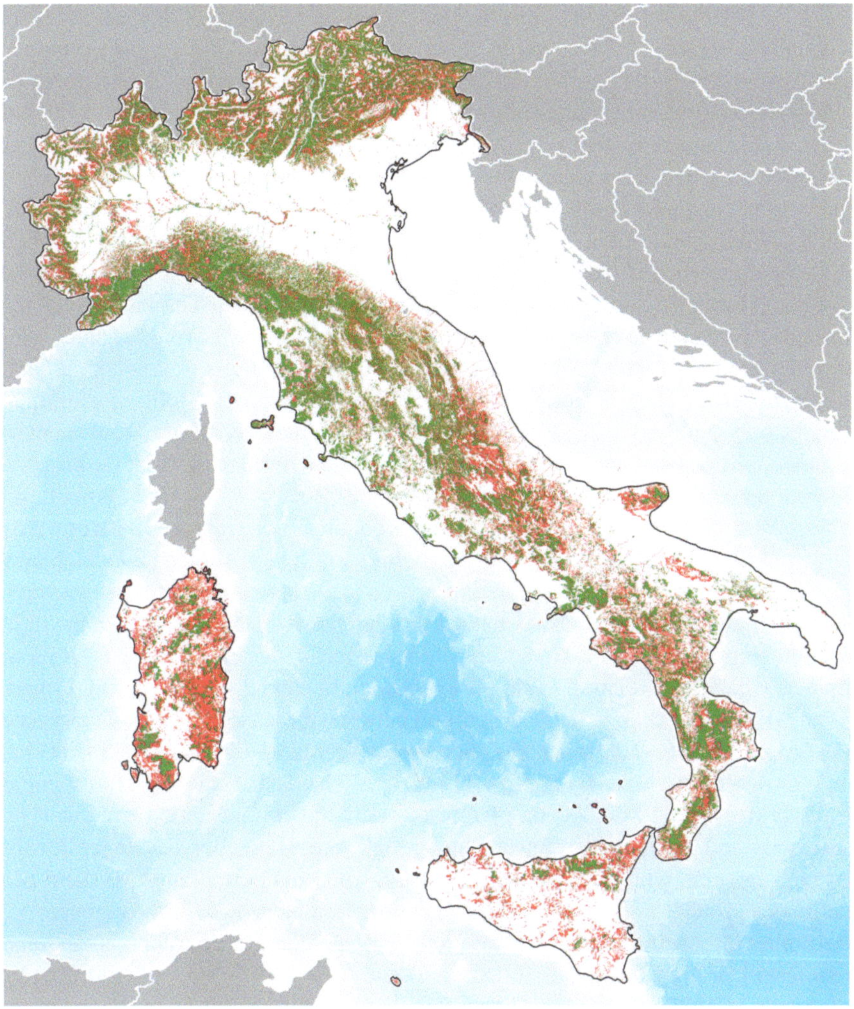

Fig. 11.2 Forests existing in 1936 (green color) and new forests grown from 1936 to 2018 (red color). The new forests are almost entirely the result of the abandonment of cultivated land and pastures. The forest has advanced at the rate of 116,000 ha/year (Agnoletti et al. 2022)

1871, even more significant if we consider the annexation of Veneto region in 1867 and Lazio in 1871, which brought about 550,000 ha of additional forests. The deforestation reported in the period between 1874 and 1912 can be estimated at around 30,000 ha per year. The reforestation that occurred in the same period, amounting to 37,595 ha, is a very modest result compared to the initially declared intentions, as we have seen in the chapter on afforestation the judgement must take into account factors extremely unfavourable to the realisation of extensive reforestation, especially in mountainous areas (Agnoletti 2002a, b).[3]

The intense deforestation of this period naturally does not exclude the possibility of equally important or even more significant phenomena occurring in previous periods, as indicated by Serpieri (1926), and which at the European level seem to be confirmed also by recent studies of forest history (Agnoletti and Anderson 2000) but which in this case must still take into account the specificity of local situations, as demonstrated for Italy by recent studies for the first decades of the 1800s (Armiero 1999). It is not possible for reasons of space to analyse the phenomenon for all regions of Italy, but it is interesting to note that this trend appears to be confirmed for Lombardy, while in Tuscany there is a continuous growth of forests from 1870 onwards (Martegani and Agnoletti 2003; Agnoletti 2002a, b). The increase in the forested area detectable between 1914 and 1929 is partly due to the forests gained with the annexation of Trentino Alto Adige and Friuli Venezia Giulia, following the First World War. However, it is not possible to explain the nearly two million additional hectares only with the acquisition of forests from the new territories, equivalent to about one million hectares (Tofani 1970), we must therefore also consider the possible different detection methods used in these years, especially in the 1929 agricultural land register, as already explained. What seems certain is that from the 1920s onwards there is a stable reversal of trend, which remains constant until today and which now sees the forest area more than doubled compared to the period between the two wars.

Another element that can be evaluated through the available statistics is the ratio between coppice (low forest) and high forest, even if the data can be even more imprecise than those relating to simple extension. As can be seen in Fig. 11.1, there seems to be a great change in the proportion between these two forms of governance between 1868 and 1911, since the high forest decreases by about 1,850,000 ha, and the coppice increases by about 1,114,000 ha. The decrease in high forest could be partly due to the 800,000 ha of forests destroyed in that period estimated by Lunardoni, and partly to the conversions of high forest into coppice. Regardless of this phenomenon, which we will address later, from this period onwards there is a continuous growth of the high forest, and a more limited growth of the coppice.

Another interesting fact is the variation in the percentage between conifers and broadleaves in high forests. Between 1870 and 1925 the former seem to double, going from 450,000 to 1,000,000 ha, while the broadleaves go from 2,556,000 to 1,000,000 ha. Probably a decrease in broadleaves can be attributed to the already mentioned process of deforestation and conversions, but it would remain to explain the increase in conifers in such a short period. The proportions between the two groups seem to remain unchanged until the second post-war period, when a slight increase in conifers and a progressive reduction in broadleaves begin, but even for this period the data generate some perplexity. In fact, although the total of reforestation up to 1950 is around 194,000 ha, those that occurred in the subsequent period (about 850,000) would bring the total reforestation carried out in Italy from 1862 to more than 1,000,000 ha, mostly composed of conifers (Romano 1986).

In this case, as in others, it would be interesting to delve into the situations at the regional level, comparing them with general trends. In the national panorama, there are certain testimonies of the increase in conifers, as demonstrated by the already described increase of a third of conifers in Trentino, the most important Italian region for timber production, and the halving of the extension of the coppice in the same region. Striking in the south is the fourfold increase in the extension of the forests in Sardinia from 1929 to today, and the general reduction of the coppice with standards, almost disappeared in Campania from 1947 onwards, with the consequent increase of the simple coppice. Sardinia is actually a very interesting case of that little adaptability of the scientific concept of forest to a reality in which grazed forests, tree pastures, bushes and pollards dominated a landscape in which the forest, understood as continuous and spatially well-defined coverage was very rare, representing however a faithful mirror of the needs of the local economy as well as of the deforestation that occurred in Sardinian history. The surveys of the 1985 national forest inventory indicate as many as 976,500 ha, but about 65% of the surfaces reported by it as "forestry" refer to shrubs, rocky and riparian formations, while only 22% are high forests and coppice forests.[2]

As we have seen, the latest statistical surveys of 2015 confirm the further advance of the forest, now covering almost 30% of Italy, this advancement affected all the regions of Italy. The size of this land use change, entirely due to socioeconomic changes, affecting more than 5,000,000 ha, is higher than the predicted land use change that might be due to the projections on global warming for the end of the century. The national statistical agency has also estimated that the comparison between the rate of urban sprawl and abandonment between 2001 and 2011 shows that abandonment prevails. It is worth noting that despite this increase Italian forests represents only about 0.21% of the forest of the world, thus, their contribution to carbon sink is quite limited at world level. At the same time, we can observe that Italy is still importing more or less the same amount of timber indicated at the end of the Nineteenth century, if not more.

The inventory of 2015 introduces two new categories, "natural forests" and "virgin forests". According to these estimates, the first would amount to 6,671,331 ha, the second to 1,485,354 ha. Considering the historical evidence presented in this volume and the fact that, for example, in the inventory carried out by the State forestry Corp in 1936 the forests were just over five million and at the time were all utilized for different purposes, it is not clear how these new categories have been detected (Agnoletti et al. 2022),. The forests regrown from then to today, more than six million hectares, even if they had never been utilized do not have a natural origin, having originated on soils previously occupied by other crops. From the ecological point of view, they should be classified as secondary successions, no matter if the resemble theoretical natural models. The justification offered, for this inclusion in the statistics, is that the classification was assigned to forests that no longer show visible traces of management carried out by man (CREA 2017), that is exactly the logic of the rules establishing the national list of Old Growth Forests. The explanation appears questionable from a scientific point of view, hinting at motivations more than anything of a political nature. In the inventory, there is no detection of forests of cultural value, which, properly considered, would also have the right to be recognised and protected. As those who work in statistical sectors and in planning well know: "if you don't measure it, it doesn't exist". As we already wrote, the situation described is the result of a multiplicity of factors that have led the forest landscape to be determined no longer by the economy, but by the perception of the landscape and the environment developed in society in recent decades.

11.2 Forests, Society and Economy

Among the factors responsible for the changes described especially for the period straddling the nineteenth and twentieth centuries, demographic growth has undoubtedly played a fundamental role. The processes in place during the period from the Unification of Italy to 1925 caused the doubling of the Italian population and gave rise to a real "assault on the mountain", a phenomenon that had no equal in modern times, affecting the entire national territory and bringing the mountain population from about 5,000,000 to 8,500,000 inhabitants in 1951. This increase did not occur uniformly in the various parts of the country, assuming different growth rates in the various areas, but had significant consequences at the forest level, which the foresters did not fail to notice. For the Apennines, there is a continuous increase in population with a slight reversal of this trend towards the 1930s in the northern part and towards the '50s in the southern part, while the Alps and Prealps show a trend of continuous growth of the resident population even in the post-war period. The need for new arable land was not solved with an increase in unit production through technological development, but rather with the extension of the cultivated area, also involving marginal areas, cultivating very steep areas and reclaiming marshy lands (Fig. 11.3).

As already seen for Trentino, it is in this period that the national dependence on imported timber and the development of a processing industry (essentially furniture), with the progressive reduced importance of the production of national forests. This also explains the attempts to increase the extension of high forests, especially those of conifers, since the most requested construction material came from the alpine

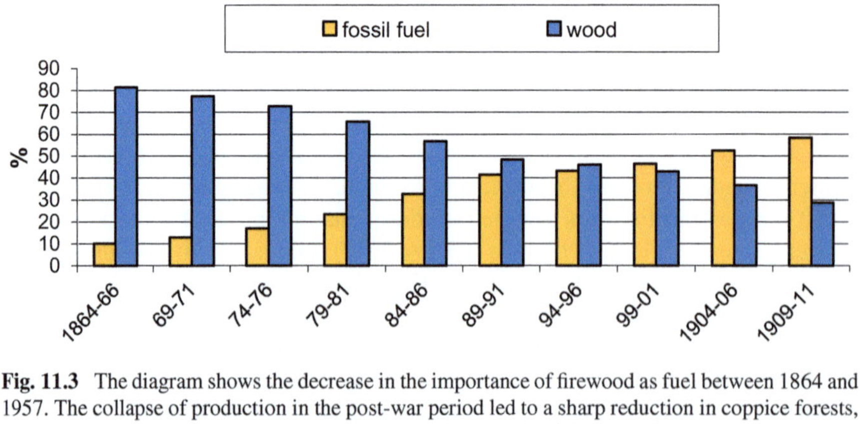

Fig. 11.3 The diagram shows the decrease in the importance of firewood as fuel between 1864 and 1957. The collapse of production in the post-war period led to a sharp reduction in coppice forests, from which a lot of firewood was extracted, and an increase in high forest, due to the conversion of the coppice into to high trunk or to the ageing of coppice woods

conifer forests. The incidence of charcoal production and construction timber had little impact on deforestation compared to the expansion of agricultural lands that were pushing higher up in the mountain, where most of the forest resources were concentrated.

Against the set of these large processes that invest Italian society, legislative measures could do little, and the modest reforestation occurred testify to this failure. The Luzzatti law of 1910 had little influence in changing the situation, more than anything else linked to the establishment of the State Forest Property and which mainly testifies to the changed attitude of the State towards the commitment in forestry matters, and to the reduced use of resources to solve the problem of the mountain. The influence of the 1923 law seems to be greater, but the feeling is that the increase in forests that occurs in the '20s, and more generally after 1950, is due to the changed relationship between population and natural resources. Certainly, the increase in population ceases to produce deforestation when the forests are no longer threatened by the extension of pastures and crops. This happens due to a complex set of factors that can also be partly attributed to technological development in agriculture, which no longer needs to extend surfaces to increase production, and because the need for agricultural goods and timber is met with imports, but also due to the lesser weight in the national economy of the rural sector recorded after the crisis of 1929 and the increasing imports of food from abroad.

As can be seen from the diagram in Fig. 11.2, in the post-war period new energy sources replace vegetable fuels, reducing pressure on the forest, but above all industry and services absorb more and more labour that this time also abandons the countryside, not just the mountain. Only when these conditions occur can the forest begin to reclaim abandoned lands. With economic progress and the increasing marginality of forest resources for the life of the populations, the social perception of the forest also changes, increasingly often seen not as a product of human work but as a natural component of the landscape.

As demonstrated by the data, the increase in the forest is specular to the reduction of agricultural surfaces, equal to more than 9,000,000 ha, a symptom of a transformation that reflects the dominant role of socio-economic factors in the transformations of the Italian landscape. This is an abandonment that advances at a speed of about 118,000 ha per year in the period considered, depending on the statistical data available. This has happened in the face of a strong increase in urbanised areas, although 90% of this occurred around already urbanised areas. As a result of this process, every Italian, who theoretically has about 5000 m^2 of land available for survival, can actually only rely on 1500 m^2 of cultivated areas for their sustenance, given that the rest is one-third forest and the rest are areas in transition or already urbanised.

Projecting this trend into the future, as is done for climate change scenarios, and adding other statistics, we have a very worrying situation. According to ISTAT, in the last ten years we have lost about 1,500,000 ha of cultivated area, which today amounts to 12,885,000 ha, and if we continue like this, by the end of the century we could have lost almost all cultivated land. Considering the annual increase in urbanisation and adding it to the current "urban poles"—a definition used in the national strategic plan for rural development that includes the presence of widespread urbanisation in addition to fully urbanised areas in the national rural development plan—we cover about 3,320,000 ha, 11% of the country. If we keep the data on the increase in forest in the same period constant, we would end up with almost 18,000,000 ha of forests, about 60% of the territory. Despite the obvious difficulty in determining which of the trends illustrated will remain constant, it is likely that we could end up with only large metropolitan areas surrounded by some peri-urban agriculture, and the rest of the territory no longer productive but in transition, with predictable economic, social and environmental consequences. The abandonment of the countryside and the increase of the forests is a wide spread phenomenon in Europe, according also to data made available by the European Environmental Agency, affecting many countries from Scandinavia to Italy, from Poland to France. It is interesting that compared these trends to the deforestation and desertification scenarios predicted as a result of global warming.

The increase in urbanised areas and that of forested areas represent the two sides of the same problem, namely abandonment. They essentially sanction an epochal transformation of the landscape, which took place in just 100 years, in which the forest has assumed the role of protagonist. Within agricultural areas, the most significant decline concerns the reduction of arable land and then that of meadows and pastures. The reduction of cultivated areas has only been partially compensated by an increase in productivity. Food dependence is also a cause of ecological imbalances as the ecological footprint of Italy is increasing also because we let other countries produce the food for us. We should also remember that the production of food outside the country contribute to the emissions of CO_2, therefore, the positive effects of the increase of the Italian forests is matched by the increased emission in other countries.

Various factors have contributed to the profound change in the rural landscape in the last eighty years. Among these, it is worth mentioning demographic movements from the mountains to the plains and the spread of innovations determined by technological progress, such as fertilisation and chemical weed control in addition to mechanisation, which have ended up favouring rather than curbing the exodus from the countryside. The use of agricultural machinery, which has significantly reduced the use of labour, together with the industrialisation of the country, has contributed to the abandonment of many cultivated areas, starting from marginal mountain and high-hill areas. This evolution has also been accompanied by the transformation of the corporate structure of agricultural companies, which have significantly reduced their number.

Essentially, a polarisation of the landscape has occurred, with large urban areas in the plains, around which are the areas of intensive agriculture. In fact, industrial agriculture in Italy occurs entirely on the plains. Nevertheless, the increase of the forests was not limited to mountains and hills but has affected all the altitudinal bands. Currently 37.7% of the forests is located between 0 and 500 m AMLS, 35.7% between 500 and 1000 m, 26.6% between 1000 and 2000 m and above. It is clear that, faced with an increase in forest there is a decrease in herbaceous species linked to meadows and pastures (Cevasco 2007) and animal species linked to cultivated habitats, as well as landscape-scale diversity. In this sense, the investigations carried out by Almo Farina are symptomatic, according to which the reduction of areas occupied by olive groves and the evolution towards greater forestation result in a decrease in avifaunal diversity (Farina 1993).

From a silvicultural and landscape point of view, it would be more desirable to have fewer but better managed forests, with a higher level of spatial diversity and control of fauna, especially predators and ungulates, which increasingly encroach on cultivated and inhabited areas in search of prey and easier food. Animal species that have natural habitats of a few hundred kilometers, as the wolf, can only come into conflict with a national density of 200 inhabitants/km^2 agricultural activities and wild grazing. In this sense, the abandonment of agriculture and forest management in the hilly and mountainous areas goes hand in hand with the concentration of the population in large urban areas and industrial agriculture, both occurring in the plains. This has created an interesting coincidence of interests, certainly not intended, between a certain type of environmentalism, agro-industry and construction industry. All of them contribute to the abandonment of rural areas, especially the marginal ones, albeit for different reasons: the first to facilitate the return of nature and wildlife, the second for to higher production costs of hills and mountains, the latter to increase the extension of built up areas.

11.3 Tuscany

Tuscany is the region of Italy with the largest extension of forests (not including scrublands), but also the first for the development of agritourism, a fundamental economic activity within the context of rural economy. More in general, landscape is considered a fundamental economic, environmental and social asset, therefore the regional government attributes an important role also to the conservation of the rural landscape and on maintaining a balance among agricultural, forest and urban areas. Tuscany was the first region of Italy to develop a landscape plan, approved in 2014, and establish a landscape observatory, according to the National code on cultural heritage of 2004. The report on landscape policies in Tuscany present the first official analysis on the landscape changes affecting also the rural territory as well as the first results of the application of the plan (Regione Toscana 2020) (Fig. 11.4).

A research we carried out in 12 study areas—equivalent to about 1% of the territory—located in different landscape types analysed the dynamics of the landscape over a period of about two centuries (1832–2012), with economic, ecological and historical investigations, the results of which can provide a point of reference on a larger scale compared to processes that have occurred in many areas of the national territory (Agnoletti 2002a, b). A strong reduction in landscape diversity was observed, amounting to about 70–75% in the study areas analysed, mainly due to the increase in forest in the mountainous and hilly areas abandoned by man. In the plains, on the other hand, a strong reduction in diversity is also observed, but this is due to the elimination of trees and the extension of monocultures, as has happened in other areas of Italy (Cazzola 1996; Martegani and Agnoletti 2003). From a quantitative

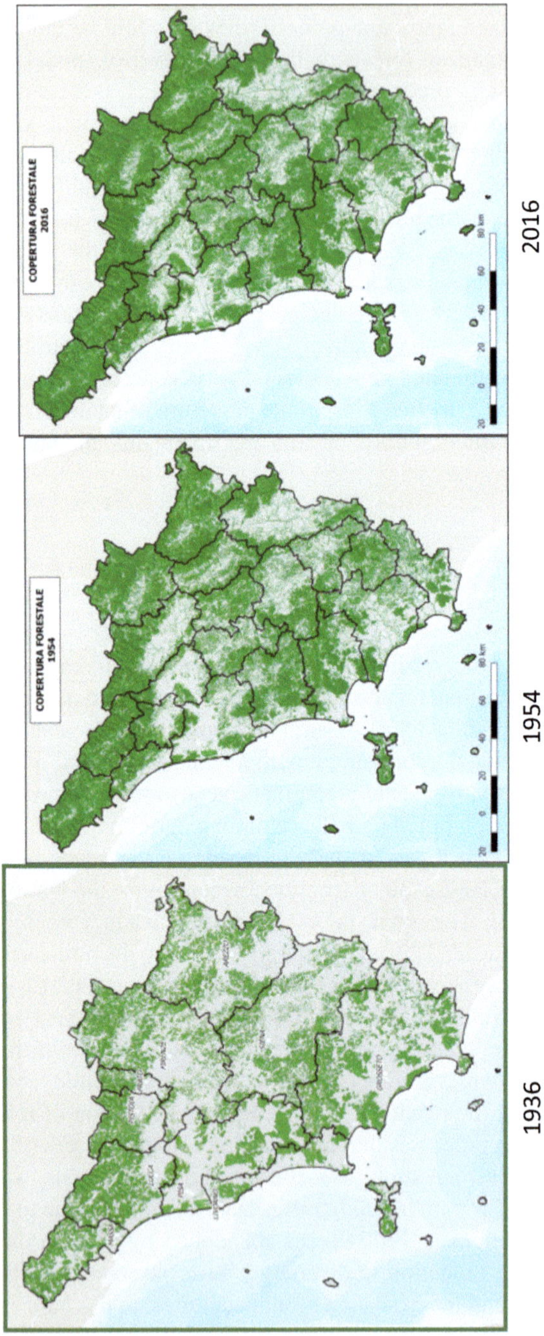

Fig. 11.4 In 1936 in Tuscany the forests had an extension of 870,735 ha; in 1954 the surface was 1,050,832 ha; in 2016 1,208,849

point of view, the major changes in the landscape concern the processes of afforestation that have affected 21% of the territory, to which is added a 6% of phenomena related to "coniferisation", due to reforestation and the development of pine tree (Pinus pinaster) forests in former pastures and fire-affected areas. The data from the Tuscan study areas confirm the trends already reported at a national scale, which identify the forest as the main dynamic factor, affect 37% of the total landscape changes recorded in the last 200 years. The growth of forest on pasture and former agricultural areas represents the process that imposes itself on others in this period. More specifically, there is a strong decrease in wooded pastures, while in the former nineteenth-century pastures, the greatest number of reforestations have also been carried out, introducing mainly artificial monocultures of black pine, as we have already observed in Chap. 7, that should have been replaced by native broadleaves, but have instead remained abandoned without silvicultural care.

The situation is more complex if we consider the transformations within the "forest" category, with the great growth of coppice forests that have almost tripled their extension since 1832, mainly at the expense of high forests, as already observed at a national level. A portion of these new coppice forests is due to the transformation of fruit chestnut groves (30%) into coppices, due to the abandonment of their cultivation and the diseases affecting this species, with an overall decrease of almost 80% of their surface, confirming the continuous loss of a significant cultural landscape. Breaking down the general data into two periods, 1832–1954 and 1954–2012, it is clearly observed that in 1954 almost all the transformation of wooded pastures into forest and chestnut grove into coppice forest had already occurred, in addition to the most important processes of intensification. In the second period (1954–2012) at least 77% of the territory remains unchanged in the main categories, with 10% of forestation processes and 8% of extensification. This demonstrates the greater importance of the first period (1832–1954) in terms of the quantity of changes that have occurred in the Tuscan landscape compared to the post-war period.

From the 1950s onwards, a small portion of the chestnut groves is still transformed into high forests and reforestations, while the recovery of the high forest mainly occurs in tree crops and abandoned crops. As expected, there is an increase in shrubs in the former cultivated areas and in the pastures, but a certain quota that develops in the former high forests should not be overlooked, a process in some cases because of fires, after which shrub vegetation and pines settle. Certainly the shrub, therefore the Mediterranean scrub in the coastal areas, has now almost completely changed its function within the landscape, going from an important element of the agro-forestry-pastoral regime, cultivated to obtain fundamental assortments for agricultural activities, to a relic element of a previous landscape, often indicated as a degradative state of the forest vegetation, not understanding its original function anymore. Within the "forest" category, the high forest has remained over time the most stable form of landscape in terms of extension, in fact about half of the previously existing surface has never changed its main destination. The most significant variations are those that see the transition from high forest to coppice, followed by the reverse transition, from coppice to high forest, followed by that from chestnut grove to coppice.

Analyzing separately the first period (1832–1954) and the second (1954–2012), it is noted that in the first 122 years, as we have seen, most of the transitions from high forest to coppice have already occurred, while in the period 1954–2012 no important transitions are recorded, confirming that the push to coppicing mainly occurs in the nineteenth century. As for the crops, between the 1800s and the post-World War II period, there has been a predominance of the reduction of mixed cultivations with trees. This confirms the transformations of agricultural techniques that see the progressive uselessness of trees in the fields and the creation of large plots, that is, mechanisable monocultures. Comparing these data with the historical maps available for all of Tuscany, it is confirmed that, despite the growth of new forests, about 380,000 ha in the last century, only half of the forests are still managed today, revealing the abandonment of many forests of historical value. Due to the continuous alteration of the balance between forest surfaces and agricultural surfaces, noted in recent years, and acknowledging the need to maintain it, the regional landscape plan has provided for the possibility of authorising the removal of woodlands developed on abandoned fields and pastures, a possibility now established by the national forest law of 2018.

11.4 Landscape Dynamics and Cultural Dynamics

The data concerning the relevance of the phenomena of abandonment and expansion of vegetation in previously cultivated areas take on a particular meaning if related to the forms of protection, an issue analysed in detail by various researches, among which the investigation for the constitution of the National Catalogue of historical rural landscapes carried out by the Ministry of Agriculture in 2010 (Agnoletti 2012). This investigation, the results of which led to the establishment of the "National Register of rural landscapes of historical interest, agricultural practices and traditional knowledge", intended to promote the protection of rural landscapes that have still preserved historical-cultural characteristics, among which some forests are also present. Analysing the areas identified by the investigation, the number of those affected by protected areas and sites of community interest (60%) appears high, but in all areas the processes of abandonment and reforestation appear more intense than outside them. These processes give rise to unmanaged forests, degrading the historical landscape, but most of all, to a number of problems due to a misleading application of the concept landscape conservation often intended as a "return to naturalness". This has little to do not only with the features of the Italian landscape but also with the need of effectively coping with climate change and biodiversity loss.

Examples of this situation can be found throughout Italy. In the Cinque Terre area, in the last thirty years, the phenomena of abandonment and reforestation have affected 70% of the terraced areas, nor has the establishment of the UNESCO site and the Park been able to change this trend. In this area, characterised by slopes that even exceed 60%, it has often been thought that the presence of the forest due to abandonment was at least a guarantee against landslides. In reality, some studies on the landslide events that occurred in 2011, due to an exceptional rain fall occurred in a very short time, show that 85% of the large landslides detected occurred on abandoned terraces covered by the forest and only 5% on terraces in good state of maintenance (Agnoletti et al 2012). The restoration of the historical rural landscape of Punta Mesco, owned by the Italian Environmental Fund (FAI) and located within the Park, started in 2013, had to deal with the fact that the dry stone terraces subject to the project, abandoned and colonised by tree and shrub vegetation, were put inside a EU protected area established after the abandonment of agricultural activities and declared "natural" by law. This has created considerable difficulties in carrying out the restoration project ending with a law suite against the director of the project.

In the forest of Sant'Antonio in Pescocostanzo inside the Maiella Park—a forest of monumental beech pollard created by centuries-old grazing activities—the Park has banned the exercise of grazing to protect the natural habitat of beech, which in scientific terms does not exist, since it is a forest entirely shaped by man. The municipality has appealed to the Aquila court which in the first ruling has recognised that the rights associated with the historical use by the local community take precedence over the impositions of the Park. On the other hand the Park appealed stating that the municipality was acting for the private interest of shepherds while the Park acted on behalf of the "supreme interest of nature". Regardless of the final outcome of the matter, this small example reveals the conflict that exists today between a perception of the landscape detached from its historical context, tied mainly to urban classes, but also to many scientific studies, where naturalistic interpretations of the landscape have found greater resonance, and the rural population. Evidence of this disconnection can be found in many interviews with farmers and woodsmen, who recall how the woods were once cultivated "like a garden" and are now abandoned (Agnoletti 2002a, b). In many protected areas, therefore, orientations are favored that facilitate, directly or indirectly, abandonment. According to these guidelines, the habitats to be protected are the natural ones, even though there are no "natural" forests in Italy, but rather forests where management has been interrupted, or forests grown on abandoned land. As we have seen in the case of Cardoso, among the "natural habitats" protected by European directives we find "mixed woods dominated by chestnut", which are nothing more than an ecological phase of transition determined by the abandonment of an original fruit chestnut grove. The constraint imposed would make a restoration intervention of the original fruit chestnut grove, the real habitat at risk of extinction, a violation of community regulations, and of landscape and environmental constraints, with consequent convictions in criminal and civil court. Other examples are the recent one of the Marganai forest, in Sardinia, where a few years ago the restoration of the coppice management—already existing at the time when the landscape constraint was imposed to that area- was interpreted as a permanent modification to the state of the places and prosecuted by law. In this case the modification of the forest management, from high forest to coppice, thus cutting the upper part of the tree and leaving the stump from which the tree regenerates, was interpreted as a permanent destruction of the forest, while it was simply a phase of the management technique adopted.

These are not only Italian situations, to be clear. The famous "natural forest" of Bialowieza in Poland, universally known for being the last natural habitat of the European bison, has been brought to the attention of the chronicles for forest cuts made by the Polish authorities considered destructive for the natural habitat, also protected by UNESCO. In reality, environmental history investigations in that forest show a continuity of anthropic influences, such as fire, grazing, reforestation, forest cuts, etc., that have been going on continuously at least since the Middle Ages, demonstrating the little naturalness of the place, and the fact that the bison has evidently adapted to anthropic influences (Samojlik et al 2016). Such conflict situations are not the rule, fortunately. The Park of the Apennine rural landscape of Moscheta in Tuscany was realised thanks to a restoration of the historical landscape while being inside a protected area, adapting the management plan. Today the area is in fact included in the National Register of Historical Landscapes, demonstrating that protected areas could be aimed at the conservation of the rural landscape. There is therefore a problem of adapting the concept of nature protection to landscapes having a cultural origin.

The phenomena of afforestation, artificial and spontaneous, together with technological development in agriculture, industrial and urban development, have triggered new processes that are not always desirable, even from an ecological point of view, and that in some way contribute to the progressive deterioration of the quality of the landscape, including forest landscape. This situation, which may seem paradoxical, is also the result of cultural dynamics and scientific approaches. The movement towards a return to nature that swept across Europe at the end of the nineteenth century, thanks to Gayer, Biolley and Gurnaud, but also promoted in the United States by thinkers like Thoreau (1817–1862), considered a precursor of nature conservation, and then by George Perkins Marsh (1801–1882) and John Miuir (1838–1914) will indeed assert itself worldwide, influencing society's view of the landscape. Within a century, the complex of values proposed by the travelers of the Grand Tour, who, crossing Italy from the fifteenth century onwards, highlighted the characteristics of a landscape shaped by man, will be almost overturned. The ecologists of the twentieth century, from Clements (1916) to Odum, from Tansley to Forman and Godron, have proposed new values to which Italian culture as a whole has adapted, first of all in urban the upper-middle classes.

It is in this historical context that the theme of the protection of the landscape and cultural heritage also appears, first with the law proposed by the Italian philosopher Benedetto Croce in 1920–1922 and then with the law of 1939. We do not know if the appearance of landscape defence in the constitution of the short-lived Weimar Republic in 1919 influenced Croce. Certainly, asides from safeguarding monuments and works of art, the interpretation of the open landscape as a "natural picture" marks a strong difference with the social and economic nature of the Italian rural landscape, highlighted half a century later by Emilio Sereni in his History of the Italian Agricultural Landscape, proposing an idea of protection from which the current concept of landscape constraint springs. On the other hand, to which study current could philosophers like Croce, could have turned to propose an alternative vision? As Sereni himself explains in the introduction to his book, the only references could perhaps be found outside Italy, may be in the French school of the Annales and in the writings of historians like Bloch (1970), or in the thinking of geographers like Carl Sauer (1925) who defined the concept of cultural landscapes in 1925. In essence, the concept of natural beauty and its protection developed in the early decades of the century appear largely influenced by the settings of art history and the protection of the artistic heritage, particularly by landscape painting. The protection of the landscape in Italy is seen as the extension to nature of the principles used for art (D'Angelo 2014). Symptomatic in this sense is the frequent absence of correct interpretations of the rural landscape in the catalogues of painting exhibitions and museums, where it is almost always considered a natural backdrop to sacred images, or to human subjects depicted in the paintings of various epochs (Agnoletti and Signorini 2011).

The post-war period is dominated by international initiatives—such as the Stockholm Declaration of 1972, the Brundtland Report of 1987, as well as the UNESCO convention of 1972 which, in addition to monuments, safeguards natural heritage—that characterise a direction of the concept of sustainability increasingly oriented towards nature and the undisturbed ecosystem, as a model of reference also for conservation policies. Not coincidentally, the landscape only appears in the operational document Agenda 21, produced during the United Nations conference on "Environment and Development" in 1992, while in the same year UNESCO defines the concept of cultural landscape. However, although many agricultural landscape are designated in this UNESCO category, the word "agriculture" does not appear in any document concerning the world heritage convention. In Italy the famous Galasso decree, which in 1985 finally introduced "environmental goods" among the categories subject to protection, including mountain peaks, coasts, rivers, bodies of water, wooded areas, did not consider the landscape in historical-cultural terms, as forests are protected as an environmental value and agricultural landscapes are not even mentioned (Santoloci 2000) (Fig. 11.5).

It can be observed that, despite Sereni, the naturalistic vision of the forest landscape is the prevalent one in protection tools. In practice, naturalistic values, associated with forests, they become part of the cultural heritage protected by law, while forests with socio-cultural characteristics were not yet protected or recognized as such. It is not surprising that the result of such a static vision of the landscape, and in particular of the forest, is automatically applied to reforestation that occurs on abandoned agricultural lands, which once reclaimed by the forest are automatically "sealed" by the system of restrictions: all this in a current context in which only 28% of our forests are still cultivated.

There are obviously scientific approaches at the base of the directions aimed at attributing naturalness to a certain vegetational system, or at orienting management towards such an objective. At the base of such directions would be geobotany, whose roots can be traced back to Alexander von Humboldt and his philosophical conviction of the unity of landscape and nature. In its developments such science identifies the plant community as a set of species that occupies a defined space, to which the theories of Clements have applied a theoretical evolutionary model aimed at achieving the stage of maximum naturalness defined as *climax*. The system determined by the

Fig. 11.5 In the drawing at the top, the order of agricultural and forestry cultivations typical of the share cropping system present in central Italy from the thirteenth century until the '80s of the post-war period. Below, the effect of the development of forest vegetation after the abandonment of agriculture and forest management (Agnoletti [ed], *The Tuscan agro-forestry landscape,* cit.)

interaction between plant species in an area corresponds to the plant association that is the subject of phytosociology, developed following the studies of the Swiss Braun-Blanquet (1884–1980) and today widely used in Europe and in the world to interpret vegetation patterns at any scale. This view assumes that a certain environment, for example a mountain slope, corresponds to a certain community of plant species, and vice versa, with the result that the plant community present at the time of observation becomes a "biological proof" of the natural characteristics of that environment. Among the various factors that influence vegetation there is also human influence, understood however as factors that "degrade" the vegetation, interrupting natural evolution or returning it to primitive stages (for example from forest to scrub with fire). Each individual vegetation survey is therefore referred to a theoretical natural model, identified as potential vegetation, that is a higher order population to which the vegetation should tend, according to repeatable models on a geographical scale, wherever equivalent environmental conditions occur.

This view unfortunately does not consider the historical processes that occurred locally, which are instead studied by forest history and historical ecology (Agnoletti 2001), a problem already addressed by Carl Sauer's critique of "environmental determinist thinking", emphasising culture as a shaping force, as opposed to something shaped by nature. The difference between the two approaches is that the phytosociological approach tends to abstract from the historical-topographical context the place studied to refer it to a theoretical model of plant association towards which to direct management. The models created by phytosociology, when they define the evolutionary or involutionary history of landscapes, through the comparative analysis of current ecosystems with potential ones, hardly manage to take into account the complexity of the historical dynamics that have affected the current ecosystems. These are historical dynamics not necessarily coinciding with the dynamic series proposed by Clements with the concept of *climax*. In other words, the concepts of "potential vegetation", "climax", "evolutionary/regressive series" and "degree of naturalness" often obscure the reality of historical processes behind theoretical schemes that in most cases are to be demonstrated, but above all deny the historical-cultural values associated with vegetation structures (Cevasco 2007). This scientific approach therefore develops interpretative and management models poorly suited to landscapes largely determined by human action, also influencing the media and cultural models. Much of the forests today classified as "natural" or "semi-natural", according to a rigorous study could instead be classified as cultural or semi-cultural.

These may seem only questions of methodological interest, in reality such classifications are aimed at defining precise areas of cultural, scientific and political influence that turn into norms sanctioned by law. It is a contrast not only between scientific methods, but between cultural models. The first aimed at directing the management and attribution of value to a landscape to the extent that it approaches its greatest state of naturalness, regardless of its ties with history and human society, often seen as factors of degradation. The second aimed at valuing the result of the interaction between culture and nature. Associated to these scientific developments there have been also policies at university level, occurred in Italy since the 80s of the last century, aimed at removing the views and the skills related to silviculture and generally to an active forest management, replacing them with an approach basically dominated by "nonintervention". Politically speaking, this has also resulted in an improved capacity to collect research funding related to nature conservation and the climatic role of the forests, the consequent strong reduction of teachings and research activities related to the active management of the forests and to a public image of forestry more in tune with the perception of forests among urban population. A very different point of view is obviously the one of the rural population or any category of workers in agriculture and forestry, but they represent small minorities in our modern societies.

11.5 Conclusion

The forest has served mankind for millennia, without its contribution we could not have developed our civilisation, and the Italian landscape would not have assumed the characteristics that are typical of a cultural landscape. Unfortunately, the historical and cultural nature of Italian forests is little known and, so far, is not subject to particular protection. This awareness is the heritage of a limited number of inhabitants and scholars, given that now 95% of the population lives in urban areas, and a large part of it has been influenced by cultural models originating outside of rural society, and often even outside of the country. Visions of naturalness combined with landscapes created by man are therefore a constant in the way many people today look at forests. This has created a progressive detachment between rural population, urban population and territorial entities responsible for the governance of the territory, which often leads to conflicts.

Europe has never really addressed the problem of conserving traditional rural landscapes, whether agricultural or forestry. The lack of a directive on landscape from the European Commission is not surprising. In fact, the European landscape convention, a treaty signed in the year 2000 and proposed by Italy, France and Spain, is in charge of the Council of Europe which deals with human rights and considers the right to landscape as an expression of democracy, not of the European Union. Not coincidentally, the project of a political unity of Europe finds in the lack of recognition of local identities, also expressed by the diversity of landscape, some of its weak points. Agricultural, forest and, more in general environmental policies, have homogenized the European landscape, rather than preserving diversity. In this respect they can be considered another aspect of globalization. Unfortunately, those who do not feel to be recognized by a centralized political decisions are quite often the ones that will claim an unrecognized identity that can lead to distancing from the common project.

The lack of attention to the historical and cultural values of forests is also characteristic of the majority of the forestry sector. In reality, the Ministerial Conference for the Protection of Forests in Europe (MCPFE) as early as April 2003 affirmed the importance of the social and cultural value of forests, defining it as the third pillar of sustainable forest management, while guidelines for the implementation where prepared for the implementation of this concept (Agnoletti et al 2007). The MCPFE Vienna declaration of 2003 recognised the relationship between people and forests and that the social and cultural dimensions of sustainable forest management are reflected in landscapes, historical sites and monuments, artistic, traditional or linguistic knowledge, values, experiences and traditional practices related to forests and uses of wood, non-wood goods and services. It also states that SFM should maintain and increase the attractiveness of the landscape by, inter alia, enhancing and preserving traditional elements of the cultural landscape; raise awareness of the contribution of traditional forest knowledge and practices in sustainable forest management for the protection of landscapes (Parrotta and Trosper 2012).

The guidelines explains that many world's "primary forests" and biodiversity "hotspots" are in regions with the highest diversity of indigenous populations who manage their natural resources based on their distinctive cultures and their associated traditional knowledge and wisdom. In other rural environments a long history of integration of forestry and agricultural activities has also created land use forms and biological diversity that is closely connected to complex landscape patterns. Cultural landscapes often show a high level of habitat diversity tighten into a versatile mosaic produced by the application of different management forms, and the introduction of a great variety of species over the years, that came to meet specific economic, social and environmental functions. Considering the future scenarios presented by environmental change and especially by global warming, the conservation of traditional woodlands and forest management practices, as well as their associated landscape-level adaptations to difficult environmental conditions, should be given priority attention. The efficacy in coping with challenging environmental conditions depends on the interactions between key factors that require careful consideration in order to understand their historical success. Many have been achieved through internal experience and logic that has rarely been formalized into formal science. In traditional rural communities, the different types of forest land, from scattered trees in the fields, to dense forest cover, provide a variety of products and environmental services. Marginal and apparently non-productive lands, such as areas with low tree cover or shrublands have been traditionally exploited providing valuable resources to local populations helping to reduce external energy inputs. Such landscapes are rapidly shrinking in Europe by lack of protection mechanisms and appropriate management.

In order to better explore the implications of these concept for biodiversity in 2014 the Florence Declaration of UNESCO and CBD (United Nations Convention on Biological Diversity) was produced, which established the biocultural character of the rural landscape, of which forests are an important part. This initiative had several goals. One is to recognize that biodiversity is a wide concept that cannot be reduced to the protection of certain species linked to the theoretical trophic chains existing in an untouched natural environment. The long relationships between man and nature have created different patterns of biodiversity, different areas of the world may have different needs for the conservation of biodiversity. In practical terms, Scandinavia may have different goals compared to southern Europe, as Europe have different goals compared to Amazonia or Siberia. We must also consider that biodiversity is a product of history, we had several mass destructions of biodiversity in the history of the earth, while the last few thousand years of human history have witnessed changes in animal and vegetal species, because the environmental and landscape features have changed. The concept of biocultural diversity has been officially received by the FAO in its programme for the conservation of agricultural heritage, called GIAHS.[4] In Italy the Ministry of Agricultural, Food and Forestry Policies mentions it among the objectives of the activities in the decree establishing the National Observatory for Rural Landscape.

The degradationist vision, with which man's action is evaluated at a global level has limited the ability to evaluate the importance of a landscape approach considered as a strategy to integrate human society and environment. There is indeed a great difference between environmental policies and landscape policies. We can propose a global environmental policy because it is mainly based on indicators related on elements such as the quality of air, quality of soil, quality of water etc. Theoretically, if all the countries achieve a certain standard we can have a good quality of environment everywhere. On the contrary, landscape quality is not based on a set of standardized indicators. Human society has interacted with the environment in different ways, creating different landscapes representing the local cultural identity as also different ways of adapting to different and changing environmental conditions. In the case of forests, it is not fundamental whether there are dense woodlands or not, if we have an untouched forest or not, to define the value of a forest landscape. This depends on the interrelationships between the environment and human society across time.

Forests in Italy were a fundamental element of the daily life of a large part of the population only up to seventy years ago. To the historians of the future the task of establishing whether the new values assigned to the forest landscape will replace the ones that for centuries have characterized it.

Notes

1. According to the national forest inventory, forests are considered forest stands with a minimum ground cover of 20%, minimum extension 0.2 ha and minimum width of 20 m (Ministry of Agriculture and Forests, National Forest Inventory 1983, Themes, Trento 1988). According to ISTAT, forests are considered forest stands with a minimum cover of 50% and a minimum extension of 0.5 ha (ISTAT, Environmental Statistics, Rome 1998). The FAO considers a forest a territory having a cover greater than 10% of trees capable of growing, at maturity, to heights greater than 5 m, in formations wider than 0.5 ha and 20 m in width (UN/ECE-FAO (2000) Forest resources of Europe, CIS, North America, Australia, Japan and New Zealand: Main Report, United Nations, New York 2000).
2. Data was also taken from Bardini's work on the population. Particular thanks go to Carlo Bardini for allowing me to view the original data used in the preparation of his work, which does not appear in his article.
3. Data from the Corine Land Cover, IV level, for Italy.
4. Globally Important Agricultural Heritage Systems (http://www.fao.org/giahs/en/).

References

Agnoletti M (2001) Fra storia e tecnica. Sviluppi e tendenze della storia forestale. In: Agnoletti M (ed) Storia e Risorse Forestali. Accademia Italiana di Scienze Forestali, Firenze

Agnoletti M (2002) Le dinamiche del paesaggio fra XIX e XX secolo. In: Agnoletti M (ed): Il paesaggio agro-forestale toscano, strumenti per l'analisi la gestione e la conservazione. ARSIA, research report, Firenze

Agnoletti M (2002) Le sistemazioni idraulico-forestali dei bacini montani dall'Unità d'Italia alla metà del XX secolo. In: Lazzarini A (ed) Diboscamento montano e politiche territoriali. Alpi e Appennini dal Settecento al Duemila, Franco Angeli, Milano

Agnoletti M (ed) (2012) Historical rural landscape. Cultural values for the environment and rural development, Springer, Dordrecht and New York

Agnoletti M, Anderson S (2000) Forest history: international studies on socioeconomic and forest ecosystem change. CABI International, Wallingford and New York
Agnoletti M, Signorini MA (2011) Il paesaggio nella Cavalcata dei Magi. Pacini, Pisa
Agnoletti M et al (2007) Guidelines for the implementation of social and cultural values in sustainable forest management. A scientific contribution to the implementation of MCPFE – Vienna resolution 3, IUFRO Occasional Paper
Agnoletti M, Emanueli F, Maggiari G, Preti F (2012) Paesaggio e dissesto idrogeologico: il disastro ambientale del 25 ottobre 2011 nelle Cinque Terre. In: Agnoletti M, Carandini A, Santagata W (eds) Florens 2012. Studi e Ricerche. Bandecchi e Vivaldi, Pisa
Agnoletti M, Piras F, Venturi M, Santoro A (2022) Cultural values and forest dynamics: the Italian forests in the last 150 years. Forest Ecol Manag Special Issue on Restoration of cultural values in forested landscapes, John Parrotta & Mauro Agnoletti (guest eds) 503:119655. ISSN 0378-1127. https://doi.org/10.1016/j.foreco.2021.119655
Arma dei Carabinieri—CREA (2021) Le Foreste Italiane. Sintesi dei risultati del terzo Inventario Forestale Nazionale, Trento
Armiero M (1999) Il territorio come risorsa. Liguori, Napoli
Bardini C (1991) L'economia energetica italiana (1863–1913): una prospettiva inconsueta per lo studio del processo di industrializzazione. Rivista di Storia economica
Bloch M (1970) French Rural History. University of California Press, Oakland
Carloni P (1926) La estensione e la distribuzione territoriale dei boschi italiani. L'Italia forestale
Cazzola F (1996) Disboscamento e riforestazione ordinata nella pianura del Po: la piantata di alberi nell'economia agraria padana, secoli XV–XIX. Storia Urbana 20(76–77):35–64
Cevasco R (2007) Memoria verde. Nuovi spazi per la geografia. Diabasis, Reggio Emilia
Clements FE (1916) Plant succession: an analysis of the development of vegetation. Carnegie Institution of Washington, Washington
Comitato nazionale per le celebrazioni del centenario dell'Unità d'Italia (1961) Cento anni di vita nazionale attraverso le statistiche delle regioni. SVIMEZ, Tip. Failli, Roma
CREA (2017) Tutela e valorizzazione del patrimonio forestale italiano. Roma
D'Angelo P (2014) Filosofia del paesaggio. Quodlibet, Macerata
Farina A (1993) Bird Fauna in the changing agricultural landscape. In: Bunce RGH, Ryszkowski L, Paoletti MG (eds) Landscape ecology and agroecosystems. Lewis Publishers, Boca Raton
ISTAT (1958) Sommario di statistiche storiche italiane, 1861–1933. Roma
Lunardoni A (1904) Vini, uva e legnami nei trattati di commercio. Cooperativa Poligrafica Editrice, Roma
Martegani V, Agnoletti M (2003) Il paesaggio forestale lombardo. Immagini e storia. Cierre edizioni, Verona
Milizia nazionale forestale (1936) Carta forestale d'Italia. Roma
Ministero di Agricoltura, Industria e Commercio (1876) Relazione intorno alle condizioni dell'agricoltura nel quinquennio 1870–1874. Direzione generale dell'agricoltura. Barbera, Roma
Ministero di Agricoltura, Industria e Commercio (1886) Annuario statistico italiano, Roma
Ministero di Agricoltura, Industria e Commercio (1894) Bollettino di notizie agrarie. Direzione generale dell'agricoltura. Tip. Eredi Botta, Roma
Ministero di Agricoltura, Industria e Commercio (1913) Superficie territoriale e superficie agraria e forestale dei comuni del Regno d'Italia al 10 gennaio 1913. Direzione generale della statistica, Ufficio di statistica agraria. Tip. Bertero, Roma
Parrotta JA, Trosper RL (eds) (2012) Traditional forest-related knowledge: sustaining communities, ecosystems and biocultural diversity. World Forest Series vol. 12. Springer, Dordrecht, 621 p
Regione Toscana, Coordinated by Mauro Agnoletti, First report on the state of landscape policies in Tuscany, Firenze, 2020
Romano D (1986) I rimboschimenti nella politica forestale italiana. Monti e Boschi 6:7–12

Samojlik T, Fedotova A, Kuijper DP (2016) Transition from traditional to modern forest management shaped the spatial extent of cattle pasturing in Białowieża Primevai Forest in the nineteenth and twentieth centuries. Ambio 45(8):904–918
Santoloci M (2000) Diritto all'ambiente. Edizioni Ambiente, Milano
Serpieri A (1926) L'ambiente economico e sociale. L'Italia forestale
Sauer CO (1925) The morphology of landscape. Univers California Publ Geogr 2:19–53
Tofani M (1970) L'ambiente economico e sociale. Accademia italiana di scienze forestali, L'Italia forestale nel centenario della fondazione della scuola di Vallombrosa. Tip. Coppini, Firenze
Valenti G (1911) L'Italia agricola dal 1861 al 1911. Accademia dei Lincei, Roma

MIX
Papier aus verantwortungsvollen Quellen
Paper from responsible sources
FSC® C105338

If you have any concerns about our products,
you can contact us on
ProductSafety@springernature.com

In case Publisher is established outside the EU,
the EU authorized representative is:
Springer Nature Customer Service Center GmbH
Europaplatz 3, 69115 Heidelberg, Germany

Printed by Libri Plureos GmbH
in Hamburg, Germany